ENGLISH PLACE-NAME SOCIETY   VOLUME LXVI

66

FOR 1988-89

THE PLACE-NAMES
OF LINCOLNSHIRE

By

KENNETH CAMERON

in collaboration with JOHN FIELD and JOHN INSLEY

PART THREE

THE WAPENTAKE OF WALSHCROFT

ENGLISH PLACE-NAME SOCIETY
1992

Published by the English Place-Name Society

ISBN: 0 904889 18 1

Printed in Great Britain
by Woolnough Bookbinding, Irthlingborough.

This volume is dedicated to the memory of

Bruce Dickins,
himself a Lincolnshire man

ACKNOWLEDGEMENTS

This collection of material, the preparation and printing of this volume have been greatly assisted by generous grants from the British Academy; for the printing and publication by gifts from two anonymous members of the Society. The collection of material and the editing of the text over the past two years were made possible by the generous award of an Emeritus Fellowship by the Leverhulme Trust.

*The Camera-ready Copy of this volume
has been produced by Mrs Esmé Pattison
on equipment proovided
by
Messrs Allied Breweries plc
and by
Messrs Advent Desktop Publishing Limited*

CONTENTS

PREFACE

The third part of *The Place-Names of Lincolnshire*, like the second, covers only one Wapentake in the North Riding of Lindsey, that of Walshcroft lying to the south of that of Yarborough. Once more I have to thank friends in the Lincolnshire Archives Office, who have constantly helped during the past three years, particularly my friends Mr Nigel Colley and Mr Peter Noon. A glance through the text of this volume will show just how much material in the Archives I have searched and the time and energy on the part of the staff it has required to put it at my disposal. In addition, I have to thank Mr Brian Dyson, University Archivist, the University of Hull and Mrs C.H. Cobbold of the Suffolk Record Office, Ipswich, who guided me through important collections of Lincolnshire documents at Hull and at Ipswich. Mr Ian George, Sites and Monuments Record Officer, the City and County Museum, Lincoln, most kindly provided me with details of Brokenback in the parish of West Rasen.

My friends Mr John Field and Dr John Insley have helped so much in the preparation of this volume that it is right and proper for their names to appear on the title page as collaborators. Mr Field prepared the first draft of each of the parish field-name lists and made numerous suggestions of etymology, based on his own considerable experience of the study of field-names, though many have been silently included in the text and only he will know the extent of my debt to him. Dr Insley has read the whole of the volume and put at my disposal his unrivalled expertise in the field of early personal names. His contribution in this area and in others is very considerable indeed and I trust I have attributed to him all his suggested etymologies. Particular attention should be drawn to the "new" etymologies he proposed for Kingerby, Orford, Tealby and Thoresway.

My grateful thanks go to Mrs Anne Tarver, who drew the map of Walshcroft Wapentake and especially my friend Mrs Jean Russell-Gebbett. She has read through the entire text at least twice and saved me from numerous errors, though what remain are entirely my responsibility. She has also accompanied me on numerous trips round the area checking on the topography of individual places and providing a photographic record of them.

Once again, the text itself has been prepared for press by our Publications Officer, Mrs Esmé Pattison, whose experience and knowledge in handling place-name material has earned the admiration of the Society's editors and members over many years. Her patience and application are endless and again I stand greatly in her debt.

University of Nottingham   Kenneth Cameron

ADDITIONS to the ABBREVIATIONS and BIBLIOGRAPHY printed in THE PLACE-NAMES OF LINCOLNSHIRE, PARTS 1 and 2

Barne	Documents in the Barne Collection in the Ipswich Record Office
BPRentals	Bishops' Possessions, BP/Rentals/2 in LAO
Drax	The Cartulary of Drax Priory, Bodleian Library, Top. Yorks c.72
Featley	The Notebook of John Featley, Provost of Lincoln Cathedral, CC 8/152941 in LAO
GDC	Documents in the Gray, Dodsworth and Cobb Collection in LAO
Hjertstedt	Ingrid Hjertstedt, *Middle English Nicknames in the Lay Subsidy Rolls for Warwickshire*, Uppsala 1987
Inv	*Probate Inventories of Lincoln Citizens, 1661-1714* (LRS 80), 1991
JEPN	*Journal of the English Place-Name Society*, in progress
Jönsjö	Jan Jönsjö, *Studies on Middle English Nicknames*, Lund 1979
MC	*Calendar of Documents relating to the family of Maxwell-Constable*, East Riding County Record Office, 1965
MC	Documents in the Maxwell-Constable Collection in the Brynmor Jones Library, University of Hull (forms in badly damaged documents quoted from MC)
Mills	A.D. Mills, *A Dictionary of English Place-Names*, Oxford 1991
ONFr	Old Northern French
RDa	The Register of Bishop Dalderby, BR 3 in LAO

(1) Following the name Walshcroft Wapentake, the parishes in the Wapentake are set out in alphabetical order.

(2) Each of the parish names is printed in bold type as a heading. Within each parish the names are arranged as follows: (i) the parish name; (ii) other major names (i.e. names of sizeable settlements and names of primary historical or linguistic interest), each treated separately in alphabetical order; (iii) all minor names (i.e. the remaining names recorded on the 1906 edition of the O.S. 6" map, as well as some names that are 'lost' or 'local', *v. infra*), again treated in alphabetical order but in a single paragraph; (iv) field-names (which include other unidentified minor names) in small type, (a) modern field-names, normally those recorded since 1750, with any older spellings of these names in brackets and printed in italics, (b) medieval and early modern field-names, i.e. those recorded before about 1750, printed in italics, the names in each group being arranged alphabetically.

(3) Place-names no longer current, those not recorded on the editions of the 1" and 6" maps are marked '(lost)'. This does not mean that the site to which the name refers is unknown. Such names are normally printed in italics when referred to elsewhere.

(4) Place-names marked '(local)' are those not recorded on the 1" and 6" O.S. maps but which are still current locally.

(5) The local and standard pronunciations of a name, when of interest and not readily suggested by the modern spelling, are given in phonetic symbols in square brackets after the name.

(6) The early spellings of each name are presented in the order 'spelling, date, source'. When, however, the head-form of a name is followed only by a 'date and source', e.g. CLAXBY MOOR, 1794 *Dixon*, 1824 O, 1828 Bry, 1830 Gre, the spelling in 1794 *Dixon*, 1824 O, 1828 Bry and 1830 Gre is the same as that of the head-form.

(7) In explaining the various place-names and field-names summary reference is often made, by printing the elements in bold type, to the analysis of elements which will appear in the final volume of the Lincolnshire County Survey, and more particularly to *English Place-Name Elements* (EPNS 25, 26) and to *Addenda and Corrigenda* to these volumes in *English Place-Name Society Journal* 1. In many of the minor names and field-names the meaning is so obvious as to need no comment or so uncertain as not to warrant it. For personal-names which are cited without authority, reference

should be made for Old English names to Redin, Searle and Feilitzen, for Old (Continental) German to Förstemann PN and Forsner, and for English surnames to Bardsley and Reaney (for details of these sources *v. Abbreviations and Bibliography in The Place-Names of Lincolnshire*, Part 1 (EPNS 58).

(8) Unprinted sources of the early spellings of place-names are indicated by printing the abbreviation for the source in italics. The abbreviation for a printed source is printed in roman type. The exact page, folio or membrane is only given where the precise identification of an entry is of special importance or value, as e.g. under ALMSHOUSES in Kingerby *infra, MiscDon 238*, 183.

(9) Where two dates are given for a spelling, e.g. Hy2 (e13), 1190 (m13), the first is the date at which the document purports to have been composed and the second the date of the copy that has come down to us (in many cases the latter is a Cartulary, ecclesiastic or lay). Sources whose dates cannot be fixed to a particular year are dated by century, e.g. 11, 12, 13, 14 etc. (often more specifically e13, m13, l13 etc., early, mid and late 13th century respectively), by regnal date, e.g. Ed1, Hy2, Eliz, Jas1 etc., or by a range of years, e.g. 1150-60, 1401-2 etc., although this last form of date may alternatively mean that the spellings belong to a particular year within the limit indicated.

(10) The sign (p) after the source indicates that the particular spelling given appears in that source as a person's surname, not primarily as a reference to a place.

(11) When a letter or letters (sometimes words or phrases) in an early place-name form are enclosed in brackets, it means that spellings with and without the enclosed letter(s), words or phrases occur. When only one part of a place-name spelling is given as a variant, preceded or followed by a hyphen, it means that the particular spelling only differs in respect of the cited part from the preceding or following spelling. Occasional spellings given in inverted commas are usually editorial translations or modernisations and whilst they have no authority linguistically they have chronologically.

(12) Cross-references to other names are given with *supra* or *infra*, the former referring to a name already dealt with, the latter to a name dealt with later in the text.

(13) Putative forms of personal names and place-name elements which appear asterisked in the concluding volume of this survey are

not always asterisked in the text, although the discussion will often make it clear which are on independent record and which are inferred.

(14) In order to save space in presenting the early spellings of a name, *et passim* and *et freq* are sometimes used to indicate that the preceding form(s) occur respectively from time to time or frequently from the date of the last quoted source to that of the following one, or to the present day.

ADDENDA and CORRIGENDA

Volume 58

11	s.n.	CLOSE WALL (lost), add *the Close-wall* 1662 *Featley*.
12	Add.	THE MASON'S YARD (lost), *the Mason's yard* 1662 *Featley*.
18-19	s.n.	BRAYFORD HEAD (lost), add *Braiford-head* 1662 *Featley*.
23	s.n.	DUNSTAN'S LOCK, add *Danston Lock* 1662 *Featley*.
24	s.n.	LITTLE and GREAT GOWTS (lost). Mr A.E.B. Owen comments "Whatever the original meaning of **gotu**, the meaning of *gowt, gote, goat* in Lincs place-names is 'sluice'. This is abundantly evidenced.... I think some form of physical check at the outfall, which a sluice represents, must always be understood. I have never in drainage records met *gowt* with the sense 'watercourse' and I am confident that this is not a valid interpretation of it in Lincs at any period".
38	s.n.	OLD EYE (lost). Mr A.E.B. Owen points out that Smith's comment in EPN s.v. ēa that *eau* is 'Lincs dialect' cannot be substantiated. He notes that the forms for the word are *ea, eay, eae, e, ee, ey, eye, eyey, egh*, etc., but never *eau* in Sewer Records. He, further, draws attention to PN C 332, where *eau* is described as a pseudo-French spelling. This interpretation is no doubt correct.
38-39	s.n.	SINCIL DYKE, add *Sinsill Ditch* 1685 Inv.
48	s.n.	ALMA TCE. lines 2-3 should read "most names in *Alma*, as here, commemorate the battle of that name in the Crimean War, 1854."
54-55	s.n.	BROADGATE, add *Ware-dike* 1661 *Featley*.
83-84	s.n.	MUCH LANE, add *Much Lane* 1682 Inv.
94	s.n.	ST LAWRENCE'S LANE (lost), add *St Lawrence lane* 1661 *Featley*.
141	s.n.	CASTLE OF CRAKE (lost). Delete "The form of the name ... remain obscure", since the name

is to be associated with the *Crake* family. The suggestion of possible Scottish influence arose because the formula *The Castle of* ... appears to be specifically Scottish today. However, Barrie Cox 'The Use of Middle English *castel* in the Names of Medieval Town-Houses', *JEPN* 25 forthcoming, provides parallels to this name, e.g. *the Castel of Croydone in Herlone* 1331 in Gloucester, PN Gl 2, 135, also derived from a family name. He demonstrates convincingly that the references are to fortified merchant town mansions, a further example from Lincoln being *castello* ... *de Tornegat*, PN 1, 158.

150	s.n.	KYME HALL (lost). Delete in line 3 "Hungate", RA 1, 312b". line 3 should read "hall "may well have been the Castle of Thorngate" Hill i, 159-60".
159	s.n.	WHITE HO (lost), add *the White house* 1662 *Featley*.
161-62	s.n.	ANGEL (lost), add *the Angel in the Bale* 1662 *Featley*.
164	s.n.	THE BULL (lost) (2), add *the Bull in East-gate* 1662 *Featley*.
178	s.n.	*the Ash(e) garthe*, add *the Ash-garth* 1662 *Featley*.
185	s.n.	(*a pingle or*) *pond(e) garth*, add *the Pond-garth* 1662 *Featley*.
186	after	*Paradyze* add *the Pin-fold* 1662 *Featley* (it was in SNich).
187	after	*in campo australi* (*Lincoln'*) add *One Mill standing and being udard* (i.e. under) *the Comon comonly called or knowne by the name of ye Spring Mill* 1703 Inv.
209-10	s.n.	SHEEPWASH GRANGE. For "Stangeways" read "Strangeways".

Volume 64/65

xx	Map	The parish of Nettleton, immediately south of Caistor, was inadvertently omitted from the map.

It has been included on that of Walshcroft Wapentake *infra.*

2-7   s.n.   LINDSEY. Mr M.S. Parker has drawn attention to the following early spellings:

> *in lindissi* c.704-14 (9) *The Earliest Life of Gregory the Great. By an Anonymous Monk of Whitby,* ed. B. Colgrave, reprinted 1968
>
> *in Lindisse* s.a. 765, s.a. 796 (12), *in Lindissi* s.a. 767 (12) all *Symeonis Monachi Opera Omnia* (RS), 2 vols., 1885.

Mr Parker points out that the form dated c.704-14 is apparently the earliest recorded reference to Lindsey.

13 etc.   s.n.   *ye acredikes.* The earliest reference to this name is *on þa æcer dik* 956 (c.1200) BCS 924 (S 606) in the bounds of Abingdon, PN Berks 735, where it is translated 'acre ditch'. A further example, *Acre Ditch* 1767, is recorded from Sheldon, PN Db 165. Its frequency in L suggested a possible ODan origin, but the Berks example clearly shows that it must be derived from the OE **æcer-dic,* though in L the forms have often been Scandinavianized.

44   s.n.   *Alkazedail.* A further example of the obscure surn. *Alkaz* has been noted in John *Alkaz de Spalding'* 1316 *RDa.*

67   s.n.   MAUSOLEUM. For "Arabella" read "Sophia".

118 and 292   s.n.   BROCKLESBY OX (INN). Through the kindness of Mr N.G. Glover, landlord of the Brocklesby Ox, Ulceby, I was able to see a copy of a document dated 1810 giving a full description of the Brocklesby Ox. The animal was bred by R. Goulton, Esq. of Bonby (cf. *Goultons Cover* in Bonby, PN L 2 58) and was the property of Messrs. Marshall, Lovitt and Co. of Hull. The weight is given as "265 Stones, of 14 pounds to the Stone", together with its dimensions in full, as well as the name of its sire and dam.

145   s.n.   *le Moldfanch(e).* Mr A.E.B. Owen points out

that this term is linked to saltmaking, known to have taken place in Habrough Marsh. He draws attention particularly to D.M. Williamson, 'Some Notes on the Medieval Manors of Fulstow', LAAS 4, 1951, pp. 1-56, especially p. 37, where there is a reference to "a holme with a saltern and two plots (*placeis*) called *Molfanges*" and to D.M. Owen, 'The Medieval Salt Industry in the Lindsey Marshland', LAAS 8, 1960, pp. 76-84, especially pp. 81-83. He notes further to the references quoted from RA v, 188, that ib, 189 Dr Kathleen Major comments "A *moldfang* is apparently an alternative word for a sandpit or place where sand could be taken, *molde* being glossed as *sabulam* in the Anglo-Saxon period. (See Bosworth-Toller A.S. Dictionary). I am indebted to Professor Stenton for this interpretation". Clearly *moldefang* is closely linked to saltmaking, and though the exact meaning is uncertain, it may be 'the place where sand is taken', as Sir Frank Stenton suggested.

191	s.n.	*Brice Willows*, add cf. Margaret *Brice* 1591 *BT* (South Kelsey).
222	s.n.	Great Limber Grange, for "Edward Gilllet" read "Edward Gillett".
277	s.n.	*Nettelbustmar*, seven lines from bottom, for "Croxhill f.ns." read "Great Limber f.ns.".
290	s.n.	*Snore Hill.* A detailed examination of the sources in which the names occur demonstrates that this is in fact to be identified with Smooth Hill in f. ns. (a), as was tentatively suggested. As a result, it has been possible to identify the site of the hill as being TF 078 169. It lies close to the minor road to Barton upon Humber, a continuation of the modern A 1077, which itself takes a right-handed bend into Wootton. Dr Margaret Gelling identified *Snore Hill* as probably being a sixth example of the name, hence the importance of trying to localise it precisely. She will eventually publish a detailed article on this name, examining

the site of each example, but a summary of her tentative conclusions may be given here. She comments that "Snower Hall Nf, *Snora* 1086 DB, is the only unequivocal evidence for an OE **snōr*, but there are now six names in which it is possible that this word is compounded with **hyll**". She points out that the name has been discussed by Dodgson in PN Ch **3** 37, s.n. Snow Hill and by Ekwall in LnStN 180-81, s.n. Snow Hill. Commenting on the latter in PN Ch **3** 37 the late Professor Mattias Löfvenberg believed that Ekwall was correct in suggesting that **snōr* meant 'something twisted or knotted', in a transferred sense 'hillock'. In an investigation of the topography of the examples which can be located, including the present one, Dr Gelling suggests that **snōr* was a specialised term for a type of hill, and that it was "used for a road which twisted in order to negotiate a small hill". Of *Snore Hill* in Thornton Curtis she notes "The road curves sinuously to follow a slight valley, thus avoiding higher ground on either side. This is quite different from the right-angled bend which the road makes further north". Personal examination confirms Dr Gelling's description. At present, it would seem that her interpretation of the name is not only plausible, but also likely and that the first el. is an OE **snōr* in this specialised sense. No doubt *Hill* was added because the meaning of **snōr* was no longer understood.

297	s.n.	*the Eae.* See the comment on OLD EYE *supra.*
299	s.n.	HOWE HILL. The meaning of *Howe* here is undoubtedly 'a mound', since this is the name of a round barrow.

Volume 66

2-3	s.n.	Binbrook. For a detailed analysis of p.ns. derived from OE **burna** and OE **brōc**, *v.* Ann Cole, '*Burna*

and *Brōc*. Problems Involved in Retrieving the Old English Usage of these Place-Name Elements', JEPN 23, 26-48. On Binbrook, Mrs Cole, in a private communication, comments "It is unusual to find a *brōc* on the edge of the chalk, but it has a muddy floor and banks and plenty of apium growing in it, so it is a fairly good example".

18   s.n.   CLAXBY MINES. Dr Rod Ambler points out that there is a reference in *Yarb* 5/2/17/1 dated 1776 to charges incurred in the trial to find coals at Claxby, where a note describes how Thomas Poundal "acted in a very disingenuous part his whole view seeming to be to get what he could for himself and men, they trifling their time away in a manner beyond imagination".

19   s.n.   TERRACE HOUSES. Dr Rod Ambler notes that there is an item in *Yarb* accounts for £1,195 spent on ten new terrace houses, together with one for the manager, at Claxby in 1871. They have now been pulled down.

74   s.n.   SMITHS COTTAGE, add *Smyth thyng* m16 *MiscDep 43* (*v.* þing 'possession, property').

76   after   *gat a furlonge* add *vna Grangie* m16 *MiscDep 43* (*v.* **grange**).

77   after   *Northinge* add *norton Howse* m16 *MiscDep 43* (no doubt from a family called *Norton*).

138   add   THE SMOOTING (local). This is dial. *smooting* 'a narrow passage between houses', *v.* EDD s.v., a word recorded only from L. EDD has also *smoot* and cf. Danish dial *smutte* 'an opening, a small passage or entrance into a place', 'a narrow passage between houses. *Smooting* is clearly a derivative of *smoot.*

       s.n.   TEALBY THORPE. The references to *Thorpe Lane* should be put separately under THORPE LANE, since this is still used locally.

141-42   Cow Lane. The forms quoted for this name should have been given under the head name COW LANE since the name is still in use locally.

143   s.n.   the Green. The forms quoted for this name should have been given under the head name THE GREEN, since this is still in use locally.

147   s.n.   Wailesby Lane. The three forms should have appeared under the head name WALESBY LANE, a name still in use locally.

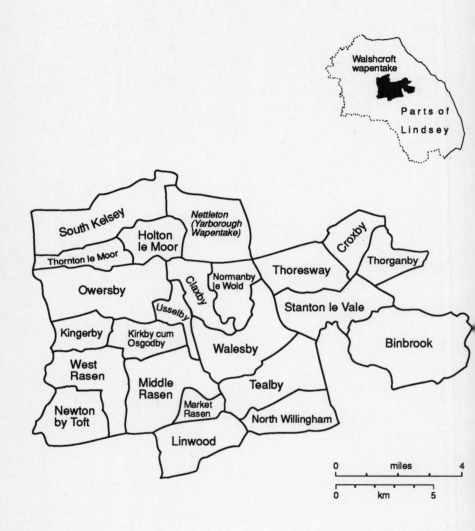

Walshcroft wapentake

Parts of Lindsey

South Kelsey

Holton le Moor

Nettleton
(Yarborough
Wapentake)

Thornton le Moor

Croxby

Thorganby

Owersby

Claxby

Normanby
le Wold

Thoresway

Usselby

Stanton le Vale

Kingerby

Kirkby cum
Osgodby

Walesby

Binbrook

West
Rasen

Middle
Rasen

Tealby

Newton
by Toft

Market
Rasen

North Willingham

Linwood

0   miles   4

0   km   5

WALSHCROFT WAPENTAKE

Based upon the 1963 Ordnance Survey four miles to one inch map, with the
permission of the controller of Her Majesty's Stationery Office. © Crown Copyright.

WALSHCROFT WAPENTAKE

   Walescros 1086 DB, 1183, 1184, 1185, 1186, 1187, 1202 P,
      Waliscros 1175 ib
   Walecros 1086 DB, 1201 P
   Walescroft c. 1115 LS, 1130 (p), 1166, 1167, 1168, 1169,
      1170 P *et passim* to 1526 Sub, *-crofte* 1562-67 LNQ v,
      1585 SC, 1601 *Terrier, Walescrof* (sic) 1198 P, *Waliscroft*
      1277 RRGr, 1291 Tax, 1342 Pat, 1428 FA
   Wallescroft 1177 P, 1373 Peace, 1723 SDL, *-crofte* 1576 LER
   Walecroft 1192, 1193, 1194 P, 1308 *LCCA, Walcroft* 1200 P
   Walscroft 1242-43 Fees, 1276 RH, 1281 QW, 1296 *Ass,* 1298
      Ass, 1338 Pat, 1393 Works, 1565 SP
   Walshecroft 1287 Ipm, 1327, 1332 *SR, Walshcroft* 1456 Pat,
      1465-66 Lanc *et passim, -crofte* 1465 Pat
   Waleschcroft 1288 Ipm, 1392 Works, *Walschcroft* 1535 VE
      iv, *Walesshcroft* 1343 NI, *Walsshcroft* 1347 Pat, 1373
      Peace, *Walchecroft* 1428 FA, 1536 LP xi

    The forms are preceded or followed by Wapentake or
Deanery, usually in a Latin form. It is once, 1373 Peace, referred
to as a Hundred.
    As suggested by Fellows-Jensen, SSNEM 162, this is probably
derived from the Scand pers.n. *Váli* or *Valr* and **cros** 'a cross', a
word borrowed from OIr and brought to England by the Vikings,
spreading rapidly as far south as Hu by the time of DB. The *cross*
presumably marked the site (now not known) of the wapentake
meeting. The name is to be compared with the lost *Walecros,*
PNDb 622, while the same first el. occurs in Walesby in Walshcroft
Wapentake, both presumably named from the same man, in which
case, *Valr* is to be preferred. The second el. was replaced by **croft**,
certainly by the early 12th century. See further DEPN s.n.
Walesby and EHN 51-52.

Binbrook

BINBROOK

Binnibroc (3x) 1086 DB
Binibroch' 1179 P (p), *Bynibroka* Hy3 (1409) Gilb
Binnabroc c. 1115 LS
Binne broke 1088-93 (14) YCh i, *Binnebroke* 1156-57 (Ed2) ib
    i, *-brok* R1 (1308) Ch, 1243-46 ChantCert, 1291, 1294,
    1295 RSu, *-broc* 1099 RA i, 1150-55, 1155-60 Dane, 1163
    RA i, 1182, 1183, 1184 *et passim* to 1314 Ch, *-broch* 1146
    RA i, 1180 P (p), 1181, 1182 ib, 1183 ChancR, c. 1200
    (e13) *NCot*, 1314 Ch
Bynnebroc 1099 RA i, 1235 Cl, 1235 Ch, 1256 *DC* (p), *-brock*
    1276 RH, *-brok*(') 1291 Tax, 1292 Pat, 1301 Fine, 1301
    Cl, 1332 *SR*
Binebroc(') 1160-66 Dane, c. 1189 LAAS v (p), 1194 CurP
    (p), 1196 ChancR, 1197, 1198, 1200 P *et passim* to 1269
    RRGr, *-brok* 1252 Ch, 1252 (1389) Pat, *-brock'* 1254
    ValNor
Bynebrok 1239-40 RRG, 1250 FF, *-broke* 1302 Abbr
Binbroka 1228-32 (1409) Gilb (p), *-brock'* 1240-50 RA v,
    -brok 1244 FF, 1276 RH, 1281 QW, 1301 Ipm, 1348
    Pap, (- *Gabrielis*, - *Marie*) 1526 Sub, 1576 Saxton, 1610
    Speed, *-broke* 1275, 1453 *Tat*, *-broc* 1261 Cl, *-brook*(*e*)
    1587, 1592 *Goulding*, 1596 *Foster*, (- St *Maries*, - S^t
    Gabriell) 1610 *Goulding*, (- *nigh Stainton Le Hole*) 1698
    MiD. 1749 *Td'E et passim*
Bynbroc 1242-43 Fees, *-brok*(') 1271-72 *Ass*, 1303 FA, 1304
    FF, 1316 FA, 1316, 1327 Pat, 1335 Cl, 1341 Pat, 1343
    Ipm *et freq* to 1428 FA, *-broke* eHy3 (1409) Gilb, 1340
    Ch, 1372 Cl, 1383 Peace, 1393 Cl *et freq* to 1566
    Goulding, (- *Gabriel*) 1558 InstBen, (- *Sancti Gabriell*)
    1566 *LCCA*, *-brocke* 1327 *SR*, - *brook*(*e*) 1343 Cl, 1676
    BRA 1125
Binnesbroch 1177 P (p), *Bynesbroke* 1302 Abbr
Bininghbroc 1237 (13) *LOC*, *Byngbroke or Byngbrukes* 1495
    Ipm

    Ekwall, DEPN s.n., suggests that the meaning is 'Bynna's
brook', from the OE pers.n. *Bynna* and **brōc** 'a brook', adding that

OE *binnan brōce* '(land) inside the brook' does not seem suitable. The latter is certainly the etymology of Binbrook Lane, PN C 44, which is translated '(land) enclosed by the brook' with the elliptical use of the OE preposition **binnan**. The village is situated on the west side of the brook, the course of which is hardly such as to make a meaning '(land) enclosed by or within the brook' topographically appropriate. We are, therefore, clearly concerned with the OE pers.n. here. The spellings *Binnibroc, Binibroch'* and *Bynibroka* seem to indicate that an -**ing**[4] variant existed beside the *-an-* gen.sg. Hence, the base would seem to be *Bynningbrōc 'the brook associated with Bynna' by the side of *Bynnanbrōc 'the brook of Bynna'. For other formations of this type *v.* EPN i, 294 and for the pers.n. *Bynna v.* Redin 61-62. The affixes *Gabriel* and *Mary* are from the dedications of two churches here, the former now "lost".

BECKFIELD (lost, approx TF 191928), *Bekfeld'* 1202 Ass (p), 1329 *Ass, -feld* 1535 VE v, *-felde* 1393 Cl, *Beckefeld* 1224-42 RA v, a1235 ib v, 1316 FA, *Bekefeld* 1219 Ass, 1240 FF, 1302 *FF, Bekesfeld* 1219 Ass, *Beckfeld* 1555 *AddCh, beckfielde* 1577-80 *Terrier, beckefield* 1638 *ib, -feild* 1638 *Foster, Beckfield* 1659 *MiD,* cf. *Bek* Hy2 (14) Dugd iii, *Bec* R1 (1308) Ch, and later *Beckfield close, - hill* 1577-80 *Terrier, Beckfeild hill* e17 *MiD, beckfeald close, - hyll* 1605, *beckfeld close, Beakfeld hill* (sic) 1612, *Beckfield(e) hedg(e)* 1634, *Beck field hedg* 1664, *Bec feid hedg* (sic) 1668, *Beckfeild hedge* 1679 all *Terrier.* The name seems earlier to have been simply *Beck* 'the stream' (*v..* **bekkr**), to which was added **feld** 'open land, land for pasture or cultivation'. It is the site of an "extinct hamlet", *v.* further DB xlix, and was named from the stream which flows beside Kirmond le Mire and which forms the boundary between the latter and Binbrook. Cf. *bekfurhill* in f.ns. (b) *infra.*

EASTHORPE FM (lost), *East-Thorpe Farme* e17, *Easthorpe* 1609, *East Thorpe* 1658 all *MiD,* 1692 *Foster, -thorp* 1662, 1664 *Terrier, Eastthorpe Farme* 1697 *MiD,* presumably a late example of the use of ODan **þorp** in some such sense as 'an outlying farmstead', east of the village. Its site is not known.

LIMBER HILL, - *hill* 1662 *Terrier*, cf. *Linbereydale* (sic) 1244 FF, *Limber hill foot* e17 *MiD*, - *hill side*, - *hill Furs* 1668, *Limberhill side* 1671, *Lymber hillside* 1679, *Limber botham* 1638, *Lymber fures* 1577-80, *limber fures* 1612 (*v.* **furh** 'a furrow' in the pl.), *Limber slacks* 1662 all *Terrier* (*v.* **slakki** 'a shallow valley, etc.'), - *crosse* 1668 *ib.* The 1244 spelling suggests that here Limber is to be compared with Great Limber, PN L 2 219, and shows a similar development. It means, therefore, 'the lime-tree hill', *v.* **lind, beorg.** A formally possible alternative is 'the hill where flax grown', *v.* **lín, beorg.**

THE SCALLOWS HALL, *Scallow Farm* 1781 *Tur, new erected Messuage or Tenement Cottage and Farm called Scallow farm* 1784 *ib, Binbrook Scallow* 1830 Gre, cf. *Scalowe way* 1577-80 *Terrier, Scallagate* e17 *MiD, Skally gate* 1662, 1679, *Scalley Gate* 1668, *Scallow gate* 1671 all *Terrier* (*v.* **gata** 'a road'), *Scalla End - Nooking* e17 *MiD, Scallow Hill* 1740 *Cragg,* - *Platts* 1782 *Tur* (*v.* **plat**^{2} 'a plot of land'). The forms are late, but it may be suggested that Scallow is to be compared with Scallow, PN Cu 407, which is taken as being probably from ON **skalli** 'a bald head', used of 'a bare hill', and **haugr** 'a hill', a Scand compound, and also with *Scalawe* in East Halton f.ns. (b), PN L 2 157. The hall is on the slope of a steep hill, and the suggested etymology seems topographically appropriate.

SPOTTLE HILL FM, *Spottle Hill B.*^{n} 1828 Bry, cf. *Spotedaleklif* 1244 FF (*v.* **clif**), *Spodill* 1577-80, *Spodall* 1612, *Spoddale* 1664, 1668, 1679 all *Terrier, Spodall bottom* e17 *MiD, Spottle or Speardale Hill* 1838 *CC, Speardale Hill* 1800 *Tur.* With only a single early form it is difficult to suggest an etymology for *Spottle.* The farm lies on the slope of a valley, so that the second el. is likely to be **dalr** 'a valley', but *Spote-* is difficult to interpret. There appears to have been an OE ***spot** 'a spot', in ME 'a small piece of ground' (*v.* MED s.v. (1), 4 (a)), while ON **spotti** 'a piece, a particle' is used in Norwegian of 'a piece of ground'. It is, however, difficult to see the significance of either in a compound with **dalr**. There is also an OE nickname *Spot* (*v.* Tengvik 335-6), which at least formally might be just possible. No certain etymology can be suggested.

ASH HOLT. BACK LANE. BINBROOK GRANGE, "a grange in the vill of" *Bynnebroke* 1302 ChronLP, *ye Grange* 1662, - *grange* 1668 both *Terrier*, cf. *Grainge Close hedge* e17, (*Le*) *Grange Close* 1625, *the grange close* 1639, *ye Grange Closes* 1679 all *MiD*, *the Grange Close* 1727 *et passim* to 1814 *Tur* and *v.* also Adam Storeys Garths Closes in f.ns. (a) *infra*. This was a **grange** of Louth Park Abbey. BINBROOK HALL, 1830 Gre, - *House* 1824 O, - *H^o* 1828 Bry, but cf. *the hall garthe* 1605 *Terrier* (*v.* **garðr** 'an enclosure'), *Hall Close end* e17 *MiD*, *ye Hall close end* 1679, - *Hall close noock* 1664, - *hall close Noock* 1668 all *Terrier*. BINBROOK HILL FM, 1833 *Tur*, *Binbrook Hill* 1830 Gre, cf. *the hill side* 1662 *Terrier*. BINBROOK MILL, 1828 Bry, *molendinum juxta Binnebroc* 1156-57 (Ed2) YCh i, *molendino de Bynbroka* Hy2 (1409) Gilb, *The Mill* 1740 Cragg, cf. *miln hill* 1612, *mill hill* 1664, *the milne dame* 1612 all *Terrier* (*v.* **dammr**), *the Mill Gate* e17 *MiD*, *Milne gate* 1668 *Terrier* (*v.* **gata**), *the Miln Close* 1703 *MiD*, *Mill Close* 1738 (1818) *Tur*, *the Mill Beck* 1740 *EnclA* (*v.* **bekkr**), *v.* **myln**. BINBROOK TOP, TOP FM, *Top B^n* 1828 Bry. BINBROOK VILLA is *The Home Farm* 1626 *Tur*, *Binbrooke Home Farm* 1833 *ib.* BINBROOK WALK HO, 1824 O, *v.* **walk**. BINGHAM'S TOP is *Millisons Top* 1828 Bry, both no doubt named from local families. BLACK HOLT. BRATS LANE, part of the boundary with Swinhope LNR. BURKINSHAW'S TOP is *Fletchers Top* 1828 Bry, named from local families, cf. John *Fletcher* 1826 White. BUTTER CROSS, now in the churchyard. CHESTNUT FM. DOVE COTE (lost), 1828 Bry, cf. *dovecoategarth* 1609 *MiD* (*v.* **garðr** 'an enclosure'), *the duffecoate close* 1697 *ib*, *Dove Coat Close* 1782 *Tur*, *Dovecoat Close* 1784 *ib*, *the dove cot Close* 1822 *ib*, self-explanatory and *v.* also Adam Storeys Garths Closes in f.ns. (a) *infra*. FIVES CROSS (lost), 1828 Bry, at the intersection of the south-east corner of the parish with Kelstern, where five roads meet. THE GREEN (local), *Atte Grene* 1327 *SR*, *atte Grene* 1332 *ib*, 1343 NI, *del Grene de Bynbroke* 1383 Peace all (p), *v.* **grēne**^{2} and note "A spreading village, the Green on the side of the hill" P&H 191. HORSESHOE PLANTATION, *Horse Shoe Clump* 1828 Bry, no doubt a reference to its shape. ISAAC PLOT PLANTATION. LODGE, *The Lodge* 1828 Bry. LONG PLANTATION. LOW FM is *Binbrook End F^m* 1828 Bry, - *Farm* 1830 Gre. LOW LANE. MANOR HO. MARKET PLACE. MARQUESS OF GRANBY, *Granby Inn* 1804 *Tur*, *Granby's Head Inn* 1805 *ib*, *Granby* 1826 White, *Marquis of*

Granby 1842 ib. NORTH HALLS, *Northall maner* 1453 *Tat,* 1453 Pat, *the North Hall* 1634, *North Hall* 1662, *The North hall* 1668 all *Terrier, Farme ... called Northalls* 1709 *Tur,* cf. *the North halls pece* 1658 *MiD,* self-explanatory. ORFORD BRIDGE, - *bridge* 1720 *Webb,* - *Briggs* 1740 *EnclA,* named from Orford in Stainton le Vale parish *infra. The Ford* is marked here 1828 Bry. PARSONAGE FM, cf. *y*^{e} *parsonage close* 1668 *Terrier, The Parsonage Close* 1692 *BT,* - *parsonage close* 1697, 1717, *parsonige Close* 1718 all *Terrier,* the reference is apparently to the church of St Mary. PLOUGH INN, *Plough and Horses* 1826 White, *Plough* 1842 ib. POCKET BRIDGE, cf. *Pocket Close* 1738 (1818), 1806, 1818 *Tur,* probably referring to 'a nook of land', cf. Pocket Patch, Field 170, and Pocket Cl in f.ns. (a) *infra.* POND HO (lost), 1828 Bry. THE POPLARS is *Bland F.*^{m} 1828 ib, named from a local family, cf. Luke *Bland* 1804 *Tur.* RECTORY FM. SWINHOPE HILL, *Swinnop hill* e17 *MiD,* cf. *Swinhop Hill Close* 1822 *Tur,* and also "the way of" *Swynehop* 1244 FF; it leads to the adjoining parish of Swinhope. TEMPERANCE HALL, "a handsome Temperance Hall, erected in 1840, at the cost of £550, including four tenements in the basement story" 1842 White. WEST HO.

Field-Names

Forms dated 1244 are FF; 1327, 1332 are *SR;* 1570, 1594, e17, 1602, 1604, 1609, 1616, 1635, 1646, 1648, 1649, 1657, 1658, 1662^1, 1666, 1667, 1676, 1679^1, 1680, 1690, 1697^1, 1698, 1700, 1703, 1722 are *MiD;* 1577-80, 1601, 1605, 1606, 1612, 1634, 1638^1, 1662^2, 1664, 1668, 1679^2, 1697^2, 1717, 1718, and 1822^2 are *Terrier;* 1587, 1589, 1614, 1638^2 are *Goulding;* 1596, 1638^3, 1641, 1692^1 are *Foster.* 1601, 1719, 1743, 1747, 1776, 1803, 1804, and 1814^1 are *Webb;* 1692^2 are *BT;* 1709, 1714, 1727, 1727 (1814), 1733, 1737, 1738 (1818), 1739, 1761, 1763, 1782, 1784, 1802, 1804, 1806^1, 1812, 1814^2, 1818, 1819, 1820, 1822^1, 1826, 1827, 1830, 1833, 1834, 1835, 1836, and 1840 are *Tur;* 1740^1 are *Cragg;* 1740^2, 1806^2 are *EnclA;* 1775, 1806^2, and 1838 are *CC 129;* 1859 are *Padley.* Other sources are noted.

(a) Bast(e) Ings 1761, 1763, Baste Ings Cl 1806^2, 1814^1 (*Bassekynges, a peace of medowe called basse kinges* 1577-80, *medowe growne lying at basse kings* 1601, *basekinges* 1612, *bayste Inges* 1638^1, *Bayes Inges* 1638^3, *Baise -* 1641, *Baists Ings* 1657, *Base ings* 1662^2, *Baseings* 1664, *Bastinges* 1668, *bastings* 1692, *Bast-Inges* 1697^2, *y*^{e} *Bast Inggs* 1717, *bast Inggs* 1718; *Basceinge Furlong* 1634, *Bastinges*

Forlong 1668 (*v.* **furlang**); *Baesling gate* e17, *Basceinge* - 1634, *Baseings* - 1662², 1664, *Baslinges* - 1668, *Baslinge* - 1679², *Bastingate* 1668, *Bassinge gate Furlonge* 1679² (*v.* **geat, furlang**), *Bassekinge heade* 1577-80, *basse kynges heades* 1605, *baskinge heades* 1612, *Baesling head* e17 (*v.* **hēafod**), *Basing thing* 1589, *cotagiu'* ... *voc' Basingthing* 1614, *Bases thing* 1662², *bayes* - 1664, *v.* **þing** 'premises, property'. Some forms have apparently been influenced by the surn. *Bayes,* cf. Christopher *bayes* 1562 BT, but the history of this name, presupposing all the early forms refer to one place, is difficult to ascertain. Dr John Insley puts forward a very tentative suggestion that the ultimate base might be a surn. deriving from the Flemish pers.n. *Basekinus,* apparently a hypocoristic form of *Basilius, v.* Tavernier-Vereecken, *Gentse naamkunde van c, 1000 tot 1253,* Tongeren, 1968, p. 153); *Beedham Cl (sic)* 1806¹, The Beedham Cl, Pocket Close, Mill Cl but more commonly called ... the Hill Cl (*sic*) 1822¹ (*v.* also Hill Cl *infra*); Binbrooke Pasture 1814¹ (*the pasture* 1577-80, 1605, - *pastures* 1612, - *Pastures* 1634, *le Pastures* 1638², *ye* - 1664 and cf. the Pasture Cl *infra*); Near Bottom 1833 (*v.* **botm**); Brathers Cl 1782, 1784, 1814², - but now called Brady Garths 1822¹, Brady Garth 1833, 1840 (*v.* **garðr** 'an enclosure, a small plot of ground', as elsewhere in this parish; the earlier name is from the *Brather* family, cf. Richard *Brather* 1574 BT, the later probably from the surn. *Brady* not been noted so far in the parish); Brooks House 1782, Brocks House (sic) 1822¹ (cf. *Brookes headland* e17, *v.* **hēafod-land**), *Brooks great piece* 1662² (*v.* **pece** 'a piece or plot of land', sometimes a consolidation of strips acquired by piecemeal enclosure; for the surn. cf. Thomas *Brooke* 1577-80, 1612, James *Brokes* 1668); the Caister Road 1806², Castor Road 1822² (*Caister gate* e17, *Caster Road* 1740¹, the road to Caistor, *v.* **gata**); Cap Cl 1776, 1803 (*le Capcase* 1638³, *Caprase* (sic) 1719, *Cap-close* 1743); Chalk pit 1820; Chapmans Garth 1782, 1822¹, Chapman's Garth 1814² (1727 (1814)) (named from the family of Thomas *Chapman* 1806 *EnclA,* with **garðr**); Cooke Cl 1833 (presumably from the surn. *Cooke*); Far Cottam, Well Cottam Flat 1833 (*v.* **flat**); the Cover Platt 1822¹ (*v.* **plat** and cf. Fox Cover *infra*); Cow Cl 1833; the Cowpasture 1763, 1776, 1818, 1819, 1830, 1834, the late - 1814², the Cow Pasture 1761, 1804, 1812, 1814², 1819, 1833, 1834, Binbrooke Cowpasture 1804, Cowpasture 1838, the common - 1806¹ (*the Cowe pasture* 1648, *Cows Pasture, Cow* - 1740², said in *EnclA* "to be held and occupied as a common Cow Pasture"); Croft Flat 1833; the East Fd 1782, 1814², (*in campo orient* 1570, *in campo orient'* 1589, *in campo orien'* 1614, *in orien campo de Binbrooke* 1680, *in oriental' campo* 1722, *Eastfeild* 1577-80, *the East feild* e17, 1612, *le East* - 1638², *yᵉ east feld* 1606, *the East Field* 1657, 1662², 1668, - *Feild* 1719, *yᵉ East Field* 1679², 1718, one of the open fields of Binbrook, apparently later combined with the South Field, cf. the south and east fields *infra*); The Eighteen Acres 1859; Five Acres (2x) 1833; Football yd 1806¹ (1738 (1818)), Foot

Ball Yd 1818, The Football - 1822[1], Football Garth 1820, 1826 (*v.* **garðr**; the alternation between *yard* and *garth* is noteworthy); The Four Acres 1806[1], - acre 1820, Four Acre Mdw 1826; Four Score Acres 1812, four score acres 1819, - Score acres 1833, Fourscore Acres 1830, 1834; Fourteen acres close called Pasture Land 1822[2]; Fox Cover 1833, The - 1859 (on bdy with Swinhope); Frimmer's House 1818 (*fryme waye* 1577-80, *frimars gate* e17, *frime gate, Fryme* - 1612, *Freemers* - 1664, *Fremers* - 1679[2]; the forms are too late to suggest a convincing etymology); Garth Cls 1784 (*v.* **garðr**); Messuage ... commonly called or known by the sign of the George 1763; Gilliatts Pingle 1822[1] (from the surn. *Gilliatt,* cf. William *Gylliatt* 1740 *BT,* with **pingel**); John Goodhands Platt 1782, - but now called ... Saintfoin Platt 1822[1], John Goodhands Platt 1814[2] (*Mr Goodhand pasture Close* e17, *John Goodhand* mentioned 1782, cf. *Mystres Goodhand*(*e*) 1577-80, John *Goodhand* 1614, *Mr Goodhand* 1612 *Terrier* and Saintfoin Cl *infra*); Grantham's Occupation Road 1806, Grantham's Platt 1820, 1826, 1859 (cf. *Grantham Lane* 1740[2], named from the family of Peter *Grantham* 1683 *BT,* John *Grantham* 1703 *ib,* Thomas *Grantham* 1804 *Tur*); Hackforths Pingle 1803 (named from the *Hackforth* family, cf. Violet *Hackforth* 1822[1] with **pingel**); the Hall Yd 1822[1] (cf. Binbrook Hall *supra*); the Heath Cl 1806[1] (*v.* **hæð**); Hill Cl, - Flatt 1833; the Holt Cl 1806[1] (*v.* **holt**); Home Cl 1820, 1826; East -, South -, West Home Walk 1833 (*v.* **walk** 'a sheep walk or pasture', frequent in this parish); Horse Cl 1833; Hund. Acres 1812 *LTR* (i.e. Hundred); Jocky Cl 1782, Tocking or Jockey - 1784 (cf. Adam Storeys Garths Closes *infra*); Kirmond Flat 1833 (- *Closes* 1740[2]), - Road 1806, 1814[2], 1822[1] ("the way of" *Keuermund* 1244, *kirmin waye* 1577-80, *Kermon gate* e17, *kirmon gat*(*t*)*e, kirmon waie* 1612, *Kirmundgayte* 1638[1], *Kirmund gate* 1664, *the two kirmund gates* 1634, *2 Kirmond gates* 1679[2], *the Lower Kirmund gate* 1634, *lower kirmund* - 1664, *the upper kirmund gate* 1634, *upper kirmo* - (sic) 1664; *middle kirmin waye* 1577-80, *kyrmane way* 1605, *mydell kyrmo* (sic) *gattes* 1605, *Midlekirmon gatte* 1612, 'the road to Kirmond le Mire' a neighbouring parish, *v.* **gata**); Long Cl 1782, 1814[2], 1822[1] (*Long Close end* e17, *the Long close* 1697[1], - *Close* 1709, 1727 (1818)); Longhill Plats 1782, - Platts 1814[2], The - 1822[1] (*The Longhill plats* 1727 (1814), *v.* **plat**[2] 'a plot, a small piece of land'); (the) Longhills 1822[1], 1835, 1836 (*longe hill* 1612, *Langhill* 1638[3], 1719, self-explanatory); Louth Road 1822[2] (1740[2]) (*louthe waie* 1577-80, *lowth gatte, kythorn lowth gate* 1612, *Louthgate Dive* 1658, *Louth gate* 1662[2], *louth* -, *lowth* - 1668, 'the road to Louth', *v.* **gata**; *kythorn* is difficult, but may reflect a compound of OE *cȳ,* nom.pl. of **cū**, and **þorn** 'a thorn bush', though it is probably recorded too late to be satisfactorily explained); The Manor Yd 1822[1], - Garth 1833 (*v.* **garðr**; again, the variation between *yard* and *garth* is noteworthy); Market Rasen Road 1806[2], 1814[2], 1822[1], the Rasen Road 1822[2], M. Rasen Road 1833 (self-explanatory); Marris's Occupation Road 1806 (cf. Thomas

Maris 1806); the Middle Walk 1822[1] (*v.* **walk**); Mill Cl 1833, - Mdw 1859 (on the northern boundary of the parish by "Water Mill", and cf. also Binbrook Mill *supra*); Newton Land Platt 1820, 1826 (perhaps for Newton Lane -, cf. Wold Newton Lane *infra*); North Cl 1820, 1826; North Hall Cls 1826 (cf. North Halls *supra*); One Acre 1833; the onset 1776 (*the Onsett* 1709, *that Close of pasture ... comonly called the Onsett* 1719, *that Close of Pasture* - 1743, possibly a reference to the infield-outfield system, the terms used in some northern counties being *Onset* and *Outset, v.* Adams 155; *onset* is common as an appellative in north L); Orford cl 1822[2] (cf. *Orford Grounds* 1740[1]), - Plantn 1833, a gate called Orford Gate 1806[2] (from Orford in Stainton le Vale *infra*); Osier Holt 1820; Far -, Near Paddock 1833; the Pall Cl 1806[1]; The Parsons Poll 1822[1]; the Pasture Cl 1776, 1804, 1840, Pasture Cls 1818, the - 1806, 1822[1] (*pasture Close* 1577-80, *the pasture close* 1601, 1612, *Pasture Close(s)* 1634, 1679[2], 1740, - *close* 1662, 1664, *pastor close* 1668, *The pasture Close* 1692[2], 1697[2], *the* - 1719, *the Pasture Close* 1709, *pasture close* 1717, - *Close* 1718, *the pasture Close* 1719, *the Pasture Closes* 1738 (1818), *the Pasture-Close* 1743 and cf. Binbrook Pasture *supra*); Pasture Platt, - Walk 1820, 1826; Plantation 1833; Pocket Cl 1818, 1822[1] (*v.* Beedham Cl *supra*); Potatoe Garth 1833 (*v.* **garðr**); the Prebend Cl 1838 (cf. *the prebend ground* 1577-80, *the* -, *le Prebend ground* 1638[3], *y*[e] *prebend* e17, *the prebend* 1634, *Prebend* 1740[2], *the prebend house* 1838[1] and *Prebendarium de Milton' et Bynebroke in ecclesia Lincolnien'* 1300 *DC, Milton Mannour Prebend Founded in the Cathedral Church of Lincoln* 1775, *that Manor Seignory or Lordship of the Prebend of Milton Manor cum Binbrooke* 1806[2], i.e. Milton in Oxfordshire, *v.* RA iii, 290); the Premian Pce 1834; the Quake Well Pingle (*v.* Wirewells Pingle *infra*); Reedham Cl 1818 (1738 (1818), and probably to be identified with *Redholm* 1244, *v.* **hrēod, holmr**); Sainfoin Cl 1833 (*v.* **sainfoin**, cf. John Goodhards Platt *supra*); Seed Walk 1820, 1826, Far -, Near - 1833 (*v.* **walk**); Seven Acres 1833; Sheephouse Cl 1776 (1740[2]), - Cls 1802 (*Sheep house close* 1697[1], - *house Close* 1719, - *House Close* 1740, 1747, *the Sheephouse Close* 1743); Shepherds House Plantn 1833; the Six Acre 1820, Six Acre Mdw 1826; South Cl 1820, 1826; the south and east fds 1776 (cf. the East Fd *supra*); the Spearing Hill 1834; Stonum cl 1822[2] (*stonam* 1577-80, 1605, 1634, 1679[2], *Stonam* 1668, 1697[2], *great* - 1717, *stoneham* 1634, 1638[2], *great* - 1718, *Stone ham* 1662[2], *Stonham Close* 1635, *Stoneham close* 1649, *stonam* - 1662[2], *stonum* 1692[2], *Stonams Close* 1717, *Stoneham* - 1727, 1733, 1737, 1739, *east stonam next y*[e] *bridge* 1606, *east Stonam* - 1718, *west stonam* 1606, 1718; the forms are late, but could represent **holmr** 'an island, a water-meadow, etc.' as frequently is the case in north L., the first being **stān** 'a stone, etc.'); Adam Storeys Garths Cls 1782, - Garth Cls 1814[2] (*v.* **garðr** and note: a Homestead in three closes and called by the names of Dove Coat Close, Grange Close and Jocky close and now called by the several names of Adam Storeys Garth Closes 1782, a Homestead formerly in three Closes called Dovecoat Close, Grange Close and Tocking or Jockey Close but then called Adam Storeys Garth Close 1784, that Homestead ... which was formerly in three closes called by the names of the dove cot Close Grange Close Tokey or Taking close (sic) 1822[1]); Adam Storey's Homestead 1814[2] (*Adam Storeys homestead* 1727 (1814)); Sutton Garden 1822[1] (from the Sutton family, cf. *Roberte Sutton* 1591 *Inv*); Swan Pool 1806[2], Swann pool 1822[1] (self-explanatory); Swinewell's Walk 1820, Swinewal's - 1826, Swinewall's - 1827 (*Swyndall* (sic) 1577-80, *swinewells* 1662, *v.* **walk**; *swinewells* is presumably self-explanatory); Ten Acres 1833, the - 1834, The - 1859; Thirty Acres 1833; Thorpe Corner 1820, 1826, Far -. Nr Thorpe Flatt 1833, Thorpe Lane 1833, Thorpe Plat 1838, Gt -, Lt Thorpe Walk 1820, 1826 (cf. *thorpe dale* 1679[2]; *thorpe fielde* 1577-80, *Thorpfield* 1638[1], *Thorpe feld side* 1612; *Thorp Long close* 1664, - *long close hedge* 1634, *Thorpe Closes* 1740[1]; *Thorp Meare* 1634, - *meare* 1664, y[e] *thorpe Meare* 1679[2] (*v.* **(ge)mǣre** 'a boundary'); *Thorpe stone* e17, *Thorp stones* 1634, *thorpe* - 1679[2]; - *stong* 1664 (*v.* **stong**), *Thorpe walk* e17 (*v.* **walk**); all named from Thorpe le Vale in Ludford LSR); Tup Paddock 1833; Turner's House 1818 (named from the *Turner* family, cf. *William Turner* 1650 *BT*); Twenty five Acres 1833; The Vicar of St Gabriel Road 1806 (cf. *Gabrill thing* 1612, *v.* **þing**, property belonging to the former church of Binbrook St Gabriel); the Warren of Binbrooke 1806[2] (*v.* **wareine**); Well Platt 1820, 1826 (*v.* **wella**, **plat**[2], cf. *ad Fontem* 1327, *atte Well* 1332 both (p)); the Wharth Cl 1806[1] (*v.* **waroð**); John Whites Platt 1782, 1822[1], - Platts 1814[2]; Wirewells Pingle 1782, 1784, 1814[2], - but now called the Quaker Well Pingle 1822[1] (*v.* **pingel**); Wold Newton Lane 1833 (*Newton Road* 1740[1], 1740[2] ('the road to Wold Newton (to the east of Binbrook)').

(b) *Abiegate, abeygate* 1577-80, *Abye* -, *Aby gatt* 1605, *Abbey gat(t)e* 1612, *the abeywaye, the abie way* 1577-80 ('the road or path to the abbey (i.e. Orford Priory in Stainton le Vale *infra*)', *v.* **abbaye, gata, weg**); *Aldeklif* 1244 FF (*v.* **ald, clif**); *le Ashgarth* 1638[3], *the Ashe garth side* 1658 (*v.* **æsc, garðr**); *Aungeres Cros* 1244 FF ('Aunger's cross', *v.* **cros**. The first el. is the ME pers.n. or surn. *Aunger,* from Norman *Ansger* (<West Frankish *Ansgêr* or *Ásgeirr*) or, when used as a byname, from the appellative ON *angr* 'grief', as Dr John Insley points out); *Axeldaile* e17, *Auxstelldale stithe* 1577-80, *Axilldale stithe* 1605, *Anxelldall sicke* 1612 ('Ansketil's portion, share of land', from the Norman pers.n. *Ansketil* > ON *Asketill* and **deill**, with **stigr** 'a path, a narrow road' and apparently **sík** 'a stream'); *Basset land* 1577-80 (from the surn. *Basset*); *bekfurhill* e17 (cf. *the becke* 1638[1]. *the beck that parts Kirmond and Stainton from Binbrook* 1740[2], *v.* **bekkr** 'a stream', and *Ourebeck'* 1327, *Ouerbeck* 1332, *at Bek de Bynbrook* 1387 Peace all (p)); *Beesby Grounds* 1740 (from the neighbouring Beesby in Hawerby cum Beesby LNR); *Belgarth* 1609, *Bellgarth* 1662, 1664, 1697[1], 1698, 1700; *Bell Close Lane* 1740[2] (either pieces of land, *v.* **garðr, clos(e)**, the rent of which was used for the upkeep of a bell in the church, or perhaps less likely from the surn. *Bell*, cf. John *Bell* 1720 *BT*); *Mr Bewley headland* e17, (from the surn. *Bewley* and **hēafod-land**, cf. Richard *Bewley* 1740 *EnclA*); *Bimundeker* c. 1237 *LOC*, *bymond Care* 1577-80 (Dr John Insley suggests that this is a compound of the proposition **bī** 'by' and a lost f.n. *Mundeker*, the latter being derived from the ON pers.n. **Mundi*, a regular hypocoristic form of ON *Asmundr*, *Ingimundr* etc. and **kjarr**, ME *ker* 'a bog, a marsh'); *Binbrook Cros* 1740 (*v.* **cros**); *Bloods* e17, *the Bloods* 1662[2], *ye bloods* 1668, *ye Bloods* 1679[2] (obscure); *blowe hill* 1577-80, *Blow hill gate* e17, *Blow-hill (gate)* 1634, *Blow hill gate* 1664, *Blowhill gate* 1679[2] (presumably self-explanatory and *v.* **gata** 'a road'); *Bolehowegate* 1244, *Bulla gate* e17, *bully waye(e)* 1577-80, *bolliegate*, *bolliegatt* 1612, *bully gate* 1662 (perhaps 'the bull mound', *v.* **bula, haugr**, with **gata** and for the development of *-howe* cf. *Langhou infra*)); *Bonkelffeld* 1554 Pat; *ye bottome* 1662 (*v.* **botm**); *Brakland* 1244, *bracke lande* 1577-80, *Bracklands* e17, *Brake lands* 1664 (perhaps 'land covered with brushwood or thickets', *v.* **bræc[1], land**); *bragdall* 1668; *Brandon Cross* e17, *Brandon Crosse* 1577-80, 1668, 1679[2], *brandon crosse* 1612, *Brandon crosse* 1662 (the first el. is presumably a surn. from Brandon, Kest, the second is **cros**); *breakdiss hill* 1577-80, *Brackdish -* e17, *breakditch -* 1612, *Bragdish -* 1662[2], *Bragdishhill* 1679[2]; *(y^e) Bulbanke* 1606, 1718 (*v.* **bula, banke**); *Bull Medow* 1718; *Burdell* 1577-80, *Bourdaile* e17, *Burdail* e17, *Burdall* 1612, *Burdale* 1634, *- Forlong* 1668, *- Furlong* 1679[2], *Birdale* 1664 (perhaps 'the share of land attached to a cottage, dwelling house' *v.* **bur[1]** (ME *bour*) and **deill**); *Burrell Raise* e17, *Burwell* 1664 (perhaps named from the *Burrell* family, cf. Mark *Burrill* 1786 *BT*); *ye Butt medowe* 1606 (*v.* **butte**); *Cok acre* e17, *Cocke Acres* 1658, *Cock-acre* 1664 (probably self-explanatory, *v.* **cocc[2], æcer**); *Cockfeild* 1641; *Cocking Close* 1697[1] (probably from a surn.); *coke myll, cocke myll dam* 1577-80, *cockmiln* 1612, *the Cocke mill, Cocke mill hill* 1662, *the Cocke milne* 1668, *cock Milne gate* 1668, *Cocke Milne* 1679[2]; *Collyn Hill* 1577-80, *Collinge hill* 1612, *Collin -* 1662[2], *Colling hill* 1668, *on Colling* (sic) 1679[2] (the first el. is probably ME *Colin*, a short form of *Nicholas*); *the comen feld* 1577-80, *the common field* 1668, *ye common Feild* 1679[2], *Common field* 1718; *the comon lane, - layne* 1577-80, *ye Common Lane* 1679[2]; *ye common Sewer* 1668, 1679[2]; *the Corne feild* 1648; *Counter pece inge* 1577-80, *counter peacelinge* (sic) 1612; *Cour fien rigge* (sic) 1662; *Crakedale* 1244 FF, *Crake daile* e17, *Crake dale* 1664 (*v.* **kráka** 'a crow, a raven', **deill**); *dandall stighe, - stythe* 1577-80, *Dandelesteght, dandeliesteght* (sic), *dundele sicke* (sic) 1612, *Dandal stight* 1662. *dandy stigh* 1668, *Dandale Stigh* 1679[2] (obscure); *Depedale* 1244, *deep daile* e17, *deepdale* 1634, 1679[2], *deep dale* 164 (*v.* **dēop, dalr**); *the*

Dinge 1612 (perhaps from **dynge** 'a dung-heap'); *Doghill* e17; *the Downehouse Garth* (sic) 1709 (*v.* **garðr**); *draughtes* 1577-80, 1612, *draught* 1612, *fare draught dall* 1577-80, *fare draghts* 1612, *Midledraughtes* 1612 (alluding to draught-roads, 'far', *v.* **feor**, and in the 'middle', *v.* **middel**, cf. Foredraught, PN Db 632, with **deill** in 1577-80); *the East end* e17 (*v.* **ēast, ende**[1], cf. *the North end, the South end, y*[e] *west End infra*); *east welles* 1577-80, *- wells* 1612, *East wells* 1662[2], *- wels* (sic) 1668, *Eastwells* 1679[2] (*v.* **ēast, wella**); *the Esk* e17, *Easke* (sic) 1662, *y*[e] *Eske* 1668, 1679[2] (from ON **eski** 'a place growing with ash-trees'); *the fallows, the Fallows, the Fallows end acre* e17, *the fallow feild* 1648; *in campo de Binnebroc* 1244, *in campis de Binnebroc* 1150-55 Dane, Hy2 (1314) Ch, *campum de Binbrok* 1276 RH, *in campis de bynbroke* 1570, *in ... campis ... de Binbrook* 1587, *in both the Fields of Binbrooke* 1690, *binbrooke Feild* 1703, *The Field* 1634, *campis de Binbrooke* 1676 (alluding to the open fields of Binbrook, evidently reduced to two by 1690); *Freforde feld* 1606; *the gate siche* 1662 (*v.* **gata, sīc**); *Gilbies thinge* 1641 (from the surn. *Gilby* with **þing**, cf. *a lande of M*[r] *Gylbies* 1577-80); *Golding thing* 1589, *Goulding -* 1638[2], *Mes(s)uagium vocat Goulding* 1676, *- Gouldinge thinge* 1679[1], *- Golding thing* 1680, *Golden farme* 1666 (from the *Goulding* family, whose collection of documents is deposited in LAO, with **þing**); *goodknapp thing* 1570 (from the surn. *Goodknapp* with **þing**); *goose close* 1577-80, *Goose wong* 1662, *a Pingle called Goose Toft* 1662, *goose Toft* 1668, *Goose toft* 1679[2] (*v.* **gōs, vangr, toft**); *Goseberdland* 1244 FF (from the ME pers.n. *Goseberd* (Continental Germanic *Gôzbert*) and **land**); *Gosehoudale* 1244 (*v.* **gōs, haugr, deill**); *Goulton Stight* 1634 (from an ancestor of Richard *Goulton* 1781 *Tur* and **stīg, stigr** 'a path'); *Hardie Sicke* 1612 (from the surn. *Hardy* and **sīk**); *Hatclif -, hatlif waie* 1577-80, *Hatcliffe gat(t)e* 1612, *- gate* 1679[2], *Hatt cliff gate* 1662[2], *hatliff gate* (sic) 1668 (*v.* **gata**), *Hatlif hill side* e17, *Hattecliffe thing* 1638[3] (*v.* **þing** 'property, possession' and no doubt the surn. *Hatcliffe*); *hauerdall furlonge, - hole, hauerdell waie* 1577-80, *Haverdalle, Haverdall gate, - hole* 1612, *Haverdayle* 1638[1], *hauerdale, - gate, - hole* 1662, 1668, *Haverdale* 1679[2], *haverdale gate, - mouth, haverdalehole* e17, *Haverdale hole* 1658 (*v.* **hafri** 'oats', **deill**, a Scand. compound); *Hauerhoc Klif* 1244 FF ('steep slope by *Haverhook*' the latter perhaps being 'the hook-land on which oats were grown', *v.* **hafri, hōc, clif**); *the hede land* 1577-80, *the headland* 1612, *headland & fellow* 1662[2] (*v.* **hēafod-land**; for *fellow*, cf. *two lands ... the headland & his fellow* in Barnetby le Wold f.ns. (a), PN L 2 12, where *fellow* is taken "perhaps in the sense 'a counterpart, a match'", *v.* NED s.v. 4b); *the High Close* 1709; *hole house* 1666; *houlto-stight* (sic) 1664 (probably an error for *Goulton supra*); *Hunters close* 1662 (presumably from the surn. *Hunter*, though the earliest reference so far noted is William *Hunter* 1842 White); *Hunybroughe* 1577-80, *hony brough* 1612, 1668, *Honybrough* 1634, *Honey-brough* 1664, *Hony Broughs* 1679[2] (probably 'the mound, hill where honey is found', v. **hunig, beorg** and for spellings in -*brough* cf. *Middle Bruf infra*); *Cotage called the house in the Field* 1657; *the house close* 1697[1]; *Ibrey Thinge* 1596 (from the surn. *Ibrey,* cf. William *ybr* a1537 *MiscDep 43,* Margaret *Ibrye* 1558 *Inv,* John *Ibre* 1562 *BT,* with **þing**); *Iuett crosse* 1577-80, *Iwetcroft* (*sic*) 1612 (from the ME fem. pers.n. *Ivet* (*v.* Reaney s.n. Ivatt) and **cros** or **croft**); *the kiln house* 1612; *ye Kings land* e17; *Kirmonde Crose* 1577-80, *kirman crose* 1605 (perhaps a boundary cross named from the neighbouring parish of Kirmond le Mire); *Kitchine, Kitchyn*(*e*) 1577-80, *Kitshen* (sic) e17, *in kichine* 1612, *in Kitchen* 1662, *ye kitching* 1668, - *Kitchin* 1679[2] (presumably from ME *kichen*(*e*)) in some undetermined sense); *Ladees Whong* e17, *lady wong* 1662 (*v.* **hlæfdige** 'a lady, Our Lady', **vangr**); *the Lamas Ground* 1709 (*v.* **lammas**); *Lamb-coats* 1664, *lam'* *cotes headland* 1668 (self-explanatory and *v.* **hēafod-land**); *langdale, Landall* *scoles, longdalliwaye* 1577-80, *long daile* e17, *longdall'* 1679[2], *langdall gate,* *Langdalscoles* 1612, *langdayle* 1638[1], *Langdale* 1664, 1679[2] (*v.* **lang, deill,** with **skáli** 'a temporary hut or shed' and **weg**); *langhou, Langhowgate grenes* 1244 ('the long mound or hill', *v.* **lang, haugr**, perhaps to be identified with this is *Lanley hill* (sic) 1577-80, *langa hill* 1668, *Langa Hill* 1679[2], with the second el. -*hou* reduced to -*a* (though the forms are late) as possibly in other names in this parish); *Langing*(*e*) 1718 (*v.* **lang, eng** 'meadow, pasture'); *Largadike* (sic) e17 (perhaps an error for *Lange-*, and so related to *Langhou supra*); *Lingam furlonge* (sic) 1577-80, *Ling a furlong* 1664, *ling^a forlong* 1668, *Linga furlonge* 1679[2] (if -*a* is a reduced form of **haugr**, this is comparable to *Langa-* above, and means 'the mound, hill where heather grows', *v.* **lyng, haugr** with **furlang**); *the linges* 1638[1] (*v.* **lyng**)*; Linghill bancke* 1638[1] ('the heather-covered hill', *v.* **lyng, hyll,** with **bankc**); *Littlehowendinges* 1244 ('the little mound or hill', *v.* **lytel, haugr,** with the pl. of **wending** 'a bend in a road or stream', perhaps developing in meaning to denote place(s) where there is a bend or bends, cf. *Wendinges* in East Halton f.ns. (b), PN L 2 158, *littel Dall* 1577-80, *little a dale* 1668, *Littleadale* 1679[2] perhaps belong here, if -*a* is a reduced form of **haugr**; *dale* is probably from **deill** 'a share, a portion of land'); *little thorne* 1577-80, *litle thorn* 1612, *little a thorne* 1664, *Little a thorne* 1668, *Litleathorne* 1679[2] (perhaps to be compared with the prec., though forms in -*a* are again late); *the lords close* 1612; *Louth bottome* e17 (*v.* **botm** and cf. Louth Rd *supra*); *Ludford waye* 1577-80, *ludfoord gatt* 1605, *Ludforth gat*(*t*)*e* 1612, *Ludford gate* e17, 1634, 1664, 1679[2], - *Road* 1740[1] ('the road to Ludford (a parish to the south of Binbrook)', *v.* **gata**); *Ludford knot* 1577-80, 1664, - *knott* 1679[2], *ludfoord knott* 1605, *Ludforth knot* 1612 ('hillock by Ludford', *v.* **knottr, cnotta**); *Lurbur -, lurbur slacke* 1577-80, *Lurborslak, lurbour* *slack* 1612, *Ludbrough slack* (sic) 1638[1], *Lurbur slack* 1662[2], *lorbor slack* 1668, *Lurbur Slacke* 1679[2] (*v.* **slakki** 'a shallow valley'; *Lurbur* is obscure); "the way called" *Ludburgstrete* 1244 ('the road to Ludborough (several miles to the east))', *v.* **strǣt**); *the mare bank* e17 ('the boundary bank', *v.* **(ge)mǣre, banke**); *Marklands* e17; *Mackrill headland* e17 (from the *Makerell* family, cf. Thomas *Makerell* 1558 *Inv*, and **hēafod-land**); *marshe thing* 1589, *Marshthing* 1614 (from the surn. *Marsh* and **þing**); *Massone rigge* 1577-80, *Mason Rigge* 1679^2 (*v.* **hrycg, hryggr** 'a ridge', possibly in the sense of 'a ridge or land in the open field'; the first el. is the surn. *Mason*); *Middle Bruf* e17, *midell broughe* 1605, *midlebrough* 1612, *Middle brough* 1634, 1664, *Middlebrough* 1679^2 ('the middle hill or mound', *v.* **middel, beorg**); *midle forlong* 1612, *middle furlong* 1634, 1664, *the middle furlong* 1662^2, *Middle Furlong* 1668, - *Furlonge* 1679^2 (*v.* **midel, furlang**); *the Low -, the Midle Mires* 1709 (*v.* **myrr**); *Nebills thinge* 1641 (named from the *Nevill* family, cf. John *de Neuill* 1250 FF, Thomas *Nevell'* a1537 *MiscDep 43*, Robert *Nebill* 1576 *MiD*, with **þing** 'premises, property'); *Nesse* 1327 (p) (*v.* **næss**); *Netherhowe* 1244 FF (*v.* **neoðera** 'lower', **haugr**); *Netherwik'* 1332 (p) (*v.* **neoðera, wic**, though this may not be a local surn.); *nettle-busk* e17 (*v.* **netle-bush** 'a nettle-bed, a nettle patch', the second el. being a Scandinavianized form; for an early example of this name *v.* *Nettelbustmar* in Stallingborough, PN L 2 277); *the North end* e17 (cf. *the East end supra*, *the South end*, *ye west End infra*); *Osmondall* 1577-80, *Osmandall* e17, *Osmendall* 1612, *Osmondale* 1662^2, 1679^2, *Osmadale (hill)* 1668 ('Osmond's portion of land', *v.* **deill**. The first el. is ME *Osmond* used as a pers.n. or surn. ME *Osmond* is derived from the OE pers.n. *Ōsmund*, Norman *Osmund* (itself of OE or Old Saxon origin) or an anglicised form of the corresponding ON *Ásmundr*); *the ox pasture* 1648; *pail Croft side* e17 ('the (park) fence croft', *v.* **pale, croft**, cf. foll.); *parcrofte (ende)* 1577-80, *Parke Crofte* 1609, *Parcroftend* 1612, *Parcrofte end, Parcroft hedg* 1662^2, *Parcroft, Par croft end* 1668, *Par Croft* 1679^2 (*v.* **parke, croft**, with **ende**^{1}); *cotag'm ... voc' patyson thing* 1589, *Cotag' ... voc Patysonthing* 1614 ('Patison's premises', from the surn. *Patison* and **þing**); *pease acre* 1577-80, *peaseackers* 1612, *pease acres* 1679^2, *Peaseacres* 1662^2, *peas(e) acres* 1668 (*v.* **pise, æcer**); *Pilkico, Pilkico banks* 1662. *pilkico, pilcico* 1668, *pilkicowe* 1679^2 (obscure); *the pitte* 1577-80, *the pyttes* 1605, *the Pittes* 1634, *ye Pitts* 1664, *ye pits* 1668 (*v.* **pytt**); *Tenement or Toft wasted commonly called Podhole* (sic) 1667 (*v.* **padde** 'toad', **hol**^{1}); *Poolelane* 1697^1; *preastlynge waye(e), prestlynge way* 1577-80, *Priskelin gate, prisk line gate, prisline gate (sic)* 1612, *priestlings gate* 1668 (*v.* **weg, gata**; *priestling* is on record as a contemptuous term for a priest, but the sense here is not clear, though there may be a reference to the prebend, cf. the Prebend Cl in (a) *supra*); *William Pulmerse Close* 1577-80; *the Rood* 1612 (perhaps 'the wayside cross', *v.* **rōd**^{1}); *the Round Close* 1709; *Royalty Walk* 1740^2 (perhaps cf. ye *Kings land supra*); *in Le Rye, in the Rye* 1327 (p) (*v.* **ryge**); *Messuage Farme of Land Meadow or Pasture ... called by the name of St John of Jerusalem* 1692^1 (presumably alluding to former ownership by the Knights Hospitallers of St John); *deep Salterdaile, Great*

Salterdail, Salterdaile dike e17, *Salterdale* 1664, *Saturdale* (sic) 1679² ('the salter's portion of land'. *v.* **saltere, deill**); *Sallter gate* e17 (*v.* **saltere, gata**); *Scabbepit dale* 1244 FF (*Scabbepit* seems clearly to be derived from ME *scab*(*be*) 'a scab, crust' and **pytt**, perhaps in the sense 'the pit with a rough surface', cf. the use of ME *scabi*(*e*) adj. MED s.v. (b), of branches etc. 'rough of the surface, scaly, scurfy'); *the Scoars* e17, *Scoares* (*bottome*) 1662², *ye Scoers* 1668, - *Scores* 1679² (this may well be from ON **skor, skora** 'a cut, a notch', said in EPN 2 126 to denote in ME 'a cut, a ditch'. MED gives as some meanings of the word 'a crack, a crevice', 'a track, a way', and 'a limit, a boundary'. Since the f.n. cannot be identified the sense here must be left open); *Scrapdall forlonge* 1577-80, *Scrapadalle hill* 1668 (obscure); *Scriver wells* e17; *Seinteleriwong* 1314 Ch ('St Hilary's garden', *v.* **vangr**); *le Sheephouse Close* 1638³; *Shefeildes Thing* 1692¹ (from the surn. *Sheffield*, cf. Jasper *Sheffield* a1537 *MiscDep 43*, with **þing**); *Short Goltho styghe* 1577-80, *shortgolth sike* (sic) 1612 (*v.* **stīg, sik**; perhaps *Goltho* is here a surn. from Goltho LSR); *Siluer pits* 1668, *Silverpitts* 1679² (the sense of *silver* is not clear); *Skell thing* 1662², *Shell thing* 1664 (from **þing** 'a possession, property' and the surn. *Shell/Skell*); *Slaphow Klif* 1244 FF (*v.* **haugr, clif**); *Daniel Sles gate, Daniel Slee gate, Daniel Slee foot* e17 (the *Daniel Slee* in question has not been identified independently, *v.* **gata, fōt**); *Smalling heads* 1662, 1668 (from **smæl** 'narrow' and **eng**); *smythes close* 1605 (from the *Smith* family, cf. John *Smith* 1612); *Southedall* 1577-80 (*v.* **sūð, deill**); *Sowter dall* 1605 (probably named from an ancestor of John *Sowter* 1823 BT, with **deill**); *the south End* e17 (cf. *the East end, the North end supra, ye west End infra*); *Sraye hannige* (sic) 1577-80; *Stainton worth close noocks* 1664 (from the adjacent parish of Stainton le Vale); *Stempson headland* e17 (from a surn. and **hēafod-land**); *Steynpittes* 1244, *the stone pittes* 1577-80, 1612, - *pyttes* 1605 (*v.* **steinn, stān, pytt**); *Streate* 1577-80, *the streat* 1605, *the street* 1612, *the Old Street* e17, *the Old Street* 1634, *ye old street* 1664, *ye ould street, owld* - 1668, *the old street* 1679² (*v.* **strǣt**); *vnum mesuagium ... voc Suddaby Thing* 1594, *vnum mesuagium voc Suddaby thinge* 1604, *mesuagium vocat Suddaby thing* 1616 (from the surn. *Suddaby* with **þing**); *Swallow gate* 1612; *great -, little Swallow, Swallow prifeild-Inges* 1648 (*pri-* is obscure, *v.* **feld, eng**; it was named from the family of John *Swallow* 1646); *Swinnop bottom* 1668, - *bottome* 1679² (*v.* **botm**), *Swynoppedall* 1577-80, *Swinopp dalle* 1612 (*v.* **deill**), *swynoppe meare* 1577-80, *Swinopp*(*e*) *meare* 1577-80, 1612, *Swinhope crook meare* 1662², *Swinhop meare banke* 1662², *Swinnop meere* 1668 (*v.* **(ge)mǣre** 'a boundary') (all alluding to Swinhope, the adjoining parish); *Swinnop gate* e17, *Swinehope Road* 1740¹, *Swinhop* - 1740², ('the road to Swinhope', *v.* **gata** and cf. Swinhope Hill *supra*); *swyll beardes* 1577-80, *Swill beards* e17, *swilbcardell* (sic) 1612 (Dr John Insley wonders whether this is from a nickname 'Swill beard'); *tabutt thorne* 1577-80, *Tanbut thorne* 1612, *Talbutt thorne hill* 1662², *Talbott Thorn* 1668, -

thorne 1679[2] (from the surn. *Talbot* with **þorn, hyll**); *little Close late John Taylors* 1709; *Claus voc tenters et Halifax* 1570, *unum clausuram vocat Tenters & hallifax* 1602, *clausum ... voc - 1609, two tenem[s] called tenters and Hallifax* 1662[1] (evidently two separately named closes were combined, the first being 'land on which cloth was stretched for finishing' (*v.* **tentour**), the second probably from the surn. *Halifax*); *Theifdail* e17, *Thief dale* 1634, *Theif - 1664* (apparently from **þēof** 'a thief' and **deill** (or perhaps **dalr**)); *Thistell Clyffe, thistill cliffe, thistle - 1577-80, Thistle' list* (sic), *Thistlelist* (sic) e17, *Thixill cliffe* (sic) 1612, *thistle cliff* 1662[2], - *Clift* 1668, *thistlecliffe* 1679[2] (*v.* **þistel, clif**); *thorn tree whong* e17 (*v.* **vangr**); *the thre furlonges* 1577-80; *Threhowes* 1244 ('the three (burial) mounds', *v.* **þrēo, haugr**); *toft dalle* 1577-80, *Toft dales* 1664, 1668, - *furlonge* 1679[2], *hither Toft dales* 1634, *Great Toft Dales* 1577-80, *great toft dale* 1605, *the great Toftdalls* 1612, *little Toft dalle* e17 (from **toft** 'a curtilage, a messuage' and **deill** 'a portion, a share of land', a Scand. compound); *towes waye* 1577-80, *Tow(e)s gate* e17, 1612, 1664, 1679[2], *Towesgate* 1638[1], *Towze gate* 1658, *Towes Road* 1740[2] ('the road to Great Tows (in Ludford LSR), *v.* **gate**); *Turfer rigge* 1577-80, *turfer rige* 1612; *two wyndinges* 1577-80, *ii windinges* 1605, *tow windinges* (sic) 1612 (*winding* in a topographical sense does not seem to be recorded in dictionaries; perhaps it is an **ing**[2] derivative of **(ge)wind**[2] 'something winding, a winding path, etc.', comparable to **wending** (cf. *Littlehowendinges supra*), perhaps denoting a place or places where there is a bend or bends); *the uley waye* 1577-80; *the Waine gate* 1612 ('the waggon road', *v.* **wægn, gata**); *Wormdall* (sic) 1612, *warme dayle* 1638[1], *Warmedale* 1662[2], *warmedale* 1668, *Warm dale* 1679[2], *Warmedale bottome* 1662[2], *warmdall ende* 1577-80, *Warmedale end* 1662[2], *warmedall hole* 1577-80, *Warrandall hole* (sic) e17, *Wamedall holle* (sic) 1612, *warmedall thorne* 1577-80, *Warmedallthorne* 1612 (perhaps 'the warm or sheltered portion of land', *v.* **wearm, deill**, with **botm, ende**[1], **hol**[1], **þorn**); *Wate gate* (sic) e17, *wayregatt* 1605 (probably 'the sluice', *v.* **wer, wǣr, geat**); *Water garth* 1662 ('the wet enclosure', *v.* **wæter, garðr**); *Wattam Closes* 1709 (from the family name, cf. *John Wattham* 1641 LPT); *wellstocke windinge* 1577-80 (probably identical with this are *Wail Stock* e17, *Wales stock* 1612, *wale stock* 1664, *walestocke* 1679[2], though the development is obscure. *Wellstocke* is presumably from **wella** 'a spring' and **stoc** 'a place' or **stocc** 'a stump'; for *windinge*, cf. *two wyndings supra*); *y[e] west End* e17 (cf. *the East end supra*); *in campo occident* 1570, 1594, *in campo occidental'* 1589, *in campo occiden'* 1614, *in campo occiden* 1616, *in occidentali Campo de Binbrooke* 1638[3], *in campo occidental* 1679[1], *in occiden campo de Binbrooke* 1680, *in occidental' campo de Binbrooke* 1722, *the west feld* 1577-80, *the West Feild* e17, *y[e] west feld* 1606, *the Westfeild* 1612, 1638[3], *le west Feild* 1638[2], *the Westfeild* 1638[3], *the West Field* 1657, *ye West - 1664*, 1718, *ye west - 1679[2]* (one of the great fields of Binbrook); *wharttmolles, -moldes, whartmoles* 1577-80, *Whartmoles* 1612, 1679[2],

Wartmole forlonge 1612, *the whart moulds, whart mowles, whart molls* 1668 (probably 'the transverse strips', from **þvert** and **molde** 'mould, soil', here in the sense of 'a land, a strip in the open field'); *whild hill side* 1662^2; *Whine gate* 1577-80 (perhaps an error for *Waine gate supra*); *Whitchroft* (*heuedland*) 1658 (*v.* **croft, hēafod-land**); *White hill* e17, *Whithill* 1612, *white hill* 1668, *White hill* 1679^2, *Whithill forlonge, whittelhill sid, wyght hill* - 1577-80, *Whitehill browe*, - *gatte* 1662^2, *white hill side*, - *forlong* 1668, *Whitthill waye* 1577-80, *whyght hill waie* 1577-80, *Whitehill gate* 1612 (*v.* **hwīt, hyll,** with **furlang, gata, weg**); *Lee Whitehouse* 1570, *Whitehouse Close* 1641, *the White House* 1657, *whitehouse* 1662, *White house* 1692^1 (self-explanatory); *wildes close* 1612 (presumably from the surn. *Wild*); *windle thinge* 1641 (from the surn. *Windle* and **þing** 'property, possession'); *the windmill waye* 1577-80, *the Wind Miln gate* 1612 (cf. perhaps Binbrook Mill *supra*); *woodgate* 1577-80, 1662, *Wudgat* 1668, *Woodgate* e17, 1679^2, *woode gate sid* 1577-80 (*v.* **wudu, gata**); *wrangdale* 1577-80, *Rangdall* e17 ('the crooked plot', *v.* **vrangr, deill**); *Wyeliam Closes* 1740^2 (no doubt for *Wyckham* -, i.e. Wykeham in Ludford LSR); *atte Yate* 1332, *ad Portam* 1327, both (p) (*v.* **geat**).

Claxby

CLAXBY
    æt Cleaxbyg 1066-68 (12) ASWills
    Clachesbi (7x) 1086 DB
    Claquesbey l11 (c.1331) *Spald i*
    Clakesbi 1150-60, 1155-60, 1160-66, 1186-1200 Dane, 1197 P
        (p), 1206-14 RA iv, -*by* 1190 (1301) Dugd vi, -*bia* c.1200
        RA ii, -*bya* R1 (1318) Ch, *Clacksbey* 1200 Dugd iii
    Clakebi 1190, 1191, 1192, 1195 P (p)
    Claxiby Hy2 (1409) Gilb
    Claxebi R1 (c.1331) *Spald i,* c.1200 RA iv, 1206 Cur, 1206
        Ass, 1212 Cur (p), 1212 Fees, 1218 Ass, Hy3 *HarlCh,*
        -*by* 1199 (1330) Ch, 1200 ChR, 1202, 1207, 1208 Cur,
        1210 FF, 1242-43 Fees *et freq* to 1343 NI, - *iuxt normanby*
        1584 *LindDep 29*
    Claxbi Hy2 (1409) Gilb, 1209 P (p), 1303, 1346 FA, -*by* 1250
        FF, 1275 RH, 1295 RSu, 1311 Ipm, 1327 *SR,* 1331 Ch,
        1332 *SR et freq* "by Normanby" 1402 Pat, 1453 Fine,
        iuxta Normanby 1474 *LCCA,* "by Rasen" 1414 Cl, -*bye* 1557
        Pat, 1576 Saxton, 1590 *Foster,* 1610 Speed, - *iuxta*
        normanbye 1597 *LindDep 29,* -*bie* "near" *Normanbie* 1558

InstBen, 1576 LER, 1629 *Foster,* - *iuxt normanbie* 1591
LindDep 29
Glacseby 1203 Ass
Clexebi 1204 P, *Clexby* 1305 Pat

'Klak's farmstead, village', from the ODan pers.n. *Klak* (*v.*
Feilitzen 305, SPNLY 172-73) and **by**, identical with Claxby and
Claxby Pluckacre LSR. The same pers.n. occurs in *Claxhow* in
Great Limber, PN L 2 227, and is recorded independently in L six
times in DB. Although this name is only sparsely attested in
Scandinavia itself, it is not so infrequent in English sources. *Clac*
appears as the name of a moneyer active from the time of Athel-
stan to that of Eadwig, and there are several examples in the 10th
century Peterborough list of sureties (ASCharters, no. 40), such as
Clac on Castre (Caster Nth) and *Clac æt Byrnewillan* (Barnwell All
Saints Nth). There is no need to assume with Fellows-Jensen,
SSNEM 41, that the first el. is an unrecorded OE appellative **clacc*
'a hill, a peak', since the pers.n. occurs in L and elsewhere in
England in independent use, while **clacc* is hardly topographically
appropriate here. It is described as near Normanby le Wold and
Rasen.

ACRE HOUSE COVER (lost, TF 115965), 1830 Gre, named from
Acre Ho in Normanby le Wold *infra.* CHURCH HILL, cf. *Church
and Church Yard* 1847 *TA.* CLAXBY MINES. According to 1842
White "Some years ago, great efforts were made to get coal here,
but the quality was so inferior, that the speculation was abandoned".
CLAXBY MOOR, 1794 *Dixon,* 1824 O, 1828 Bry, 1830 Gre, *inter
moram* e13 *HarlCh, in* - eHy3, Hy3 *ib, mora de Claxby* 1358 AD,
Claxby More Hy7 Lanc, *ye moore* 1674 *Terrier,* - *Moor* 1724 *ib,
Claxby* ... *moores* 1679 *MiD,* cf. *le Morewell'* Hy3 *HarlCh* (*v.* **wella**),
Common moore, the south more e17 *AddRoll 37691, South more*
1717 *LPE,* - *Moor* 1794 *Dixon,* 1832 *Yarb,* 1847 *TA, the out Moor*
1717 *LPE,* self-explanatory, *v.* **mōr**[1]. CLAXBY PARK. CLAXBY
PLATTS is *Furze* 1828 Bry. For *Platts v.* **plat**[2] 'a plot, a small
piece of ground', as elsewhere in this parish. CLAXBY STEW
PONDS. CLAXBY WOOD, 1824 O, 1828 Bry, *The parish ... has a
noted fox-cover, called Claxby Wood* 1842 White, cf. *Wdedaile*
1150-60 Dane (*v.* **deill** 'a share of land'), *wood close* e17 *AddR*

37691, 1601 *Terrier,* 1717 *LPE, wode close* 1625, 1638 *ib, Wood Close* 1832 *Yarb,* 1847 *TA, wode end* 1625 *Terrier,* the Wood is in Normanby le Wold parish. FOX COVER, 1824 O; this is in the extreme north-west corner of the parish on the opposite side from Croxby Wood. THE GRANGE, apparently an example of the later dial. use of **grange** 'a homestead, small mansion or farm-house, esp. one standing by itself remote from others', a sense quoted from L, *v.* EDD s.v. 2. HEATHFIELD. LANGHAM FM, cf. *a Close called Langham* 1717 *LPE, Langholmes* 1779 *Yarb, Langholme Close* 1794 *Dixon, Langham Close* 1768 *Yarb,* 1832 *ib,* 1847 *TA, v.* **lang, holmr** 'an island of land, a water-meadow, etc.'. THE RECTORY, 1842 White, *the parsonage mansion house* 1625 *Terrier, The Parsonage House* 1762 *ib.* TERRACE HOUSES, self-explanatory and note "In the village ... two plain, townish brick terraces of cottages, built for iron workers in 1868" P&H 216. WASHDYKE BRIDGE, on the boundary with Walesby.

Field-Names

Forms dated 1150-60 are Dane; those dated e13, m13, l13, c.1250, eHy3, Hy3 are *HarlCh,* 1327, 1332 are *SR;* 1589 are *LindDep 29;* 1590, 1627, 1629, 1635, and 1636 are *Foster;* e17 are *AddR;* 1601, 1625, 1638, 1664, 1671, 1674, 1697, 1700, 1703, 1707, 1724, and 1762 are *Terrier;* 1587, 1768, 1774, 1779, 1831, 1832, and c.1837 are *Yarb;* 1717 are *LPE;* 1794 are *Dixon,* and 1847 are *TA.*

(a) Acrehouse Cl 1832, Part of Acrehouse Cl 1847 (*Acard house -, - howse close, - growndes* (sic) e17, from Acre Ho in Normanby le Wold parish *infra*); Ammingham Hill, - Plot 1768, Amringham Hill (*sic*) 1794, Hammingham Hill 1832, Part of Hammingham Hill(s) 1847 (*Emingham Close* 1717, the first el. being presumably a surn.); Ash Holt 1847 (*v.* **æsc, holt,** cf. Pingle *infra*); Banks Cl alias Kew Cl 1762 (*Banks Close* 1697, 1700, 1707, 1724, *Bank Close* 1703, named frolm the **.***Banks* family, cf. Martin *Bankes* 1638 *BT* and for the alternative name *v.* Cue Cl *infra*); Black Lane 1832; Blacksmith's Shop and Paddock 1847; Bondale Nook 1768, Bondle Nook Cl 1832 (cf. *Baundedaile* eHy3, *Boundedaile* Hy3, *boundaile, bondaile* 1590, *boundaile close* e17, *Bownedaile* 1601, *Bowndaile* 1625, *Bowndall* 1638, 'the bean allotment', *v.* **baun, deill,** a Scand. compound); Bratt Cl 1768, 1794, 1832, 1847, Gt - 1768, 1832 (*v.* **brot** 'a small piece of land'); a Close called also Brathing 1762, Breathing (Cl) 1768, - Cls 1794, 1832, - Cl 1847 (cf. *brathinge, Brathinge Furlonge, middle brathinge Furlonge, brathing close* e17, *A close commonly called by ye name of Brathin* 1664, *Brathin* 1671, *Brathing* 1674, - *Closes* 1697, *One close called upper Brading, lower Brading Close* 1700, *lower -, upper Brathing* 1703, *Great -, Little Braithing* 1717, *ye upper Brathing* 1724 *a close called Brathing* 1707), the lower -, the upper Brathing Glebe Cl 1762 (*lower -, upper Brading Gleab Close* 1700, *the higher -, the lower Brathing Glebe Close* 1703, *the lower -, the upper Brathing Gleab close* 1707, *One Close commonly called ye Upper Brathing Gleab* 1724, *ye Lower Brathing Gleab close* 1724), Brathing Lane 1762 (1707), Breathing Lane 1832 (1717, *Brathin Lane* 1664, *Brathing -* 1707), Breathing Mdw 1768, - Mead 1794 (perhaps 'the broad meadow', a Scand. compound from **breiðr** and **eng**, with **clos(e), furlang, glebe, lane**); The Brett Cl 1768, Brett Cl 1832, 1847, Brett's - 1847 (named from the *Brett* family, cf. Peter *Brett* 1639 *BT*); Brick kiln Cl 1768, - cl 1794, Brick Kiln Cl 1832, 1847, Brick Yd &c. 1847; Buildings, Fold Yd & Stack Yd 1847; Butt Cl 1832; Caverley Cl, - Mdw 1768, 1832, 1847, Calvelrey Mdw (*sic*) 1794 (named from the *Caverley* family, cf. Robert *Caverley* 1609 *BT*); Chapel 1847; Claxby Hall 1768, 1779; Claxby Paddock 1832; Clay cl 1794, Clay Hills 1768, Clay Hill Cl 1832, 1847 (cf. *Claye hill' Furlonge* e17, *Clay hill plat* 1717, *v.* **clæg, hyll, furlang, plat**^{2}); Cliff Cl (*a close called the Cliff* 1697, *Clife Close* 1717 and cf. *Common Cliffe, the Common Clife* 1717, *v.* **clif** and cf. North Cliffe and South Cliff Pingle *infra*); Corn Cl 1768, 1832, 1847 (1700, *ye corn close* 1671, *Corne close* 1717, *Corn Close or North Gleab Close* 1700, *the North Gleab Close sometimes called the corn close* 1697, cf. the North Glebe Cl *infra*); Cottage Moor cls, - Cls 1768, Cottage Moor (Cl) 1794, 1832, - Moor Platts 1832 (*v.* **plat**^{2}), Cottagers Moor 1832, Part of Cottager's - 1847 (occupied by the cottagers); Cowpasture 1847 (cf. *Cowclose* 1717); The Croft 1768, Croft 1832, 1847 (*v.* **croft**); Crow Wd 1768, 1794; Cue Cl 1768, 1794, 1832, 1847 (named from the *Kew* or *Cue* family, cf. Marmaduke *Kue* 1706 *BT* and cf. Banks Cl *supra*); (Lt) Deepdale Cl 1768, Lt Deep dale 1832, Deep dale Cl 1794, 1832, Little deep dale - 1832, Deepdales, (Lt) Deep Dale 1847 (*vnder deepe daile* e17, *deepdaile* 1601, *deep daile* 1625, *deepdall* 1638, *v.* **dēop, dalr**, with some forms from **deill**; it was on the boundary with Normanby le Wold, *v.* Far -, Nr Deepdale Cliff in Normanby f.ns. (a)); The Demesne cl 1768; Dovebeck Cl 1768, Dove Beck Cl 1794, Dove Beck, Dovebeck Cl 1832 (*Dowebeck* 1601, *Dowbeck* 1625, self-explanatory, *v.* **dūfe, bekkr**); Nether -, Upr Dovecoat Cl 1768, the Dovecoat Cl 1774, the Dove cote Cl 1779, Upr Dove cot Cl 1794, Nether -, Upr Dovecote 1832, (Nether -, Upr) Dove Cote Cl 1847 (*Dovecoat Close* 1697, *ye upper Dovecoat Clos* 1700, *the upper Dovecoat close* 1707, *ye Dovecoat close* 1724, *v.* **douve-cote**); Duck Dykes 1768, Duck dyke Cl 1794, Duck dike 1832, Part of Duck Dike (Cl) 1847 (presumably self-explanatory, *v.* **dūcc, dīk**); Ellis Plat 1762, 1847 (1700, 1703, 1707, - *plat* 1697, - *Plat* 1717, - *Platt* 1724), Far Ellis plott 1768, Ellis Plot 1832 (named from the *Ellis* family, cf.

Thomas *Ellis* 1696 *BT*, Richard *Ellis* 1768); Elvingarth 1768, Elving Garth 1794, Elvin - 1832, 1847 (*Eluin Garth* 1717, v. **garðr** 'an enclosure', as elsewhere in this parish); Far moor Plott 1768, Part of Far Moor 1847 (cf. *Furmer hill' Furlonge* e17, *Furmore hill* 1601, *furmore* - 1625, perhaps from ME *foremer*(*e*) etc., in the sense given by MED s.v. 2 (a) 'one who makes or shapes (tiles)' or (3) 'one who informs, a teacher', here probably a surn. Far Moor would then be due to popular etymology); Foster Cl 1768, 1794, Fosters - 1832, Foster Cl and Buildings 1847 (named from the *Foster* family, cf. Thomas *Foster* 1604 *BT*, 1620 *Inv*); Garden 1847; Gawber (House and) Cl 1768, Gawber Cl 1794, 1832, Gauber - 1847 (probably from the surn. *Gawber*); yᵉ Hale Cls, Hales cl 1768, - Cl 1847 (v. **halh** or the derived surn. *Hale*); (House, Buildings and) Home Cl 1847 (*the Home Close* 1697, 1703, 1707); Howsor Hill Cl 1768, Howser Hill 1832, 1847, Hawser - c.1837 (probably from the surn. *Houser*); Hundich Leys 1768, Hundrick Ley (sic) 1832 (cf. *hundle lea close, hundle leaze* 1590, *Close called Hundeleay* e17, *hundaile leaze* 1601, *hundell leas* 1625, *Hundale lea Close* 1636, *Hundell leas* 1638, perhaps from the ON pers.n. *Hundólfr* and **lēah** 'a glade', though the forms are late); Intake 1794, Intack 1832, Part of Intake 1847 (v. **inntak** 'a piece of land taken in and enclosed'); a Close called Kirking 1762, Kirk Ing Cl 1768, Kirk Ings Cl 1832, 1847 (*a Close called Kerking* 1697, *Kirking Close* 1703, *a Close called Kirking* 1707, - *called* yᵉ *Kirking* 1724, *Kirkinmoat Close* (sic) 1717, v. **kirkja, eng** 'meadow, pasture', a Scand. compound); Lane 1847; Gt Laram Plott, Laram Cl 1768, Laram Cl (1636), - Plot 1794, 1832, Saram plat (*sic*) c.1837, - Plat 1847 (*Laram Furlonge* e17, *illud Clausum prati vel pasture ... vocat Laram* 1627, *Larum* 1625, 1707, *Laram Close* 1636, *Great Larum* 1717, perhaps from ME *larum*, a shortened form of *alarm* (cf. MED s.v. *larum-belle*), though the significance is unclear); the Lower Cl 1774; Marshalls Road 1832 (named from the surname of Benjamin *Marshall* 1832, or one of his ancestors, cf. Richard *Marshall* 1639 *BT*); (Gt -, Lt) Meadow Cliff 1832, - Cliffe 1847 (cf. Cliff Cl *supra* and the same name in Normanby le Wold f.ns. (a)); The Meat Cl (sic) 1768, Moat Cl 1794, 1832; Middle Plat 1762 (1707, *a Close called* - 1697, *Battley Middle plat* 1700, *Thomas Thixtons middle plat* 1703, *Middle Plat* 1717, v. **plat²** 'a small plot of ground', as elsewhere in the parish; the references are to the same piece of land; *Battley* is presumably a surn., and cf. *thos. Thixtons rushy Cliffe* in (b) *infra*); Middle Wood 1794; Milkendale Plott 1768, Milkin dale Plot 1832, - plat c.1837, Milking Dale Plat 1847 (v. **plat²**); Mill Fd c.1837, 1847 (cf. *molendina de Claxebi* Hy3, v. **myln**); Moor Cl 1832, 1847 (*more close, Brats more Close* 1717), Moor intake 1768, - Intack 1832, - Intake 1847, Part of - 1847 (v. **inntak**), Moor New Pce 1832, 1847, New Moor Pce, Moor Road 1832 (all referring to Claxby Moor *supra*); North Cliffe 1794 (yᵉ *northe clyffe* 1589, yᵉ *north Clyffe* 1590, *Northcliffe* 1664, *North Clyffe meadow, the lower north Clyffe* e17, cf. Cliff Cl *supra*); the little North Cl 1831 (*the North Close* 1703); the North Glebe Cl 1762 (*the North Gleab Close* 1707, cf. Corn Cl *supra* and y^e *Home Gleab close* 1724); North Ing Mdw, - cl 1768, North Ings 1794, 1832, - Plot 1832, (Part of) North Ings, North Ing Plat 1847 (*Northinge, northinge close hedge* e17, *Northing* 1601, *Northings* 1625, *northinge* 1638, *Great Northing* 1717, *v.* **norð, eng;** the same name, presumably referring to the same meadow, occurs in Normanby le Wold f.ns. (b) *infra*); Old Pleasure Grds 1832; (Paddock and) Orchard 1847; Owersby Road 1832; Ozier Holt 1832; Paddock 1847; the Parsonage Home Cl 1762 (*the personage grounde, the personage grownd of Normanbie* (i.e. Normanby le Wold), *the personage Land* e17, *the parsonaige, the parsonaige land* 1601, cf. The Rectory *supra*); Pingle, or Ash Holt 1794 (*Pingle* 1717, *v.* **pingel** and Ash Holt *supra*); Plantation 1847; Plats (Wd) 1847 (*v.* **plat²**); Pond dale Cl c.1837, - Dale Cl 1847; Rasen Road 1832 (*Rasen road* 1703); Railway 1847; Riseholme Cl 1847 (perhaps from a surn.); Rushby Cliff (sic) 1762 (*Rushy Cliffe* 1707, 1724, *v.* **risc, -ig, clif**); Rye Cl 1768 (*v.* **ryge**); Ryeholmes Cl 1768, Rye Holme Cl 1832, Gt Rye Holm 1794, Gt Rye Holme 1832, 1847 (*v.* **ryge, holmr**, and perhaps to be identified with *Rime Close* 1717, though the reading is uncertain); Saram plat (*sic*) (evidently an error for *Laram*, *v.* Laram Cl *supra*); School House and Yard 1847; Far -, Nr Snapedale Cl 1768, Far -, Nr snape dale 1794, Far -, Nr - 1832, Nr Snape Dale 1847 (*Snap* 1150-60, *v.* **snæp** 'a boggy piece of land', **dalr;** the same name, referring to the same feature, occurs in Normanby le Wold f.ns. (a) *infra*); South Cliff Pingle 1768 (*v.* **pingel** 'a small enclosure', as elsewhere in the parish, and for the same name *v.* the South Cliff in Normanby le Wolds f.ns. (a)); Stockhill Cl 1768, 1832, Stock hill cl 1794, Stock Hill 1847 (*stockhill', Stockhill' Furlonge* e17, *v.* **stocc** 'a tree-stump', **hyll, furlang**); Stonebridge Lane, Lt Stonebridg Cl 1768, Lt Stone bridge cl, - bridge Lane cl 1794, Lt Stone Bridge, Stone Bridge Lane 1847 (1717 and cf. Gt Stonebridge Cl in Normanby le Wold f.ns. (a) *infra*); Syzer cl 1768, Siser cl 1794, Sizer Cl 1832, 1847 (named from the *Siser* family, cf. Richard *Siser* 1641 LPR); Three Cornered pce 1832, - Corner Pce 1847; Tonder Plott Cl 1768, Toder plot - (*sic*) 1794, Tonder Plot 1832 (named from the *Tander* family, cf. William *Tannder* 1623 BT); the great Town End Cl 1762, Townend Cl 1768, Town end Cl 1832, - End Cl 1847 (*Towne end furlonge* e17, *the great Townend Close* 1697, *the greater Townend Close* 1703, *the great Town end Close* 1707, y^e *great Townend Close* 1724); Town Street 1832; Tuphills 1832, Tup Hills c.1837, 1847; a Close called Walmsgate 1762, Wormsgate Cl (sic) 1768, 1832, Walmsgate c.1837, Walmsgate 1847 (*Wormsgate* 1717; Walmsgate and Wormsgate presumably refer to the same feature, but it is impossible to suggest an etymology); Water Cl 1768, 1832; West Cl 1768, 1832, 1847 (1717); Wheelwright's Shop and Yard 1847; Wilfursyke Cl 1768, Welfur Sike 1832, Wilfur - 1832, - Syke 1847 (*Willfer Sike* 1717); Willow Row Cl 1768, 1794,

Willow row Cl 1779, 1832 (*willow Row close yeat* 1601, *Willow row* 1717); Woodnook Cl 1768, Wood Nook Cl 1832, 1847 (cf. Claxby Wood *supra*); Yard and Garden 1847.

(b) *abbie hill* 1587 (*v.* **hyll**); *Aikeholm* Hy3 (from the ON pers.n. *Eykr* (*v.* SPNLY 77) and **holmr**, or alternatively ON *eikiholmr* 'the raised land in marsh, etc., where oak-trees grow', both being Scand. compounds); *selionem Alduse* Hy3 ('Aldus's selion, strip of land in the open field' from the ME fem. pers.n. *Aldus*, a characteristically north-east Midland formation (*v.* Reaney s.n. *Aldis* and PN YW 5 45) and presumably **land**; for the same name *v. toftum Alduse* in the neighbouring parish of Normanby le Wold f.ns. (b) *infra*); *the Back Yard* 1717; *Bacock Close* 1717; *bare acre* 1601 (*v.* **bere**, **æcer**); *Barley Cliff* 1700; *Billcliffes Brathing* 1724 (cf. a Close called also Brathing in (a) *supra*; *Billcliffes* is probably from a surn., cf. Joseph *Bilcliffe* (curate) 1739 *BT*); *bow-acre* 1625, *bow acre* 1638 (*v.* **boga** 'a bow', 'something curved or bent', **æcer**, the same n. occurring in f.ns. (b) of the neighbouring parish of Normanby le Wold *infra*); *Bramebusche* 1150-60, *Brambusc* e13, *Brambusc* e13 *CottCh*, *le Brambusk* Hy3, *Brangbusk*, *brangbusc* (sic), *brambusc* Hy3 (Ed1) *Newh*, *Brambusk* m13 ('the broom bush', *v.* **bröm**, **busc**, most of the forms of the second el. are from Scand. **buskr**); *Brandreth Daile* e17, *Brandred daile* 1601, *Brandred dale* 1625 (this is almost certainly from a local family and **deill** 'a share, a portion of land'. *Brandreth* has not been noted so far in the sources searched, but occurs for example in the neighbouring parish of Kirkby cum Osgodby, cf. ... *Brandrethe* 1543 *Inv* 12/186; the forename is illegible. This same name, referring to the same piece of land, occurs twice in f.ns. (b) of the neighbouring plarish of Normanby le Wold); *brathorne Close* e17 (perhaps the same as a Close called also Brathing in (a) *supra*); *breythemar* ?1227 Hy3 (Ed1) *Newh* ('broad boundary', *v.* **breiðr**, **(ge)mære**); *Brotherdik* l13 (the first el. may be the ODan pers.n. *Brothir*, though ME *brother* 'a member of a religious order, one who has become an associate or benefactor of a religious house' is perhaps more likely; the second el. is **dik**); *bull meadowe*, *the bull' meadowe* e17; *le Buskes* Hy3 ('the bushes', from ON **buskr** or a Scandinavianized form of OE **busc**); *in campis de Clakesbi* 1150-60 Dane, *In campis -* 1160-66 ib, *in campo de Claxebi* e13, *in campis de Claxeby* e13 (13) *Alv*, *- de Claxebi*, *- de claxebi* e13 *CottCh*, *- de clackesbi* e13, *- de Clakesbi* 1206-14 RA iv, *in Campis de -* Hy3, *in campis de clakesbi* m13, *Claxby feilde* 1587 (the open fields of the village, *v.* **feld**); *uia qua itur ad Castriam* 1150-60, *Caister -*, *Caiester gate* e17, *the Caister Road* 1703 (self-explanatory); *semita de Carewell'* e13 (*v.* further *Karewelbec* in Normanby le Wold f.ns. (b)); *Carfe peec* 1601, *Calf peece* 1625, 1638; *uiam de Carlehou* e13 (from the ON pers.n. *Karli* rather than the gen.pl. of **karl**, and **haugr** 'a (burial) mound, a hill', a Scand. compound); *the Cocher Plat, Coucher* - 1717 (from dial. *cotcher* 'a cottager' and **plat**[2]); *the commons* 1601, 1625, *the Commonds* 1638, *the common* 1700; *Connie thorne* e17, *Conny thorne* 1601, *Cony thorne* 1625 (*v.* **coni** 'a rabbit', **þorn**); *the Corner Platt* 1717 (*v.* **plat**[2]); *Crumdic* c.1200 RA iv ('crooked ditch or dike', *v.* **crumb, dīc, dīk**); *a street called Davie Layne* e17, *Davie lane* 1629 (named from the *Davie* family, cf. Marie *Davie* 1620 *BT*); *the Firr closes* 1717; *le Folledayle* Hy3 (perhaps from the ME byname or surn. *Folle,* Reaney s.n. *Foll,* and **deill**); *four Garth* (sic) 1717 (*v.* **garðr** 'an enclosure'); *Froske hill' furlonge* e17, *Frosk hill* 1601, 1625, *Frorsk hill* 1601, *frosk hill* 1638 (*v.* **frosc** 'frog', **hyll, furlang**); *Gaithou* e13 (perhaps 'hill where goats are found', from **geit** and **haugr,** a Scand. compound); *gose pingle in the North Feild of Claxbie, gosepingle Close* e17 (*v.* **gōs** 'a goose', **pingel, clos(e)**); *Gosewelle furlang* c.1200 RA iv (*v.* **gōs, wella, furlang**); *Great Ganbos* (sic) 1717; *hallemilne* e12 (*v.* **hall, myln**), *myllyn' de Claxby* 1503 *Monson, Corn Mill, Mill Close Bratts* 1717 (*v.* **brot**); *hally day hill* 1601, *halliday hill* 1625, 1638; *Hargrave Yard* 1717 (no doubt from the surn. *Hargrave*); *Longe Harpe Furlonge* e17 (possibly named from its shape, *v.* **lang**[1], **hearp, furlang**); *Hengdaile* 1150-60 (*v.* **eng** 'a meadow, pasture', **deill,** a Scand compound; forms with *H-* are fairly common in names in **eng**); *Hill Close* 1717; *pratum Hospitalariorum* m13 ('meadow of the Knights Hospitallers', cf. *Prior Hospitalis tenet di. f. in eiusdem villis* (i.e. Claxby) *et Templi quondam tenuit* 1346 FA); *hou* e13 *CottCh, Howe Furrs Furlonge* e17, *Hou dales* 1717 (*v.* **haugr**); *howestag* Hy3 (Ed1) *Newh* (the first el. may well be the ME pers.n. *Howe* (*Hugh*), the second being obscure, though there is an ON *tág* 'rootstock'); *Huwerhau* eHy3, *Huuerhou* Hy3, *Vuerhoubeck* 113 (*v.* **uferra** 'upper, higher', **haugr,** with **bekkr** and cf. *Netherhougate infra*); *Joneys close* 1717; *the Kinges hye waie, the Kinges street* e17; *Laming Platt* 1717 (presumably from the surn. *Laming* and **plat**[2]); *Lankynne thing* 1590 (probably from the surn. *Lambkin,* with **þing** 'property, possession'); *Lintoft* eHy3, Hy3, *Lynetoftes Furlonge* e17 ('plots on which flax was grown', *v.* **lin, toft**); *Litlebec* c.1200 RA iv (*v.* **lytel, litill, bekkr,** perhaps a Scand. compound); *Little Close* 1717; *the little holes Closes* 1717; *Longe Furrowes* e17, *long Fowres* 1601, *longe fowrwes* (sic) 1625, - *fowrowes* 1638 (*v.* **lang**[1], **furh** in the pl.); *Long Garth* 1717 (*v.* **garðr**); *Claxbie Lowe fielde,* - *north Lowe field* 1601, *the north lowe feild, Southe Lowe feild of Claxbie* e17, *the North* -, *the South Lowe feild* 1635 (one of the open fields of the village); *the low gate* 1625 (*v.* **gata**); *the lower Furlonge* e17 (cf. *the vpper Furlonge infra*); *Mar* c.1200 RA iv (*v.* **(ge)mǽre** 'a boundary'); *Marfurlang* c.1200 RA iv ('the furlong by the boundary', *v.* **(ge)mǽre, furlang**); *Middle Furlong, middle Furlonge aboue the gate* e17; *misham oxgange* 1590 (*v.* **ox-gang**; *misham* may be a surn.); *morriley thinge* 1590 (*v.* **þing,** the property of the local *Morriley* family, cf. Alan *morele* 1541 *Inv,* Oliver *Morriley de Kyngarby* 1584 *LindDep 29/1,* Hamond *Morriley* 1592 *Inv*); *le*

Netegate Hy3 ('the cattle road', v. **nēat, gata**); *Nether Furlonge* e17 (v. **neoðera, furlang**); *Netherhougate* 113 (v. **gata**), *Nether How furrhill' Furlonge* e17 (v. **neoðera** 'lower' with **haugr, furlang,** cf. *Huwerhau supra*); *in aquilonali campo* Hy3, *in campo borial' de Claxby* Hy3, - *bor' de Claxby* Hy3 (Ed1) *Newh,* y^e *northe feld of Claxbye* 1590, *Northe feild* 1601, *the Northfield* 1625, *the North feilde* 1638 (one of the open fields of the village); *the North Gleab close* 1707, y^e - 1724 (cf. Corn Cl in (a) *supra*); *Opeland* c.1200 RA iv (probably 'the open land', from ME *ope, Oppe* etc. < *open* adj., and **land**); *Le oping* c.1250, (*del*) *hoping* e13 ('the hop garden', v. **hopping**); *pece* ... *called over measure* e17; *the Oxgange* e17 (v. **ox-gang**); *Paid hole Furlonge* e17 (v. **padc** 'a toad', **hol**[1]**, furlang**); *Peart Garth* 1717 (presumably from a surn. and **garðr**); *persone pittes* e17 (v. **persone** 'a parson', **pytt,** cf. the *Parsonage Home* Cl in (a), *supra*); *Pinkenhou* 1150-60, *Pinkenhou* e13 (the same name appears as *Pinkenhou,* etc. in Nettleton f.ns. (b), PN L **2** 247, and should, no doubt, as Dr John Insley suggested be interpreted as 'the mound, hill where finches, chaffinches are found', v. **pinca, haugr.** An OE **Pinca, *Pynca* need not be considered. It was on the boundary of Claxby and Nettleton); *the pittes* 1601, *Pitts* 1628 (v. **pytt**); *pottebut(tes)* Hy3 (Ed1) *Newh* ('the strip(s) of land abutting on a boundary at a deep hole or pit', v. **potte, butte**); *Rasen wath (furlonge)* e17 (this is presumably 'the ford leading to (Market) Rasen', v. **vað**); *Roudegathe, Raudhegate in Toftbeck* 1150-60, *Raudegate eHy3 Roudgate* Hy3 (from the ON pers.n. *Rauðr* or ON **rauðr** 'red' and **gata,** in either case a Scand. compound); *Round Garth* 1717 (v. **garðr**); y^e *Rundle* 1589 (v. **rynel** 'a small stream'; for intrusive -*d*- v. NED s.v. *rundle*[2] and for its use as an appellative *a rundell* 1601 *Terrier* (Thornton le Moor)); *Sandie gate* e17 (v. **sandig, gata**); *le Segbuskes* Hy3, 113, *le sigbuskes* Hy3 (Ed1) *Newh* (v. **secg** 'sedge', **busc,** both els. here Scandinavianized); *segdaile* e13 (v. **secg, deill** and cf. the prec.); *Slecherhungate* (sic) Hy3 (Ed1) *Newh* (for -*hungate* v. Hungate PN L **1** 75, the first el. is obscure); *in Australi campo de Claxebi* Hy3 *HarlCh, the S. feild* (sic) 1587, y^e *southe feld of claxby* 1589, *Southe feild* 1601, *the Southfield* 1625, *the South feilde* 1638 (v. **sūð, feld;** one of the open fields of Claxby, cf. y^e *northe feld of Claxbye supra*); *Southiby* 1327, 1332 both (p) ('south in the village', v. **sūð, ī, bȳ**); *Sydegate* eHy3 *HarlCh, Sidegate* Hy3 *ib* ('the long road', v. **sīd, gata**); *Tho: Thixtons rushy Cliffe* 1703 (v. **risc, -ig, clif;** named from a member of the local *Thixton* family, cf. Thomas *thixton* 1601 *BT* and cf. Middle Plat in (a) *supra*); *Thorseway Acre* 1636, 1652, 1653 *Rad* (v. **æcer,** held by *Richard Brockesby of Thoresway); de spina* c.1250, *juxta le þorn ultra hou* e13 *CottCh, le þorn* Hy3 (Ed1) *Newh, the thorn close* 1674, (v. **þorn**); *the three ferme pece* e17 (v. **þrēo, ferme, pece;** *pece* here, as elsewhere in the area, evidently having the sense of 'enclosed land consolidating several holdings'); *Tophtbed, Toftbeck* 1150-60, *Toftbecce* e13, *Toftbecke* Hy3, 113, *Toftebec* Hy3 (Ed1) *Newh, Toft becsike* e13

CottCh, Toftbecsike Hy3 (Ed1) *Newh* (*v.* **toft** 'a curtilage, a messuage', **bekkr, sik**; the same name occurs once in the f.ns. (b) of the neighbouring parish of Normanby le Wold); *peece called trencher* e17 (perhaps from its shape; a *trencher* is a wooden platter, often circular); *Tussels Home close* 1703 (from the surn. *Tussel*); the *vpper Furlange* e17 (cf. the *lower Furlonge supra*); *ware hill* 1625, 1638; *Waterfalle* e13 (self-explanatory, *v.* **wæter-(ge)fall**); *Claxby Watry Lane* 1669 *MiD*; *Wellebec* eHy3, Hy3 (cf. *at Fontem* 1327 (p), *ad fontem* 1332 (p), *v.* **wella, bekkr,** cf. Welbeck Spring in Melton Ross, PN L 2 234); *West Bech* m13 (*v.* **west, bekkr**); the *West close* 1717; *Wiscartdaile* e13, *Wiskeredale, -daile* Hy3, *Wiskardside* Hy3, 113 (from the ME pers.n. *Wischard*, from Old Northern French *Wisc(h)ard*, for which *v.* Reaney 387 s.n. *Wishart*, with **deill** 'a portion of land' and **sik** 'a ditch', 'a small stream'); *Wlfarsic* c.1200 RA iv (from the OE pers.n. *Wulfheard* or the ON pers.n. *Ulfarr* and **sik**); *Wrangelandes* e13, Hy3, *Wrongelandes* e13, *Wrang(e)landes* Hy3 (Ed1) *Newh* ('crooked strips' *v.* **vrangr, land**).

Croxby

CROXBY (now added to Thoresway parish)

> *Crosbi* (3x) 1086 DB, *Crosseby* 1226-28 Fees, 1547 Pat
> *Crocsbi* (4x) 1086 DB
> *Crochesbi* c.1115 LS, 1142-53, eHy2 Dane (p), 1163 RA i
> *Crokesby* m12 (l13) *Stix*, 1226 FF, 1232 Cur, *-bi* 1154-55 RA iv, eHy2 Dane (p), 1155-60, 1160-66 Dane, c.1164 RA iv, c.1165 *HarlCh* (p), 1166 RBE (p) *et passim* to a1224 RA iv, *Crockisbia* 1177 ChancR (p), *Crokxebi* 1202 Ass (p), *Crokkesby* 1370 Pat
> *Croxebi* c.1189 LAAS v, 1202 Ass, 1212 Fees, eHy3 RA iv, 1218, 1219 Ass *-by* 1202 FF, 1208-9 Ass, 1219 FF, 1219 Welles, 1220-24 RA iv, 1225 Pat *et freq* to 1282 Ipm
> *Crocsebi* 1193, 1194 P (p)
> *Croxby* c.1200 (1409) Gilb, 1226, 1268 FF, 1279 RRGr, 1287 RSu, 1303 FA, 1327 *SR*, 1330 Orig, 1332 *SR et freq*, *-bi* 1206 Ass, 1303, 1346 FA, *-bey* 1563 *BT*, *-bie* 1576 LER, 1611 *BT*, 1611 *Terrier*, *-bye* 1601 *ib*, 1610 Speed
> *Crochebi* Hy2 Dane (p)

'Croc's farmstead, village', from the ODan pers.n. *Krōk* and **by**. For a full discussion of this pers.n. *v.* Croxton, PN L 2.

CROXBY HALL, cf. *Hall Garth* 1833 *Yarb.* v. **hall, garðr** and P&H 225. CROXBY POND, 1824 O, 1828 Bry, 1830 Gre, *Pond* 1833 *Yarb*, 1837 *TA*, cf. *Pond Top,* - *Walk* 1805 *Yarb*, 1837 *TA, Great -, Little pond close* 1814 *MiD, Fish pond* 1814 *ib.* CROXBY POND PLANTATION is *Thorganby Plantation* 1828 Bry, from the neighbouring parish of Thorganby. CROXBY TOP is *Croxby Cover* 1824 O. CROXBY WARREN WOLD (lost), 1828 Bry, the name of the area west of the road from Croxby to Beelsby and the parish boundary. FOX COVERT, 1828 Bry and is *Gorse Cover* 1814 O. LINGS FM, *Linges* eHy3, 1231-40 RA iv, cf. *the linge hill* 1601, - *Linge hill* 1611, - *ling hill* 1619, *y*^{e} *Linge hill* 1666-68, *the ling Hill* 1692, *Ling hill* 1694-95, *Lyng Hill* 1707, 1724, - *hill* 1709, *the linge closse* 1601, *great Linge closse* 1611, *the ling close* 1619, *y*^{e} *lings close* 1666-68, *Lingclose* 1671 all *Terrier, Linges* must mean 'the place where ling or heather grows' v. **lyng**. The farm is actually in Rothwell parish, but all the forms are from Croxby documents. RECTORY, *Rectory House and Paddock* 1837 *TA*.

Field-Names

Forms dated eHy3 and 1231-40 are RA iv; 1601, 1611, 1619, 1666-8, 1671, 1692, 1694-5, 1700, 1709, and 1724 are *Terrier*, 1805 and 1833 are *Yarb*; 1814 are *MiD*; 1837 are *TA* 108.

(a) High -, Low Church Cl 1814 (*the church close* 1601, 1692, *the Church close* 1619, 1709, *y*^{e} - 1671, 1724, *y*^{e} *church Close* 1666-68, *Church* - 1694-95, 1707, 1709), Church lands 1814, Church and Church Yd 1837; Colley Stable 1814 (probably from the surn. *Colley*); Cottage Walk 1805 (v. **walk**); Cottagers Plat 1833, 1837 (v. **plat**^{2}); Cow Cl 1833, 1837; Croxby Fm 1814; Croxby Lt Sheep walk 1814 (v. **shepe-walk**); Decoy Walk 1805, Decoy 1814, 1833, 1837 (a duck decoy, common in south L, v. also **walk**); East of the Road 1837; Fold Cl 1814; Furze Cover 1814; Garborough Garth 1814, 1833, 1837 (possibly from the surn. *Garborough*, with **garðr**); Garden Cl 1814, 1833, - cl 1837, Garden and Moat 1837; Grains Garth 1833 (perhaps for Grange Garth, cf. foll. entry); Grange Garth 1814 (cf. *atte Graunge* "of" *Croxby* 14 AD, *ad Grangiam* 1332 *SR* (p), *atte Graunge of Croxby* 1385 Cl (p), *Croxbye Grange* 1535-37 LDRH, v. **grange, garðr**; it was a grange of Louth Park); Grimsdale 1814 (1619, 1666-68, 1671, 1692, 1694-5, 1707, 1709, *Grimsdalle side* 1601, *Grimsdaile side* 1611, from the ON pers.n. *Grimr* or the ME surn. *Grim(m)* and **deill**, with **side**); Hall Cl 1814 (v.

Croxby Hall *supra*); Holt 1833, 1837; Home Cl 1833, 1837; Horse Park 1814, - Walk 1833, 1837 (*v.* **walk**); Far Ings 1814, Ings 1833, 1837 (cf. *hither Ingges* 1611, *midle Inggs* 1611, *the ing close* 1692, *v.* **eng**); Far New Pce 1833 (cf. *the newe close* in (b) *infra*); North Cl 1814; North Walk 1814 (*v.* **walk**); Oat Cl 1814; Parsonage Cl 1814 (cf. *the parsonage lathe* 1601, in (b) *infra*); Plantation 1833, 1837, Part of Plantation and Paddock 1837; Saintfoin Cl 1814 (*v.* **sainfoin**); Scroggs 1805 (*v.* **scrogge** 'a bush, brushwood'); Shrubbery 1837; Smithfield Cl 1814, Smith Fd 1833, Part of - 1837 (*smithfeild* 1601, *Smithfeild close* 1611, *Smithfield* 1619, 1671, 1694-95, 1707, 1709, 1724, *Smithfeild,* y^e *valley butting upon Smithfeild* 1666-88, *Smyth feild* 1692, probably from the surn. *Smith* and **feld**, cf. John & Robert *Smithe* 1621 *Inv* and *John Smiths farm* in (b) *infra*); Stainton Btm 1814, 1833, 1837 (from the adjoining parish of Stainton le Vale and **botm**); Walk 1805, Walk Cls 1814 (*v.* **walk**); Warren 1805, The - 1814 (*v.* **wareine**); Watering place 1814; Whin Bed 1805 (*v.* **hvin**); Willow Holt 1833, 1837 (*v.* **holt**); Wood Hole 1814; Nr Yard and Stables 1837.

(b) *Assefrumdale* eHy3, *Assfrondale* 1231-40 (possibly from the medieval surn. *Assefrun, Assfron,* with **deill**); *Barnards garth* 1601, *Barnard yeate* 1611 (named from the *Barnard* family, cf. Edmund *Barnard* 1583 *BT,* John *Barnard* 1608 *Inv*, with **garðr** and **geat**); *Beelsby gate* 1666-88, 1671, 1692, 1694-95, 1707, 1709, 1724 ('the road to Beelsby (an adjoining parish)', *v.* **gata**); *Beledale* eHy3, 1231-40, *Beledal'* eHy3 (*v.* **deill**; the first el. is possibly the medieval surn. *Bele, v.* Reaney s.n. *Beal*); *Borg* 1231-40, *Retro Borg* 1231-40 (variant spellings for *Bure infra*); *in a bottome* 1601, *In a Bottome* 1611, *bottom* 1619, y^e *bottome* 1671, - *bottom* 1707, 1709, 1724 (*v.* **botm**), *Bure, Bihindeburc* eHy3 (perhaps from **burh**, in the ME sense 'manor house'; the forms in *Borg supra* show the influence of the cognate ON **borg**[1]); *the Bushe, the Bushe furlonge* 1611 (*v.* **busc**, **furlang**); *viam de Cast'* eHy3, *viam de Castre* 1231-40, *the way leading to Caster* 1601, *caster gate side* 1601, *Caistergate, - gaite* 1611, *Castergate* 1619, 1666-8, *Caster Gate* 1671, *Caster gate* 1692, *Caister -* 1694-95, 1707, 1724, *- Gate* 1709, 'the road to Caistor', *v.* **gata**); *Chisteldale* eHy3, 1231-40, *Paruuchistedale* eHy3 (*C-, -c-* are no doubt errors for *T-, -t-, v.* *Thystedaile infra*); *Chorgerhenge* eHy3, *Corgerhenge* 1231-40 (*v.* **eng**; the first el. is uncertain); *Clif* eHy3, *super montem* 1231-40 (*v.* **clif**); *Collyngarth* 1519 Wills i (from the pers.n. or surn. *Collin* and **garðr**); *comon laine* 1611 (*v.* **lane**); y^e *commons* 1611; y^e *corne hades* 1601, *the Corne Heades* 1611, y^e *Corne heads* 1666-88, 1692, y^e *Corn -* 1694-95, 1707, 1709 (*v.* **corn**[1] 'corn, grain', **hēafod**, a headland in the common field); *the Cow forth* 1619, y^e *Cowforth* 1671, *the -* 1692, y^e *Cow foard* 1666-88, *Cowforth* 1694-95, 1707, 1709, 1724 (*v.* **cū**, **ford**); *Croxby wang -- Firma Grang* Hy8 NotLud, *Crosbywange -- Firma grangiæ* 1536-7 Dugd v (*v.* **vangr** 'an infield' and Grange Garth *supra*); *sub Cruce* eHy3,

1231-40 (*v.* **cros**); *Danegate* eHy3, 1231-40 (probably self-explanatory *v.* **Danir, gata**); *Dannseleland* eHy3, *Dansselelandes* 1231-40, *Dannselehfdlandes* 1231-40 (checked from original charters) (the equivalent forms in the copies in (c.1330) *R* are *Damiseleland* eHy3, *Damiselelandes, Damiselehevedlandes* 1231-40, while *Dannselehevedland* eHy3, the reading in the printed text, should almost certainly be *Damiselehevedland.* The readings -*nn*- from the original charters are interpretations of four minims and it is clear that those of the 14th century copies are more trustworthy *and* meaningful. The first el. is ME *damisele* 'a maiden' or the derived surn. (*v.* Reaney s.n. *Damsell*), the second being **land** 'a selion, a strip of arable land in the common field' and **hēafod-land** 'the head of a strip for turning the plough'); *Duuemilne* 1155-60, 1160-40 Dane, *Dunemulne* R1 (1318) Ch (*n* = *u*), *Duuemilnecroft* eHy3 (possibly *v.* **dūfe** 'a dove' with **myln, croft**); *East & South East field* 1700; *campis de Crokesbi* 1200-23 *HarlCh, the field* 1619, y^e - 1671 (*v.* **feld**); *iuxta forarium Godrici* c.1200 (1400) Gilb ('Godric's headland', and cf. *terram Godrici* eHy3); *Foxholes* eHy3, 1231-40 (*v.* **fox-hol**); *the furlong* 1619, y^e - 1707, 1709 (*v.* **furlang**); *the gate side* 1619, y^e - 1671 (*v.* **gata**); *Greneclif* eHy3, 1231-40 (*v.* **grēne**^{1}, **clif**); *Grenefelt* 1231-40 (this is an error in the original charter for *Greneclif,* as the form in the copy (c.1330) *R* shows); *Greshou* eHy3, *Groshou* (sic) 1231-40 ('grass mound', *v.* **grēs, haugr**); *Grimsbie gate* 1611 ('the road to Grimsby', *v.* **gata**); *Hakechorne* eHy3 (*c* = *t*), *Hakethorne* eHy3, 1231-40 (alluding to some kind of thorn, *v.* **haca-þorn** and cf. Hawthorn Lane in Barton upon Humber, PN L 2 38-39); *Harevange* eHy3, *Harewange* eHy3, *Harewang* 1231-40 ('the infield frequented by hares', *v.* **hara, vangr**); *the hill* 1619, 1692, y^e - 1671, 1694-5, y^e *hill side* 1724; *Biewesten Holegate* eHy3, 1231-40, *Bywestenholegate* eHy3, *Ex occidentale parte de Holegate* 1231-40 (*v.* **bī, westan** 'to the west of'; *Holegate* is no doubt 'the road running in a hollow, the sunken road', *v.* **hol**^{2}, **gata**); *Joldale, Biwestan Joldale* eHy3, 1231-40, *Biester Joldale* eHy3, *Yoldale, ex orientale parte de Yoldale* 1231-40 ('to the east of', 'to the west of', *v.* **bī, ēasterra, westan;** *Joldale* is possibly from the ON pers.n. *Jólfr* and **deill**); *Jordmilne* eHy3, 1231-40, *Yordmilne* 1231-40 ('the mill with a fence or with an enclosure', *v.* **geard, myln**); *Kukewaldemare, Kukewaldmare* eHy3, *Cocowaldmare* 1231-40 (*v.* **(ge)mǣre**), *coksewold gate* 1601, *Cuxwold* - 1611, *Coxwold* - 1619, *Coxwould* - 1671, 1692, 1694-95, *Cuxwould* - 1707, 1709, *Cuckswould* - 1724, *Cuxwold* - 1666-68 (*v.* **gata**), *Cokswold hedge corner* 1601, *Coxwold hedge* 1619, *Cuxwold* - 1666-88, *Coxwould* - 1671, 1692, 1694-95, 1709, *Cuxwould* - 1709, - *corner* 1707 (all named from the neighbouring parish of Cuxwold); *Lambecotes* eHy3, 1231-40 (*v.* **lamb, cot**); *lanhea hous* 1519 Wills i; *Lindale* eHy3, 1231-40 (probably *v.* **līn** 'flax', **deill**); *Liteldale* eHy3, 1231-40 (*v.* **lytel, deill**); *Madhou* eHy3, 1231-40 (*v.* **haugr**); *Maidenaker, Maydenaker* eHy3, *Maidenacre* 1231-40, *Maiden acres* 1611, *Mayden acres* 1671, *Maiden-Acres* 1666-88, *mayden Acres* 1692,

Mayden - 1694-95 (*v.* **mægden** 'a maiden', but in what sense is uncertain, **æcer**); *super Mare* eHy3, *super diuisam* 1231-40 (*v.* **(ge)mǽre** 'a boundary'); *Mayson laine end* 1611 (from the family name, cf. William *Mayson* 1611); y^e *middle of* y^e *Feild* 1666-88, *the middle of the field* 1692, *Middle field* 1707, 1709, y^e *middle field* 1724; *the Middle furlonge* 1611, y^e *middle furlong* 1671; *the newe close* 1601, *the new* - 1619, 1671, 1692, *New close* 1666-88, y^e *New close* 1694-95, *New close side* 1707, 1709; y^e *North furlong* 1694-95; *norwest feild* 1611, y^e *North West fields* 1707, *North & North West field* 1700, *North & West fields* 1724 (cf. *The North & East Field* 1709); *the parsonage lathe* 1601, *the Parsonage barne* 1619, *the parsonage barn* 1671, *Parsonage barn* 1694-95 (the change from *lathe*, ON **hlaða** 'a barn', to *barne* itself is worthy of note, and cf. Rectory *supra*); *Peart furlong* 1611, *peart* - 1619, 1671, 1692, 1694-95, *Pert Furlong* 1666-88, - *furlong* 1707 (possibly from the surn. *Peart* with **furlang**); *Pesfurlanges* c.1200 (1400) Gilb, *Huuerpesefurlang*, *Overpesefurlange* eHy3, *Howerepeseforlanges* 1231-40 (*v.* **uferra** 'higher', **pise** 'pease', **furlang**); *Rigdales* c.1200 (1400) Gilb, *Rigdailes* eHy3, 1231-40, *Huuerrigdail'* eHy3, *Huuerrigdailes* eHy3, *Howrerigedailes* 1231-40, *Neccherrigdail'* (with -c- for -t-), *Netherrigdailes, Netherrigedayles* all eHy3, *Netherrigedailes, Nederigedailes* 1231-40 (from **hryggr** 'a ridge', in ME 'a cultivated strip of ground' and **deill** 'a portion of land', with **uferra** 'higher' and **neoðera** 'lower'); *the sandes* 1611, - *sands* 1619, 1692, y^e *sands* 1671, y^e *Sands* 1666-68. 1694-95 (*v.* **sand**); *Sefurlanges* c.1200 (1400) Gilb (*v.* **furlang**; the first el. is apparently **sǽ** or **sǽge** probably with the sense 'pond, marsh'); *John Smiths farm* 1619, *John Smyths farme* 1692, *Jo. Smith farm* 1694-95, *John Smiths farm* w^{ch} *was so callid of old* 1707; *the Southeast feild* 1611, *South East Fields* 1707; Wm *Stamps close* 1619, 1694-95, Wm *Stampes* - 1671, 1692; y^e *stone wharfe* 1601 (the reading is doubtful); *Suneker* eHy3, *Symeker* 1231-40 (both have been checked from the original charters, and the -un- of the first is an interpretation of four minims. The form of the second shows that it should nonetheless be interpreted as -im-. The first el. is probably the ME pers.n. *Simm, Sime*, or the derived surn. (*v.* Reaney s.n. *Sim*), the second being ME **ker** (ON **kjarr**) 'a marsh'); *Swynestie* (*with a henn chamber over it*) 1611 (self-explanatory); *Thysteldale, per medium Thystheldale* 1231-40, *Thistledaile bottom* 1611 (*v.* **þistel**, **deill**; *Chisteldale supra* is an error for this); *thre thornes* 1601; *Utfurlang,* - *forlange* eHy3, 1231-40, *Howtforlanges* 1231-40 (*v.* **ūt** 'out, outer', **furlang**); *Waingate, Weyngate* eHy3, *Waynegate* 1231-40 ('the cart or waggon road', *v.* **wægn**, **gata**); *the Water Milne* 1611; *the west feild* 1601 (possibly identical with *norwest feild supra*); Wm *Wilys close* 1666-68; *Wrafurlanges* c.1200 (1400) Gilb ('furlongs in a nook or corner of land', *v.* **vrá**, **furlang**).

Holton le Moor

HOLTON LE MOOR

    Hoctun(e) (3x) 1086 DB, *-ton'* 1168, 1170, 1177 P (p), *Hotton'*
    1184 ib (p) (with *-tt-* for *-ct-*)

    Houtuna c.1115 LS, *-ton'* 1181, 1182, 1183 P (p), 1196
    ChancR, c.1200 RA iv, 1204 P, 1218 Ass, 1226 FF
    et passim to 1461 Pat, *Heuton'* 1185 Templar (with
    -eu- for *-ou-*), *Houton in Mora* 1327 *SR*, *- in mora* 1332
    ib, - "in the More" 1328 Banco, 1414 Fine, 1431 FA,
    - in la More 1331 Ch, *- in le More* 1375 Pat, *- in le
    Mor* 1387 Peace

    Howeton' 1202 Ass, 1408 Fine, 1415 Cl, - "in the More"
    1401-2 FA, *Howton* 1402, 1428 FA, 1576 Saxton, 1610
    Speed, - "in the Moor" 1472 WillsPCC, *- in le Marche*
    1509 Ipm, *- over ye More* 1539 LP xiv, *- ye Moor* 1695
    Morden

    Holton 1556 CA, 1600-01 Lanc, 1657 *Rad*, *- by more* 1653
    ParlSurv, *- in the Moore* 1653 WillsPCC, *- in le Moore*
    1657 *BRA 1597*, 1676 *MiD*, *- in le More* 1679 *ib*, *- le
    Moor* 1824 O

    Houghton otherwise called Holton in le More 1610 *MiD*, *- als
    Holton in le Moore* 1698 *ib*

    Howlton in the more 1649 *Asw*

    Houlton in the Moore 1673 *MiD*, *- in le Moore* 1683 *ib*

'The farmstead or village on a hill-spur', *v.* **hōh, tūn**, the
modern village lying on the eastern slope of a definite spur of
land. The affix is from **mōr**[1] 'a moor', though note that it is once
described as *in le Marche*.

BESTOE HO (local), named from a family recorded locally from at
least the 17th cent., cf. William *Bestoe* 1649 *Asw*, Charles *Bestoe*
1653 WillsPCC, Richard *Bestoe* 1660 *MiD*. DAISY HILL FM, said
to be *Holton-le-Moor Farm* 1875 *Dixon*. HALTON HALL (local),
Hall Farm 1782 *LTR*, *The Hall Farm* 1785 *Dixon*, *Hall* 1838 *ib*,
Holton Hall Farm 1871 *ib*. HOLTON MOOR (lost), 1828 Bry, *-
Moore* 1677, 1678, 1683, 1698, *- moore* 1679 all *MiD*; this is the
same as *Holton Warren infra*. HOLTON PARK, *Park* 1838 *TA*.

HOLTON WARREN (lost), 1824 O, 1830 Gre, *that Free warren comonly called or knowne by the name of Holton le Moore Warren* 1679 *MiD, that Free Warren or Chase* 1698 *ib, the warren* 1782 *LTR, Holton Common and Free Warren* 1821 *Dixon*; this is the name of the whole area east of the village, called in 1828 Bry *Holton Moor, v. supra.* HOLTON LE MOOR GRANGE, apparently a late use of **grange**, for which *v.* EDD s.v. 2 "a homestead, small mansion or farm-house, esp. one standing by itself remote from others", a sense quoted from L. MOUNT PLEASANT, 1824 O, 1828 Bry, 1830 Gre, no doubt a complimentary nickname. NOBLE'S WOOD, cf. *Noble's Cabbage Patch* 1821 *Dixon, Nobles Cow Close, - House and Garden, - Home Close* 1828 *TA, Noble's Cow Close, - Home Close* 1838 *Dixon*, named from a local family, cf. Samuel *Noble* 1817 *Dixon,* 1838 *TA.* SAND PIT WOOD. SANDS PLANTATION, *Sand Plantation* 1838 *Dixon,* 1838 *TA*; there is reference to the *blowing Sands and Barren Moors* of Holton le Moor 1763 *Red* 2/4/4. WATERLANE PLANTATION. YEWFIELD, *the Ewe Field* 1669, *(the) Ewfeild* 1679, 1683 all *MiD, Ewe-field* 1817 *Dixon, Ewefield* 1824 O, 1830 Gre, *Ewe Field F.^{m}* 1828 Bry, *Ewefield Farm Yard* 1838 *Dixon,* 1838 *TA,* self-explanatory, *v.* **eowu** 'a ewe', **feld**, the modern form showing confusion with *yew*, found elsewhere in p.ns.

Field-Names

Forms dated 1327 and 1332 are *SR*; those dated 1600-1 are Lanc; 1649 are *Asw*; 1660, 1661, 1666, 1669, c.1670, 1673, 1675, 1676, 1677, 1678, 1679, 1683, and 1698 are *MiD*; 1817, 1821, 1838[1], and 1875 are *Dixon*; 1838[2] are *TA* 178; forms marked 1838[3] are in both 1838 *Dixon* and 1838 *TA.*

(a) Angle Dale 1838[3] (a reference to shape, the field forming a very distinct triangle, *v.* **deill** 'a portion, a share of land'); Anthony Cl 1821, 1838[2], 1875, Anthony's Cl 1838[2], - Lane 1838[3] (from the surn. *Anthony,* not so far traced in this parish); Atra's Home fd, - House and Garden 1838[2], - Iron Stone Cl 1838[3] (William *Atra* was the occupier 1838 *TA*); East -, West Barn Platt 1821, - Plat 1838[3] (*v.* **plat**[2] 'a plot of land'); Late Betts Gdn 1838[2] (named from the family of Robert *Betts* 1784 *LTR*); Blacksmiths Shop 1838[2]; Low -, Maddison's Blow Cl, Low -, Top Blow Cl 1838[3], Blow Close Lane 1838[1] (*the Blow close* 1661, 1666, *Blowe close* 1675, *Blow Close* 1676); Brices - 1821 (*v.*

infra), Mattison's - 1821 (*v.* Mattison's Mdw *infra*), Noble's Breamers 1821 (cf. Noble's Wood *supra*), Middle -, North -, South Bramer 1838[3] (*Bramer* 1661, 1666,, *Braymoore* 1669, *Braymore* 1675, 1679, 1683, *Bramore* 1679, *Bray moore lane* 1676; the fields so named lie on the boundary of Holton, so that the name probably means 'the broad boundary, boundary land', from ON **breiðr** and **(ge)mǣre**, the latter often spelt *mo(o)re* in 17th cent. documents in north L); Late Brice's House etc. 1838[2] (named from Charles *Brice* 1817, or his family); Brick Kiln Fd 1838[3], 1875; Broughton's Cottage 1838[3], Broughtons Cow Cl, - Home Cl, - Middle Cl 1838[2] (named from the *Broughton* family, cf. Benjamin *Broughton* 1782 *LTR*); Calf Cl 1838[3], - and Farm Yd 1838[2]; Two Carltofts 1821, East -, West Carltoft 1838[3] (*Carle-Toft* 1669, *Carle Toft* 1679, 1683; the first el. may be the ON **karl** 'a freeman of the lower class', later in ME 'a husbandman, a free peasant, etc.', but frequently it is a pers.n. or surn. in compounds with **toft** 'a building plot, a curtilage, etc.' In this case, the first el. would be from the ON, ODan pers.n. *Karl(i)* found independently in L); Carpenter's Cl 1821 (from the surn. *Carpenter*, not so far traced in this parish); Carr Heads 1821 (*v.* **kjarr, hēafod**); Church and Yard 1838[2] (*ad ecclesiam* 1327, *atte Kirke* 1332, *Atte Kirk de Houton* 1343 NI all (p), *v.* **kirkja**); Claytons House and Garden 1838[3] (named from William *Clayton* 1837 *PR* or his family); Clover Cl 1821, 1838[3]; Corn Platt, - Sykes 1821 (*v.* **corn**[1], **plat**[2], **sik**); Cottagers Pasture 1821 (pasture allotted to the cottagers, fairly common in f.ns. in north L); Cow Cl 1838[3] (*the* - 1669, 1677, 1678); Middle -, North -, South Cow Fd 1838[3]; Cow Sykes 1821 (*v.* **cū, sik**); Credlands 1838[1], - Gdn 1838[1] (from the *Credland* family, cf. George *Credland* 1838[1]); Crow Croft 1838[3] (*v.* **crāwe, croft**); Dam Cl (*v.* Ley, Cow and Dam Cls *infra*); Far Cl 1838[1], Farr - 1838[2]; Farm Yd 1838[2]; Favill's Yd 1838[1], Favells House and Yd 1838[2] (Richard *Favell* was the occupier in 1838 *TA*, and note Robert *Favill* 1820 *PR*); Fish Pond 1838[3]; Fourteen Acres 1838[2], 14 acres 1838[1]; Glew and Dauber's Cottages 1821 (named from William *Dawber* 1808 *Dixon*, or his family), Glews Cl 1838[3], Glew's House and Gdn 1838[3] (named from Alexander *Glew* 1838 *TA*, the occupier, cf. also *Elwick Glue* 1819 *PR*); Grass Garth 1821 (*v.* **gærs, garðr**); Gravel Hill 1821; Great Cl 1838[3] (cf. Upper Cl *infra*); Green Cl, East -, West Green Cl Bottom, Green Cl Hill 1838[3] (*v.* **grēne**[1], **hyll, botm**); Hall Offices etc. 1838[2]; Hewits Garden and Orchard 1821, (Thomas) Hewitts Cottage, Thomas Hewitts Old Grd 1838[2], (Th[s]) Hewitt's Cottage 1838[1] (named from Thomas *Hewitt*, whose family has been noted earlier in the 19th century, cf. Robert *Hewitt* 1814 *PR*); Hewson's Home Cl 1821 (from the family name, cf. John *Hewson* 1641 *LPR*); Hills Cl 1838[3], North Hills 1838[3], South Hills 1838[2], South Hills Cl 1838[1]; Holt (Cl) 1838[3], 1875 (*v.* **holt**); Holton or Warren Cl 1821, Holton Cl 1838[3] (*v. Holton Warren supra*); Home Cl 1817, 1821, 1838[2], 1875; House Cl 1838[3]; Iron Stone Cl 1838[1] (cf. Atra's Iron Stone

Cl *supra*; the name is self-explanatory); Johnny Holme and Pingle 1821 (*Joney hole* 1669, 1679, 1683, - *Holle* 1669, 1677, - *holle* 1678, *Jone Hole* 1679, perhaps from a surn. derived from the pers.n. *John*); Laming(')s House and Gdn 1838[3] (William *Laming* was the occupier 1838 *TA*, cf. William *Laming* 1829 *PR*); Lane 1838[2]; Ley, Cow (cf. Cow Cl *supra*) and Dam Cls 1821 (*Dam Close* 1678, *the* - 1698, cf. *Dam Hill* c.1670, 1677, 1678, self-explanatory, *v.* **dammr**, ME **damme** 'a pond'); Little Cl 1821, 1838[3], 1875; Little Fd 1821; Long Leys 1821; Low Cl 1838[3], 1875 (*the low Close* 1673); Mattison's Mdw 1821, Maddison's Gdn, - Paddock 1838[2], - North Cl, - South Cl 1838[2] (Henry *Maddison* 1838 *TA*, occupier, and cf. Jonathan *Maddison* 1817 *Dixon*); the Marsh 1821; Gt Maze 1838[3], 1875 (*the great Mase* c.1670, 1678, y^e *greate Mase* 1677, *Great* - 1683), Jonathans - 1838[3]. Lt - 1838[3], Taylor - 1821 (*v.* Taylor Home Cl *infra*), Wiles Maze 1821 (from the family of Michael *wiles* 1743 LTR), Maze Cl 1838[3] (*the Mayses* 1649, *Middle Mayes* 1661, - *mayes* 1666, *the Middle mase, Middle Maze, Middlemaies* 1679, *Middlemase* 1683, *the Middle Mase close* 1669, *the litle Maze* 1698; there are six fields called Maze Cl, nos. 116-17, 124-6 and 11 on the *TAMap*; the last is in the village itself, the other five are adjoining fields east of the road to Moortown and south of Noble's Wood, and north of the road running north-east from the village. Local enquiry by Mrs Diana Cottingham confirms that there is nothing here which could physically be described as a *maze,* but the area was rough uncultivated land with few trees and much gorse. Mr Philip Gibbons, whose family has farmed here for many years, thought that perhaps the name had been given because it was possible to get lost there. The name may, of course, only superficially look like *maze*); Middle Cl 1821, 1838[3] (*the Midle Close* 1698); Miles Cl 1821; Moor Cl 1821, Low -, Top High Moor 1821, Moor Grd 1838[2], Low Moors 1838[3] (named from *Holton Moor supra*); Mount Fm 1821, Mount Lane 1838[1] (perhaps the same as Mount Pleasant *supra*); New Cl 1821, 1838[3]; North Cl 1838[1] (cf. South Cl *infra*); Old Mdw 1821; Paddock (2x) 1838[2]; Paradise and Slightings 1821, Paradise 1838[3] (1679, 1683, *a place called Paradice* 1669, *v.* **paradise** often a complimentary nickname, but ME **paradis(e)** denoted 'a garden', cf. Gt Sleightens *infra*); Pingle 1817, - Cl 1821, Gt -, Lt Pingles 1838[3] (*v.* **pingel**); Plantation 1821, 1838[3] (freq), 1875; Platts 1821 (*v.* **plat**[2]); East -, West Pond 1838[2], - Pond Cl 1838[1], Pond Cl 1838[3] (*v.* **ponde**); Punch Garth 1821 (*v.* **garðr**); Que Cl 1817, East - 1838[3], West Quee Cl 1838[2] (the first el. may be ON **kví** 'a pen, a fold'); Reed Cl 1838[3], - and Intake 1821 (*v.* **hrēod, inntak**); Rush Cl 1821, 1838[3] (adjoins West Pond *supra*); Rye Cl 1821 (*the Rye Close* 1673, *v.* **ryge**); Sand Sykes 1821 (*v.* **sand, sík,** and cf. Sands Plantation *supra*); Seed Walk 1821 (*v.* **sǣd, walk**); Sindersons Gdn 1838[2], - house and Garden 1838[1] (Thomas *Sinderson* was the occupier 1838 *TA*); East -, West Slack Cl 1838[1] (*the Slack Close* 1673, but perhaps the same as East -, West

Stack Cl *infra*); Gt -, Lt Sleightens (sic) 1817, Gt - (2x), Lt Sleightings 1838[3] (the fields lie in the extreme south-west corner of the parish and the name is probably a Scand. compound meaning 'the level meadows', from **sléttr** and **eng**); South Cl 1838[1]; Home -, Middle -, North -, South Square Cl 1817, Square Cl 1838[1], East - 1838[3], Far -, Gt Square Cl 1838[2] (named from the shape); Stack Cl 1821, East -, West Stack Cl 1838[2], Stack Yd 1838[3], Stack Yard Cl 1875 (*the Stack Close* 1698); Stalks Cl 1838[3], Stalk's - 1875 (from the surn. *Stalk*, not so far traced in this parish); Sunday School 1838[2]; Taylor Home Cl 1821, (Top) Taylors Yd 1838[1], (Top) Taylor's - 1838[2] (named from the *Taylor* family, cf. Edmund *Tayler* 1641 LPR, Jonathan *Taylor* 1785 *Dixon*); Ten Acres 1838[3] (named from its actual area); Townend Fd 1821 ('at the end of the village', *v.* **tūn**, **ende**[1], or from the surn. *Townend*); Truant Cl 1821, Truants - 1838[1], Truant's - 1838[2] (*Trowant Close* 1698, from the surn. *Truant*, cf. Vincent *Trowan* (sic) 1641 LPR); Upper Fd or Great Cl 1821 (cf. Great Cl *supra*); Gt Walk 1817, 1838[3] (2x) (*v.* **walk**); Walkers Cl 1838[1], Walker's - 1838[2] (named from the *Walker* family, cf. William *Walker* 1838 *TA*, Benjamin *Walker* 1817 *Dixon*); Weather Cl 1817, 1838[3], Wedder - 1821, Wether Cl 1838[3] (cf. *wether feild* 1660, *the weather Field* 1675, *Wetherfeild, the wether feild, the Weatherfeild* 1676, *Wether Sicke plott* 1669, *- sike plott* c.1670, *- Sike plott* 1677, *Water Sike Plott* (sic) 1683 (*v.* **weðer** 'a wether', **sik**, **plot**); Well dale Fd 1821, Weldale Fd 1838[3] (*v.* **wella**, **deill**); Westfields Paddock 1838[3]; Whites Houses and Yd 1838[2] (Thomas *White* is named as the occupier).

(b) *Eight pound Close* c.1670, *the eight pound Close* 1677, *the Eight pound Close* 1678, *Eight ponds Close* (sic) 1683 (cf. *twelve pounds close infra*); *Good closes* 1661, *Good-closes* 1666, *two Closes of pasture called ... Good Closes* 1676, *Goods close* 1669, *- Close* 1676, *- closes* 1675, *- Closes* 1683, *Goodes -* 1679 (named from the *Good* family, cf. Richard *Good* 1659 *MiD*); *the highfeild close* 1661, *the High-field close* 1666, *the High Field* 1675, *the high -* 1676; *Low -, Upper Hinderbarfes* 1649 ('hills further to the rear', from OE *hinder* 'back, lying or situated at the back' and **beorg, berg** 'a hill, a mound', cf. dial. *barf*); *the great Lodge house & the Little Lodge* 1679; *pratum de Houton* 1280-85 *Foster* ('Holton meadow'); *The Middle-feild close* 1661, *the Middlefeild close* 1666, *the Middle Field* 1675, *- feild* 1676 (*v.* **middel**, **feld**); *the New plott* c.1670, *the new plott, - platt* 1677, *the New platt* 1678, *new Plot* 1683 (*v.* **nīwe, plat**[2], **plot**); *the Nineteen Leas* 1673 ('nineteen unit-plots of grassland', *ley, lea* (OE **lēah**), often in the pl., being used of grassland units of tenure corresponding to **land** similarly used of arable); *Southfeild* 1600-1 (*v.* **sūð**, **feld**); *South Lane* 1669, *- lane* 1677, 1679, *the South Lane* 1678, 1683 (*v.* **sūð, lane**); *the new Spring close, the Ould Spring Close* 1673 (*v.* **spring**); *those severall closes ... called ... by the name of the twelve pound close*

1661, *twelve pounds close* 1666, *the Twelve pound close* 1675, - *Close* 1676 (alluding to rent or sale-value of the land, cf. *Eight pound Close supra*); *Wells Corne feilds* (sic) 1669, - *feild* 1679, *the Wells Corner Feild* 1683 (named from the *Wells* family, cf. Thomas *Wells* 1676 *MiD*).

South Kelsey

SOUTH KELSEY

   Sudkeleseia 1177, 1204 P, *-kelesia* 1199 ib, 1218 Ass, - *Keles'* 1206 ib, *-keles* 1236-37 RRG, *-kelsey* eHy3 (1409) Gilb

   Sudkelleshay 1271-72 *Ass*, *-kellesey* 1276 Ipm

   Suthkeleseye 1190 (1301) Dugd vi, a1218 RA iv, 1236 Pat, 1280 RSu *et passim* to 1345 Pat, *-kelesey* 1226 Fees, *-keleseie* 1196 ChancR, *-kelesia* 1219 Ass, *-kelese* 1265 Pat, *-kelesey* 1317-18 RA x, 1378 Pat, 1461 Fine, *-kelesheye* 1276 RH

   Sutkelesey 1242-43 Fees, *-kelesay* 1270 Misc

   Suthkellis' 1219 Fees

   Suthkelleseye 1262 Ipm, 1264 RRGr, 1271 Pat, 1272 FF *et passim* to 1391 Pat, - *Kelleseye* 1263 FF, 1276 Cl, 1295 RSu, 1298 Ipm *et passim* to 1328 Banco, - *Kellesey* 1303 Cl, *-kellesey* 1392 Pat, *-kellesay* 1273 RRGr, 1295 *Ass*, 1343 NI, *-aye* 1332 *SR*, - *Kellesay* 1346 FA

   Suth Celeseya 1275 RH

   South Kelleseye 1326 Pat, 1331 Cl, 1332, 1349, 1358, 1380, 1382 Pat, *-kelleseye* 1337 Fine, 1348 Cl, 1353 Pat, 1353 Orig, 1355 Pat *et passim* to 1408-9 RRep, *Sowthekellsey* 1542-43 *AD*, *South Kellsey* 1657 *Rad*

   Southkelseye 1327 *SR*, 1353 Ipm, 1361 Pat, 1362 BPR, 1384 Fine, 1400 Pat, - *Kelsey* 1303 FA, 1317 Inqaqd, 1402, 1428 FA, (*alias Southkellsay*) 1495 Pat, 1610 Speed *et passim*, *-kelsey* 1353 Inqaqd, 1400 Pat, 1498 Ipm, 1547 *Monson*, *Sowthkelsey* 1554, 1566 Pat, - *Kelsey* 1634 VisitN, *Southkelsay* 1332 *SR*, 1350 *Cor*, 1362 BPR, 1373 Cl, 1383 Peace, 1389 Cl, 1416 Pat, *-aye* 1539-40 Dugd iv, - *Kelsay* 1424 Pat, 1431 FA, 1472 WillsPCC, *Sowthekelsay* 1542 *Monson*

For further forms and for a discussion of the etymology of

Kelsey, *v.* North Kelsey PN L 2 180.

WINGALE PRIORY
    Wingeham (2x) 1086 DB
    Wiungle (sic) 1100-4 France
    Guiungle (sic) 1103-4 France
    Wighala (sic) c.1115 LS
    Winghall' 1224 Welles, *Wynghalle* 1361 Pat
    Wenghal 1241-42 RRG, 1275 RH, *-hal'* 1298 Ass, *-hale* 1254
        ValNor, 1281 QW, 1291 Tax, 1322 Ipm, 1342 Fine *et*
        passim to 1407 Pat, *-ale* 1348, 1352, 1414, 1423 Fine,
        1441, 1443, 1461 Pat
    Wynghale 1254 ValNor, 1276 RH, 1337 Cl, 1372 Fine, 1378,
        1382 Pat, 1382 Misc, 1382 Cl, 1384 Fine, *-al'* 1276
        RRGr, *-ale* 1351, 1374, 1390, 1391, 1392, 1400 Pat
    Wyngall 1435 Pat, 1503, 1591 *Monson,* 1576 Saxton, 1610
        Speed
    Wingall a1567 LNQ v, 1577 Harrison, 1624 *Terrier,* 1645
        Holywell, 1665, 1707 *Monson*
    Winghall' 1503 *Monson, Winghall* 1824, 1830 Bry
    Wenghall' 1506 *Monson*
    Wingle Farm 1847 *TAMap*

Dr John Insley suggests that this is a "dat.sg. (i.e. used as a
locative) OE *Wigingahale* 'at the **halh** of Wīga's followers', a
formation apparently comparable to that of Rippingale, L Kest". A
parallel is provided by Wingham K, (*UUigincgga ham* 825-32
(Original) (S 1268), *Wingeham* 1086 DB, *v.* KPN 156, 158, DEPN
s.n.). OE *Wiga* is a regular short form of such names as OE
Wiglaf, Wigmund, etc., and corresponds to Old High German, Old
Saxon *Wigo.* The exact sense of **halh** 'a nook, a corner of land'
here is uncertain, but Wingale is situated in an isolated position on
the edge of a low spur overlooking the R. Ancholme, so that 'raised
land in marsh' would be appropriate, there being carr-land between
Wingale itself and the river.

The forms recorded in France above are clearly to be
discounted, but the DB spelling, repeated twice, might possibly
suggest that the name was originally identical with Wingham K and
that the second el. was changed from **hām** 'a homestead' to the

dat.sg. *hale* of OE **halh** post Domesday. There can be no certainty, however, since the DB form is not supported by any subsequent spellings.

ASH HOLT. BECK END LOCK, - PLANTATION, *v.* South Kelsey Drain *infra.* CLAY LANE, 1847 *TAMap*, cf. *clay dales* 1652 *Rad, v.* **deill** in the pl. COACH RD. COLLEGE FM, commemorating the holdings of Trinity College, Cambridge, cf. *the medow of Trinitye Colledge in Cambridge* 1578 *Terrier, prat Collegii sce Trinitatis in Academia Cantabr'* 1588 *Monson, the Lease of Trenity Colledge in Cambridge* 1652 *Rad.* CROW HOLT. DEAL BECK. GIPSEY LANE. HOLME HILL FM, cf. *Holm Hill* 1580 *Monson,* 1830 Gre, *Home hill* 1652 *Rad, Holme Hill* 1771 *MiscDep 149,* 1824 O, *v.* **holmr, hyll.** JERVIS BRIDGE, *Jarvis Bridge* 1828 Bry, named from a local family, cf. William *Jarvis* 1797 *LTR.* KELK'S FM, named from the surn. *Kelk; John Kelk of Glamford Brigge* was given land in the adjacent parish of Thornton le Moor 1797 *Dixon.* MANOR FM, cf. *the mannour house* 1591 *Monson* and South Kelsey Hall *infra.* MILL RD, cf. *the myll gate* 1578, *ye Milgate, millgate* 1612, *Mill-gate* 1634, *the ould milne gate* 1671, *ye old milne gate, ye way to ye milne* 1697, *Old Mill-gate* 1762 all *Terrier, the Mill gate* 1652 *Rad, v.* **myln, gata** 'a road'. MOORTOWN, *Moor Town otherwise Moreton* 1808 *Td'E, Moortown* 1813, 1814 *BT,* 1830 Gre, *Morton* 1824 O, *the hamlet of Moorton, or Riverhead* 1842 White, cf. *River Head* 1828 Bry, 1830 Gre, 1838 *TAMap* (Holton le Moor); this is a comparatively modern name, though the *moor* itself is recorded earlier as *moram de Suth Kelsey* 1220-40 *Foster, Kelsey Moore* 1667, 1678, 1683, - *moore* 1679, *South Kelsey More* 1669 all *MiD,* - *Moor* 1772 *Nelthorpe.* MOORTOWN HO, *Moorton House* 1842 White. MOORTOWN PARK is *Moortown Hill* 1824 O, 1828 Bry, 1830 Gre, *Moorton Hill* 1833 *Tur,* 1837 *PT.* NEW BRIDGE. NORTH END, *Kelsey North-end* 1824 O and is *South End* 1828 Bry. NUNNERY FM, commemorating the holdings of the Nuns of the Priory of St Leonard of Grimsby, cf. *a Lea ... sometyme belonging to the Nunns of Grymsbye* 1578 *Terrier.* PINGLE WOOD, cf. *the pingle att the parke side* 1591 *Monson* (described as a wood in the text), *the Pingle End* 1652 *Rad, y^e* -, *the Pingle* 1708, 1724, 1762 *Terrier,* and *Close of Wood ground comonly called or knowne by the name of Pingle* 1667 *Monson, v.* **pingel** 'a small enclosure'. POVERTY DRAIN. RASPBERRY

PLANTATION. RECTORY (St Mary's) is *The Parsonage House* 1624, - *house* 1697, *Parsonage-House* 1724 all *Terrier, one Parsonage House* 1707 *Monson.* SAND LANE, cf. *sandyegate hill* 1580 *Monson, Sand gate furlonge* 1601, *the Sandes* 1611, 1625 all *Terrier, Sandgate hilles* 1653 *Monson,* (*the*) *Sandgate* 1671, 1703 *Terrier, v.* **sand(ig), gata.** SOUTH KELSEY CARRS, 1824, 1830 Gre, *the Carres* 1580 *Monson, the Common Carrs of South Kelsey* 1652 *Rad, the Carrs* 1591 *Monson* (Plan), 1771 *MiscDep 149, le karre* 1550 *Monson, le Carre* 1560 *ib, the Carr* 1579, 1762 *Terrier, South Kelsey carr* 1591 *Monson, - Carr* 1718 *MiscDon 108,* 1767 *Stubbs,* 1790 *Monson,* 1822 *Dixon, the Car* 1612, 1624 *Terrier, the Carre* 1624 *ib,* cf. *Northker* 1226-28 Fees, *Northkerre* m17 *Monson,* North Carr 1664 *ib* and *tenne stonges of Carreground* 1581 *ib, Carr groundes* 1663 *ib, v.* **kjarr** 'brushwood', ME *ker* 'a marsh, especially one overgrown with brushwood, etc.' All the parishes bordering on the R. Ancholme have their *Carrs.* SOUTH KELSEY DRAIN is earlier *Kelsey Beck* 1768 (1791) *LindDep Plans, the Becke* 1579, - *Beck* 1611, - *beck* all *Terrier,* 1653 *Monson,* 1671 *Terrier, le Beck* 1645 *Monson, y^e -* 1697 *Terrier,* cf. *atte Bek', - Beck'* 1332 *SR* (p), *the beck furlong* 1579, *ye becke furlong* 1612, *the beck furrelonge* 1624 all *Terrier, One land between Beck and Bank* 1703 *ib, v.* **bekkr** 'a stream, a beck'; an alternative name appears to have been *South Kelsey Dike* 1635 (c.1900) *LindDep 78, - dike* 1663 *Monson.* The earlier name survives in Beck End Lock, - Plantation *supra.* The following references seemingly refer to a different stream or streams: *ye beckes* 1601, 1612, *the Beckes* 1625, 1671, *y^e -* 1697, - *Becks* 1762 all *Terrier.* SOUTH KELSEY HALL, *att Hall'* 1503 *Monson, The Hall* 1708, 1724, 1726 *Terrier, Kelsey Hall* 1824 O, 1828 Bry, 1830 Gre; the manor of South Kelsey is called *halgarth* 1520, *Halgarth* 1527 both *Monson, v.* **hall, gar$\eth$r** 'an enclosure' and cf. Manor Fm *supra.* SOUTH KELSEY MILL. SOUTH KELSEY PARK, *The parke of South Kelsey* 1591 *Monson, the Parke of South Kelseye* 1645 *Holywell, - Southkelsey* 1656, 1678, *the parke of South Kelsy* 1672, *the parke -* 1679 all *Monson, The great Park* 1708, *y^e Great Park* 1762 both *Terrier, the little Parke* 1651 *Rad, Park, - Close, - Hill* 1822, 1840 *Dixon.* SOUTH WOOD, 1824 O, *the south woode* 1591 *Monson.* TAYLOR'S PLANTATION, named from a local family recorded at least from the 16th century, cf. Robert *Talor* 1554 *Monson.* WALKER LANE PLANTATION, named from a local family, cf. Daniel *Walker* 1743 *LindDep 82.*

WATERLAND, 1824 O, 1830 Gre. WATERMILL FM.
WESTFIELD FM, cf. *The west feilde* 1579 *Terrier, in occident campo de Southkelsey* 1588 *Monson, in campo occidn'* 1590 *ib, ye west feild* 1601, *the west feilde* 1611, *The Westfield* 1612 all *Terrier,* - *West feild* 1617 *Haigh et freq,* self-explanatory; it was one of the great fields of the village. The farm is actually in North Kelsey parish, but is *west* of South Kelsey. WILLOW LOCK. WINGALE FERRY (lost, TF 019 971), *Winghall Ferry* 1824 O, *Wingale Ferry Toll Bridge* 1830 Gre; it is marked on the Old River Ancholme, just below the modern Tollgate Bridge.

Field-Names

Forms in (a) are asterisked if they occur in both 1822[1] and 1840 *Dixon,* and are marked ** if also found in 1762 *Terrier,* but are dated if they are in only one of these. Spellings dated 1327, 1332 are *SR;* those dated 1503, 1506, 1520, 1528, 1538, 1546, 1554, 1558, 1559, 1560, 1564, 1565, 1570, 1573, 1574, 1580, 1588, 1590, 1591, 1598, 1599, c.1600, 1619, 1625, 1630, 1631, 1634, 1641, 1644. 1647, m17, 1653, 1663, 1664, 1677, 1688, 1705, 1707, 1719, and 1790 are *Monson;* 1579, 1601, 1611, 1612, 1624, 1625, 1634, 1671, 1697, 1703, 1708, 1724, 1762, and 1822[2] are *Terrier;* 1635 (c.1900) are *LindDep 78;* 1652 are *Rad;* 1718 are *MiscDon 108;* 1743 are *LindDep 82,* 1767 are *Stubbs;* 1771 are *MiscDep 149,* 1808 are *Td'E;* 1847 are *TA* 195.

(a) ye land call'd Atkinson 1762 (*Atkinson Lane* 1707, 1708, *Atkinsons lane* 1724, named from the *Atkinson* family, cf. Thomas *Atkinson* 1591 *BT,* Richard *Atkinson* 1641 LPR); Bank* (*les bankes* 1573, v. **banke**); the Bays Garth 1808, Bays Garth* (*the close cald Bayes* 1624, the same name seems to occur as Baysgarth in Barton upon Humber and East Halton and as Base Garth a f.n. in Brocklesby, PN L 2 32, 150, 68, but here the first el. is almost certainly the ME surn. *Bay,* which occurs early in the parish in Hugh *Bay* 1332 *SR,* the second being **garðr** 'an enclosure', as elsewhere); upon Blithe 1762 (*Blithe furlonge* 1579, *Blyethe nooke* 1601, *Blyth hill* 1624, *on Blyth* 1625, *on Blithe* 1634, 1653, *Blith, Blyth Nook* 1671, *on blithe* 1697, *Blith Nook* 1703, possibly named from the *Blyth* family, cf. William *Blythe* 1507 *Monson;* the references to *on Blyth, upon Blithe* etc., however, suggest that this is an old name, perhaps a stream name as in Blyth(e) Nb, Nt, Sf, St and Wa. The situation is not known, but if it were originally the name of a stream it would be derived from OE **bliðe** in some such sense as 'the gentle stream'; v. also the Hill Fm *infra*); the Brick Kiln Cl

1808, Brick Yard (Cl) 1822[1], Brick Yd, Brick-yard Cl 1840; ye Bull Marfare, - Mdw 1762, Meadow belonging to the Town for Keeping a Bull 1767 (*bull meddowe marfur* 1624, *the bull marfar* 1671, cf. *Bull Common als the Bull peece* 1664, *the Bull Common* 1635 (c.1900), v. **marfur** 'a boundary furrow', as elsewhere in this parish and cf. the appellative use in *a common Marfore or meerland* 1634); (high) Burgate 1762 (*Burgate* 1579, 1601, 1611, 1612, 1624, 1653, 1703, *Burre gate* 1634, *burgate* 1671, 1697 - *furlonge* 1579, - *stighe* 1611 (v. **stīg, stigr**), *burgat*(*e*), *burgatestigh* 1625, *burgate* 1671, 1697, *High Burgate* 1724; this is presumably **burh-geat** 'a town gate, a manor gate', found in Barton upon Humber PN L 2 36 and in many parts of England, v. EPN s.v.); Canal Bank*; Causey-Hedge 1762, Short Causey 1767 (*le Cawsydike* 1565, *the Cawsye next the medow of Trinitie Colledge in Cambridge* 1579, y^e *Causey* 1612, *Casey hedge*, - *hill* (sic) 1652, y^e *Causie hedge* 1697, v. **causie, dīk**); a little close at Wingall cal'd and known by ye name of ye Churchyard 1762 (*a little Close called the church Yard of Wingall* 1707, - *cal'd the Church-Yard at Wingall* 1708, *a little-close at Wingall cal'd & known by the name of the Church-yard* 1724, cf. *atte Kyrke* 1327 (p), v. **kirkja**); Coal Yard Bank*; ye comon 1762 (*super lee commons* 1580, *the commons* 1611, *the comons* 1625, 1653, 1671, *the Commons* 1671, y^e -, *the Comon* 1697, *the Common* 1724); Corn Cl 1822 (*Corn close* 1671, *corne close nooke* 1634, *the Corne Close side* 1652, *Corneclosegate* 1697); Cow Cl** (*Cow Close* 1707, - close 1708, *Cow-close* 1724); Croft*; Cross Swaithe 1767 (v. **swǣ ð** 'a strip of grassland'); The Dawkins Cl 1808 (no doubt from the surn. *Dawkin*, not so far noted in the sources searched); Deep-dale 1762 (*deepdale* (*mere furr*) (v. **marfur**), *deepdayle* 1579, *Deepdale furlong*, *deepedale* 1612, *deepdal* 1624, *in deepe dale* 1634, *Deepe Dale* 1652, *deepedaile* 1663, *deep dayell* 1671, *deep daile* 1677, *Deepdale* 1697, probably self-explanatory, v. **dēop, dæl, dalr**, but spellings in -*dayle*, etc., suggest at least the influence of **deill** 'a share of land', a word found elsewhere in the parish); Dickett 1762 (*dyckytt* 1574, *dicket* 1634, *the peece of common or wast commonly called the Dickett* 1644, *Dyckett Merefur* 1579, *dickett*, *dickitt* 1671, *Dicket merefur* 1612, *dicket marfur* 1624, *Dicket Mear* 1697, *dickitt gate* 1611, *Dickit* - 1625, *the top of Dickett* 1652, *Dickett flash*, - *Hill Stigh*, - *well* 1652, v. **marfur, (ge)mǣre, flasshe, hyll, stīg** or **stigr** and **wella**; *Dicket*, etc. is probably ON **dīk** with the ME (OFr) diminutive suffix -*et*, hence 'the little ditch'); Middle -, Upr Dinters 1767, 1790, Dinters Nooking 1771, Lr -, Upr Dinters*; Dog-kennel & Plantation*; the Great East cl 1762 (*the Great East close* 1707, *The great East Close* 1708, *the great-East close* 1724); East Cl* (*the easte close* 1591); East Fd*, East-Field 1847 *TA Map* (*the East Field, The east feild* 1579, *the easte feilde* 1591, *the easte feyldes of Sowth Kelsey* 1599, y^e *East feild* 1601, *The eastfield* 1612, *the East field* 1624, 1671, - *feild* 1625, 1652, 1718, *the Eastfield* 1634, - *feild* 1652, y^e *East feeld* 1697, *the East-field* 1708, 1724, *in orient'*

campo de South Kelsey 1580, *in campo orienti* (sic) 1590, one of the open fields
of South Kelsey); Fish Pond*; Green Cl 1771; Green Lane 1808 ('a track giving
access to individual plots in the open field'); Greens* (*Attegrene* 1327 (p), *atte
grene* 1332 (p), *v.* **grēne²**); Greetham-Stigh 1762 (*Greetham Stigh* 1652, *greetham
stee* 1671, *greetam steegh* 1697, named from the *Greetham* family, cf. John
Gretham 1525 *Monson*, Thomas *greathame* 1570 *BT*, with **stīg, stīgr**); Guizle-gate
1762 (*Gysell gate* 1579, *gyselgate* 1612, 1624, *Gizell gate* 1652, *gisell* - 1671,
gislegate 1697, possibly 'Gisl's road', from the ON pers.n. *Gisl*(*i*) and **gata**, as
suggested by Dr J. Insley); the Hill Fm 1762 (*cultur' voc' Hill* 1580, apparently
held at one time by members of the *Blyth* family, cf. William *Blythe de Hyll* 1570
Monson); the Holmes (Flg) 1771 (*Holme Hedge* 1718, *v.* **holmr**); Home Cl*;
Horsedale (Cls) 1771; Housham-Stigh 1762 (*Howsham Stigh* (sic) 1652, *Howsome
Stighe* 1634, *Horsham Nooke* (sic) 1652, *Howsam gate* 1671, *housam steghe* 1697,
named from the *Housham* family, cf. Thomas *Howsam* 1573 *BT*, George *Housam*
1612); Hunger hill 1762 (1652, *hungrell* 1579, *hungar hill* 1601, *Hungerhill* 1611,
hunger hill 1625, 1653, *hongger hill* 1641, *hunger hill* 1697, a common derogatory
term for infertile land, *v.* **hungor, hyll**); ye Ing furrow 1762 (y^e *ynge furr, le Ingge
furre* 1580, *yngfurr* 1590, *the Yngg furr* 1611, *the Ingfur* 1612, *the Ing fur'* 1624, *the
Ingfurre* 1625, *the Ing-furre* 1634, *the Ing furrow* 1653, *Ing furr* 1671, y^e - 1697, *v.*
eng, furh, with *furr* a regular form of *furrow* in L); Inglands 1762, 1767 (1743,
yngland furlong 1590, *overwhart Englandes* 1652, *v.* **eng** 'meadow, pasture' (as
elsewhere in this parish), **land**; *overwhart* doubtless applies to land ploughed at a
right-angle to the adjoining piece, *v.* **ofer³, þvert**); The lea cl 1762 (*the ley close*
1598, *the Lea Close* 1625, *the Lea-Close* 1707, *One Close cal'd the Lea Close* 1708,
- *called the Lea-close* 1724, from ModE **lea** (OE **lēah**) 'grassland, meadow'); little
dale 1762 (*little dale* 1612, 1624, 1634, 1697); Love-dale 1762 (*Lovedayle* 1579,
Lovedale 1579, 1612, 1703, *Louedale mouthe*, - *bottom* 1611, *lovedale* 1624,
lovedale (*mouth*) 1625, *Lovedale* 1634, *Love-dale Commons* 1671, *Lovedale Comon*
1697); Low Hole 1767; Mill Cl*, - Dam*, Mill Farm Yd*, Mill Hill* (cf. *the Mill
Mar-foure* 1652 (*v.* **marfur**) and Mill Rd *supra*); Richard Milsons lane end 1762;
ye Moor Banks 1762, Moor* (*v.* **mōr¹, banke**); ye New-intack 1762 (*the new
Intack* 1707, *the New intack* 1708, *the New Intack* 1724, *v.* **inntak**); Nordale Cl
1822¹, Nor Dale - 1840 (*Northedale furs* 1579, *the north dale* 1579, *lee north daille*
1580, y^e *North dale* 1601, *the North-dale close* 1611, *the Northdale* 1611, *the
northdale close, nordale* 1625, *North Daile comons* 1653, *north dale* 1681 *BT*, *in
Nordale* 1703, y^e *North dale* 1719, the situation of the field is not certainly
known); Orchard*; Paddock 1822¹ (several), Lr Paddock 1840 (cf. *a close Called
Paddockes* 1591); Park Ings 1762, 1847 (*the Park-Ings Adjoyning to Nettleton-Ings*
1708, *the Park Ing's* (sic) 1724, *v.* **park, eng**); 1 parcel called Parsons dike 1767
(Glebe belonging to St Mary's Church) (*dyke cald parsons dike* 1624, *the Parson*

dike 1634, *Parson dike* 1671, *v.* **persone, dīk**); Plantation* (numerous); Pleasure Grd & Garden 1822[1], Plant[n] & Pleasure Grd 1840; y[e] Rack cls 1762 (*the Racke Close* 1625, *the two Rack-closes* 1708, *Two Closes called the Rack-Closes* 1724, from the surn. *Rack*, cf. *William Rack* 1759 *BT*, William *Rack* 1800 *LTR*); A close cal'd Rougham 1762, Ruffam Longleys 1771, Ruffans* (sic) (*Roughame* 1612, *Rougham* 1634, *Rougham closse* 1624, *Ruffam Close* 1668, 1705, 1707, *Rougholme close* 1697, *a Close called Rougham* 1724; *v.* **rūh** 'rough', but it is impossible to be sure of the second el., though it is probably from **holmr**); the Sand Cl 1822[2]; One close cal'd Several 1762, Severals* (*one close called seuerall* 1625, *one Close Call'd Severall, One Close cal'd Several* 1707, - *called Several* 1724, 'land in private ownership, enclosed land cultivated separately from the open fields', *v.* **several**); Short headland Stigh 1762 (*shorthead Land stigh* 1612, *short headland Stigh* 1624, 1652, - *stee* 1671, *short head land steegh* 1697, *v.* **sceort, hēafod-land**, with **stīg, stīgr**); Smithfield Cl*; Sourback Nook 1771; South Ing Cl 1771 (*South engys* 1506, *the Sowth inges* c.1600, *v.* **sūð, eng**); Spetch lane 1762 (*Spetch Lane* 1707, *Spetch-lane* 1708, 1724, named from the *Spetch* family, cf. Peter *Spetch* 1618 *BT*, 1628 *Inv*); Spring Well Dale* (*Spryng well daylles* 1559, *v.* **spring, wella, deill**); Stack Yd*; Syke (*the Syke* 1579, *the syke* 1612, cf. *the Syke close* 1591, *v.* **sīk**); y[e] great thorn, the little - 1762 (cf. *in spinis* 1332 (p)); Thorn Cl* (*Thorne closes* 1625, *the great & Little thorn Closes* 1707, *The great & Little Thorn-Closes* 1708, *The Little-Thorn close, Great-Thorn-close* 1724); Thorn Stea-gate 1762 (*Thorn styghe gate, Thornesty gate stight* 1579, *Thorne stye gayte* 1590, *Thorne-stighe gate* 1611, *Thorne stigh gate* 1612, *thornstighgate* 1624, 1625, *Thornstie gate* 1653, *Thornsty* - 1671, - *Gate* 1702, *thornsteeghgate* 1697, 'the gate to *Thornsty* ('the path to the thorn (tree)' *v.* **þorn, stīg, stīgr, geat**; the context of the 1612 form makes it clear that *gate* is a gate and not a road; cf. ye great thorn *supra*); the Toft Flg 1771 (*cultur' voc' Tofte* 1580, *v.* **toft**, cf. Toft Hill flg in North Kelsey f.ns. (a) PN L 2 191); Town End Cl* (y[e] *towne End* 1697, 'land at the end of the village', *v.* **tūn, ende**[1]); the Town Street 1822[2]; y[e] Warlots 1762, the Warlots 1767, Warletts, Worletts 1771, Warlott's Cl 1790 (*warlot dick end* 1601, 1671, *Warlott dike* - 1611, - *dykeend* 1625, - *Dike End* 1703, *Warlotts* Chas1 *Dixon*, *Warlottes* 1631, *commons in warlotts* 1634, *Warlatts* (sic) 1634, *Warletts* 1635 (c.1900) *LindDep* 78, 1699 *Foster*, *Warlots* 1697, *War-latt-Carr* 1708, *A certain place called Warlotts* 1719, *a place called the War-lot's* 1724, *Woarlott* 1743, *v.* **warlot** and note its appellative use in *commons in warlatts* 1634; for a full discussion *v.* PN L 2 67); y[e] Warren 1762, Warren Cl* (*warennam in Suth Celseya* 1275 RH, - *in Suth Kelesbey* (sic) 1276 ib, y[e] *Warren close* 1671, *v.* **wareine**); West Cl* (*west Close* 1647); the Great West Land or Westholme Closes, Westlands Close Gate 1761 (*the two westlandes closes* 1625, *v.* **west, land, holmr**, with **clos(e), geat** or **gata**); y[e] west street lea 1767; y[e] white Pits 1762 (*Whitpits* 1612, 1624, *the*

White-pits bridge 1724); Willow(s)-gates 1762, the Willows 1771 (*willowes gate* 1579, *willowes gate furlong* 1611, *Willowes* - 1625, *Willowes gate* 1612, *willowegate* 1624, *willowegate* 1624, *willows gate* 1671, *willowes gate* 1697, *the willowes nooke* 1611, 1634, 1671, cf. the Willows in North Kelsey f.ns. (a) PN L 2 191); One Close Called Windsor 1762 (*a close called Wyndesore close* 1591, *One close call'd Wind-sore* 1707, - *cal'd Windsore* 1708, - *called Windsore* 1724); Lr -, Upr Winesome*; Wingall fore-garth 1762 (1708, 1724, *v.* **fore** 'in front', **garðr** 'an enclosure' and Wingale *supra*); wood close Nook 1762 (*the wode Nooke* 1579, y^e *wood noke furlong* 1601, y^e *Wood nooke* 1612, *wood nooke* 1634, *the* - 1671, y^e *wood Close* 1697, *v.* **wudu, nōk**); the new Wood 1808, Wood Pasture*.

(b) *Acklam hyll* 1579, - *hill* 1612, - *Hill nooke* 1652 (named from the *Acklam* family, cf. Robert *Aclam* 1553 *Monson, v.* **hyll, nōk**); *le acredike* 1558, *le acredyke inter Thornton & Kelsey* 1559, *Acredike* 1560 (*v.* **æcer, dīk** and the same name in Barnetby le Wold f.ns. (b) in PN L 2 13); *in arbrough Furlong* 1590 (*v.* **hereheorg,** ME *herber{3}e* 'a shelter (for travellers etc.)'); *the Ashgarth* 1625 (*v.* **æsc, garðr**); *Bartons headland* 1611 (William *Barton* is named in the same document, *v.* **hēafod-land**); *Beament lane end* 1634, *Bent Lane* (sic) 1703 (named from the family of Gilbert *Beamond* 1570 *Monson*); *Beangarth nooke* 1599, *Beane garth nooke* 1652 (*v.* **bēan, garðr, nōk**); *le beregate* 1503 (the same name is recorded twice in the e16th century Owersby f.ns. (b), less than three miles to the south-east, so the three references may well be to the same road; the meaning is perhaps 'the road along which barley is carried', *v.* **bere, gata**, and note that forms in *bere* are recorded in MED a1500, s.v. *ber(e)* n. (2a). It does not seem to be the same as (high) Burgate *supra*); *loco voc' Blakmyls* (sic) 1564, *Blacke Mills* 1652 (possibly alluding to land with black soil, *v.* **blæc, mylde**); *blind Laine (end)* 1652 ('(the end of) the cul-de-sac', *v.* **blind, lane, ende**[1]); *blynde well* 1580 ('a spring or well concealed by vegetation', *v.* **blind, wella**); *iuxta Bull parte* (sic) 1590; *In campo ... de Suth Keles'* a1218 RA iv, *in campis de Southe Kelsey* 1580 (*v.* **feld**); *Carr Close Nooke* 1652 (*v.* **kjarr, clos(e), nōk**); *lee caves* 1580; *Cheesmans lane* 1524 (named from the *Cheesman* family, cf. John *Cheisman* 1548 *Monson*); *the Church Streete* 1634; *Cladayels* (sic) 1671 (*v.* **clæg, deill**); *Cobham headland* 1652, *Cobheadland* 1697, *cobman Hill furlong* 1590; *The wood called the conygre* 1591 (*v.* **coninger** 'a rabbit warren'); *Cotager dale* 1630 ('an allotment of land occupied by a cottager', *v.* **deill**); *le Cowe furthe* 1554, *the Cowe Fyrthe* 1580 (*v.* **cū, fyhrðe**, here possibly in the sense 'fenland overgrown with brushwood'); *the cow gate* 1579, *the Cow gate* 1579, 1671, *the cowe gate* 1599, 1624, y^e *Cowgate* 1612, 1697, *Cow-gates* 1634 ('track(s) used by cattle', *v.* **cū, gata**, though the sense 'pasture for an individual animal' (*v.* **cow-gate**) is also possible); *crossehill* 1579; *the Crossland Close* 1635; *Crowston*

Close 1590 (named from the *Crowston* family, cf. John *Crowston* 1681 *BT*); *davye thing* 1523 Wills i (from the family of John *Davye* 1580 and þing 'a possession, property'); *dent lane* 1671, *Dent Laine end* 1652 (named from the *Dent* family, cf. W. *Dent* 1622 *Dixon*); *One peice of meadowe ground ... called the Dentures being deuided into foure partitions* 1591, *the meadow ground in the Carrs inclosed called the Dentures* 1591 (Plan) (this is unlikely to be from dial. *denshire* 'land which has been pared and burnt', v. EDD and Field 62 s.n. Dencher field. It is probably the pl. of ME *denture* (MED s.v.) 'an indenture', presumably referring here to meadow divided into four by a written formal agreement); *lez donghylles* 1520 (i.e. 'dung-hills'); *Dyson lane ende* 1599 (no doubt from the surn. *Dyson*, not so far noted in the parish); *frost holes* 1652, - *hoals* 1671 *Frost-holes* 1697 (probably from **hol**[1] 'a hollow', and the surn. *Frost*, cf. Thomas *Frost* c.1520 *Monson*); *the fures syde* 1579 (probably 'the furrows side'; for *fure* cf. ye Ing furrow in (a) *supra*); *ye Gayres* 1601, *the Gaires* 1611, 1703, *gaires* 1625, *the Garres* 1671, *gares steegh* 1697 (v. **geiri** 'a triangle of land', with **stig, stigr**); *Gamsholme* 1579, 1612, *gamesholme* 1612, 1624, *Gamsonne Close* 1652, *gamson* -, *gamsam Close, gamson lease* 1671, *Gamson* 1697 (the earliest forms, though late, suggest that this is derived from the ODan pers.n. *Gamall* and **holmr** 'a raised piece of land amidst marsh', a Scand. compound, cf. Gamston, PN Nt 51, for early ModE forms showing a similar reduction of the first el.); *atte Gatende* 1327 (p), *atte Gatende of Suth Kelseye* 1328 Banco (p) (v. **gata** 'a road', **ende**[1]); *the Graues* 1652 ('the trenches', v. **græf**); *atte Graunge* 1332 (p) (v. **grange**); *gravyll pyttes* 1580 (self-explanatory); *the close heretofore called Haule croft and now Barkers close* 1591 (it was in the tenure of John *Barker* 1591); *Haverdale* 1652 ('a plot on which oats were grown', v. **hafri, deill**, cf. Haverdale Leys in North Kelsey f.ns. (a) PN L **2** 188); *Hellpitt* 1601 (possibly a derogatory name); *the hie steet* (sic) 1611; *the high mere* 1611, *the High Meere* 1652 (v. **hēah, (ge)mǣre** 'a boundary, land on a boundary'); *long houle gate leyes* 1601, *holgate lees* 1611, *Houlgate leies* 1625, *Houlgate* 1653, *Holgate long leys* 1671, *Holegate Lays* 1703 ('the road running in a hollow', v. **hol**[1], **gata**, with **lees, leys** (from the pl. of **lēah**) 'pieces of meadow'); *The Hoppyarde* 1591 (self-explanatory, v. **hoppe, geard**); *Housewell Close* 1718; *Hove dayle* 1590; *Mrs Hunton Lane End* 1671 (*Hunton* is freq. in *BT*s, cf. Daniel *Hunton* 1635 *BT*); *the hy furs* 1579 (v. **furh** 'a furrow', in the pl.); *Kerberg* a.1218 RA iv, *Cowburght furlonge* (sic) 1579, *long Carre-barre* 1634, *Carbrough* 1653, *long Carbar, Carbor* 1671, 1697 ('a mound in or near a marsh', v. **kjarr, beorg, berg**); *the Kilnhouse Close* 1634 (v. **cyln, hūs**); *the kirke hill* 1579, *ye kirke hill* 1601, *north -, south kirkehill* 1625, *the North Kirk-hill, ye South Kirke-hill* 1671, *Kirkhill gate, the North Kirkhill* 1611, *North Kirk hill* 1703, *South Kirk hill* 1703, *the Church hill* 1652 (the frequent occurrence of *kirk(e)*, from Scand. **kirkja**, is noteworthy); *Kyngesbery garth* 1503 (no doubt from the surn. *Kingsbery* and **garðr**); *the Lady close* 1591 (perhaps a close dedicated to Our Lady); *Leifsideil* a1218 RA iv, *Laceby Dale*, - *Dayle* 1652 (probably from the Scand. pers.n. *Leifr*, in the a1218 form showing confusion with ME *Lefsi* (OE *Lēofsige*), with **deill**; the 1652 forms, if they belong here at all, are probably due to popular etymology); *the little close* 1634; *the Longe bancke end* 1652; *Luchull'* a1218 RA iv, *Luckhyll, Lurkhill* (sic) 1579, *Luckhill* 1612, 1624, 1653, 1703, *Luckhill gate* 1625 (obscure); *the Malt kill ('in eodem Clauso aedificat')* 1631 (cf. *the Kilnhouse Close supra*); *Mawdmylne hole* (sic) 1528; *meadowe cloose* 1591; *the middle land* 1624 (*v.* **middel, land**); *the middle styghe* 1579, y^e *middle stigh* 1612, 1624 (*v.* **middel, stig**); y^e *middle syke close* 1612, *the middle syke closse* 1624 (*v.* **middel, sik**); *Naylors Lane end* 1703 (named from the *Naylor* family, cf. Thomas *Nailor* 1657 *BT*); *parco vocat' the Night fold* (sic) 1647 (probably demesne land on which tenants were required to fold their sheep overnight – a frequent cause for complaint that their pasture was consumed so that the lord's fields might be manured); *northbeck* 1546 (*v.* **norð, bekkr**, cf. *the Southbeck infra*); *North dyke* 1506, *le north dike* 1574 (*v.* **norð, dík**); *Northiby* 1332 (p) ('north in the village', *v.* **norð, í, bý**); *the Oate cloose* 1591; *Ouerdale* 1327 (p) (*v.* **ofer²** or **ofer³, deill**); *the oxpasture* 1591, *oxe pasture* 1647; *the parson stackes* 1579, y^e *parsons* - 1612, *the parson-stacks* 1634 (from ON **stakkr** 'a stack', as in *Stackgarth* etc.); *the Parsonage House* 1634, - *house* 1697, *parsonage lane end* 1634 (cf. Rectory *supra*); *persons thyng* 1503 (*v.* **persone, þing**); y^e *pasture banke* 1697; *the Pingle* 1625 (*v.* **pingel**); *plummer lane end* 1634 (from the surn. *Plummer*, cf. William *Plumber* 1589 *BT*); *potter Styghte, potters Streyte* (sic) 1546, *le potter Stight* 1560 (named from the *Potter* family, cf. Margaret *Potter* 1595 *BT*; *stight* is common in north L for *sty* 'a path', *v.* **stigr**); *Purley cloose* 1599 (perhaps 'land on the route of a perambulation of a parish or a forest', *v.* **purlieu**); *lee prates de south Kelsey* 1580 ('South Kelsey meadows', *prates* being an Anglicization of Lat *prata* (pl. of *pratum* 'meadow'), cf. ME **pre(y)** 'a meadow'); *Rawlett oxgange* 1590 (named from the *Rawlett* family, cf. John *Rowlat* m16 *Inv. v.* **oxgang**); *sandyegate hill* 1580 (*v.* **sandig, gata**); *Sharpe warfe* 1579, - *Wathe* 1611, *Sharpwath* 1625, *Sharpe wath* 1652, *Sharpe-wath* 1671 (named from the *Sharpe* family, cf. Francis *Sharpe* 1683 *BT*; William *Sharpe* 1706 *BT*; John *Sharp* 1767 *Stubbs*, with a doubtful second el.; *warfe* suggests **waroð** 'streamside land', but *wath* ON **vað** 'a ford')); *Shorles lane* 1599 (*Shorles* is no doubt a surn.); *Sickle Close gate* 1652 (unless it is a variant of *Sike* (*v.* foll.), *sickle* may be from the diminutive **sícel** 'a small stream'); *Sike Close gate* 1697 (*v.* **sík**); *Soerling gate* 1652; *Sorry Lesse Lane, Sorrtles Lane* 1591, *Surrilusse Lane* 1619 (obscure); *the Southbeck* 1656, *lee south beckes* 1580 (*v.* **súð, bekkr**); *the south feilde of South Kelsy* 1591 (one of the open fields of South Kelsey); *Sowter dayle merefurr*, - *daile mere furr* 1579, *Sawter dale(s) merefur* 1612, *Sauterdale marfur* 1624 (from the occup. n. or surn.

Sowter, cf. William *Souter* 1332 *SR*, with **deill** and **marfur**); *Staynhusland'* a1218 RA iv ('(the strip) by the stone house', *v.* **steinn, hūs**, with **land**); *the stighe* 1611, - *stigh* 1625, *the neither* -, *the upper styghe* 1579, y^e *nether* -, y^e *upper stigh* 1612, *the upp & nether* - 1624, *the crooking of* y^e *stigh* 1579, y^e *crooking of the stigh* 1612, - *of* y^e *stigh* 1624, *the stigh gate* 1625 (*v.* **stīg** 'a path'); *ii landes on stoupland* 1641, *Stopland(e)s* 1652 ('strips by a post or way-marker', *v.* **stolpi, land**); *betwixt the streetes* 1634; *syx swaithes called tow stonges or two acres* 1579 (a name offering a local equation of these units of area, *v.* **swǣð, stǫng, æcer**); *toft' Johīs Sutton'* 1503, *Sutton gares* 1579 (referring to the *Sutton* family, cf. also Humfrey *Sutton*, William *Sutton* 115 Monson, *v.* **toft** 'a building plot, a curtilage' and **gara** 'a gore, a triangular piece of ground', or the cognate Scand. **geiri**); *Tenn Foote Swaith* 1743 (*v.* **swǣð**); *the townesgate* 1624; *Twede thyng* 1503 (named from the family of Robert *Twead* 1589, *v.* **þing** 'property, premises'); *the water Furres* 1579, *the Waterfurs* 1612, *Waterfurs* 1624, *waterfurres* 1634, *the water furrs* 1652, *waterfurs* 1697, *a furlong cald Waterfurswead* 1589 (*v.* **wæter, furh**); *les Waterlades, le Waterlate* 1538; *Water Laine end* 1652; *James Watson close* 1611; *the west Becks* 1703 (*v.* **bekkr**); *the weste woode* 1591; *the Wheate close* 1591; *Windinges* 1590 (this is found as an appellative in *windinges* c.1580 *Terrier* (Nettleton in Yarborough Wapentake) and *v.* *two windinges* in Binbrook f.ns. (b) *supra*); *the wood stigh* 1634 (*v.* **wudu, stīg**); *Wynghall sik* 1588, *wingall headland* 1625 (*v.* **sik, hēafod-land**; named from Wingale Priory *supra*); *Wykast hyll* (sic) 1546 (obscure).

Kingerby

KINGERBY (now included in Kirkby cum Osgodby parish)
    Chenebi (2x) 1986 DB
    Chimerebi c.1115 LS (with -*m*- for -*ni*-)
    Chinierby 1139-40 AC
    Kyinerbeia 1135-64 RA iv (with -*in*- for -*ni*-), *Kinierbia* 1212
        Fees, *Kynyerby* 1276 RH, 1291 Tax, 1343 NI, 1431 FA,
        1530 Wills ii, 1537 *AOMB 210*
    Kinierebia c.1162 RA iii, -*by* c.1230 (14) *VC* (p)
    Kinnerbi m12 Dugd vi, l12 ib vi, 1270 RRGr, -*by sive*
        Kingerby 1609 *Foster*, -*bie* 1576 LER
    Kynerebi 1163 RA i, 1231 Welles
    Kinerbi c.1160 Dugd vi, 1194 Abbr, 1194 CurP, 1230-50
        Foster, -*by* 1203 Cur, 1203 Ass
    Kenerbi 1168 P, 1169 ib (p)

Kynerbi 1190, 1191, 1192, 1193, 1194, 1195 P, *-by* 1230-50
  Foster, 1242-43 Fees, 1254 ValNor, 1275 RH, 1277 Dugd vi,
  1303 FA, 1404 *Foster*, 1535 VE iv, *-bie* 1554 PrState
Kinarby 1203 Abbr, *Kynarby* 1273 RA iv
Cunehereby 1208 FF
Kinardbi 1216 OblR, 1346 FA, *Kynardby* 1293 Ipm, e14 AD ii,
  1343 Cl, 1357 *Cor*, 1374, 1377, 1398 *Foster*, *Kynnardby* 1293
  Ass, 1364 *Foster*
Kynnerdby 1261 RRGr, *Kynardby* 1416 Cl, 1424 *Foster*
Kynerdeby 1272 *Ass*, 1332 *SR*
Kiniardby 1287, 1295 RSu, *Kynyardby* 1218 Pat, 1368 *FF*
Kynierdby 1293 Ipm, *Kynyerdby* 1322 Cl, 1322 Ipm, 1331,
  1392, 1395 Pat
Kyniardeby 1305 FF, *Kynyardeby* 1328 Ch, 1342 Fine, 1347
  Pat
Kynyerdeby 1327 *SR*, 1351 Peace (p), 1412 FA
Kyngardeby 1327 Pat
Kyngerdby 1428 FA
Kyngerby 1428 AASR xxix, 1428 FA, 1445 AASR xxix,
  Kynggerbye 1577 *Terrier*, *Kingerbye* 1576 Saxton, 1610
  Speed, *-by* 1595 Bodl, 1612 *MiscDep 332*, 1635 *Yarb et
  passim*, *-bie* 1635 *ib*

Ekwall, DEPN s.n., and Fellows-Jensen, SSNEM 57, agree
that this is 'Cynehere's farmstead or village', from the OE pers.n.
Cynehere and ODan *by*, as may be suggested by the earliest forms.
Starting from this premise, then medial *-d-* in spellings from the
13th to the 15th century is merely intrusive. Forms in *-ier-*, *-yer-*
and *-aird-*, however, can hardly be explained as developments of
Cynehere, and the first el. may well be an OE pers.n. **Cynegeard*.
Although this name is not on record, the el. *-geard* is attested in the
OE pers.ns. *Ælfgeard*, *Ōsgeard* and **Sægeard*, v. Feilitzen s.n. Later
forms in *-g-* are presumably due to association with ME *king*, cf.
perhaps Kingsbury, PN Wa 16, the first el. of which is the OE
pers.n. *Cyne*. Spellings in *Ch-* are due to AN influence. The first
el. here does not have the normal OE gen.sg. ending *-es*, but there
are a number of forms with medial *re-* and *de-*. Ekwall, IPN 69-70,
interprets *-e-* in spellings of this kind as standing for the ODan
gen.sg. in *-a* from *-ar*. In view of their relatively sporadic
appearance, however, they are perhaps better explained as

svarabhakti vowels between -r- or -d- and -b-.

ALMSHOUSES, 1824 O, 1930 Gre. Thomas Bell, in his will, instructed that his heirs *shall build or cause to be builded and erected, an Almshouse to contain six convenient rooms ... for the free dwelling of six poor Men and Women* 1675 (1725) *MiscDon 238,* 183; the original will has not so far been found. THE CHASE, 1842 *TA.* CROFT FM, cf. *Croft* 1842 *ib;* the farm is marked, but not named on *TAMap* itself. THE DAWDLES, cf. *close ... called Dawdales* 1612 *MiscDep 332, great -, little dawdales* 1657 *Red, Dawdles* 1842 *TA,* the second el. being probably the pl. of **deill** 'a share of land', earlier forms being needed to determine the etymology of the first. It could be ME **dawe** 'a jack-daw' or the surn. *Dawe,* which is derived from the bird-name, or from a hypocoristic form of the pers.n. *David.* JESMOND FM, cf. *Jesmound* 1842 *TA,* perhaps a transferred name from Jesmond Nb, but since it is not on a boundary it can hardly be a nickname of remoteness. KINGERBY BECK, 1707 *Terrier,* 1721 *Monson,* 1768 (1791) *LindDep Plans,* 1822, *the beck* 1703, *ye Beck* a1605 all *Terrier, v.* **bekkr.** KINGERBY GRANGE, cf. *Grange Field* 1842 *TA;* for the late use of **grange** in L, *v.* The Grange in Claxby *supra.* KINGERBY MANOR, *Mannor howse* 1612 *MiscDep 332, Hall* 1824 O, 1832 Gre, and cf. *ad Aulam* 1327, 1332 *SR* (p), *Hallfield* m18 *Monson, Hall Park* 1842 TA and *v.* P&H 286, s.n. Kingerby Hall. KINGERBY SPA, 1824 O, 1830 Gre, the site of a chalybeate Spring. KINGERBY VALE FM. KINGERBY WOOD, 1824 O, 1830 Gre, *Wood* 1842 *TA, Wood Plantation* 1828 Bry. THE LODGE, cf. *Lodge Field* 1842 *TA.* PINFOLD. REDHOUSE FM, *Red House* 1824 *O,* 1830 *Gre, - Ho* 1828 *Bry.* VICARAGE, cf. *Vycarage closes* 1577, *ye Virage Close* (sic) both *Terrier.* YOUNG'S WOOD, cf. *Mr Young's Swarrs* 1822 *Terrier, Youngsbury* 1842 *TA* (for *Swarrs, v.* Swards in f.ns. (a) *infra).* It was named from a local family, cf. Isaac *Young* 1789 *Monson* and note James *Younge* was lord of the manor 1842 White.

Field-Names

Principal forms in (a) are 1842 *TA* 197; Spellings dated 1230-50, 1284, 1364, and 1398 are *Foster,* 1577, 1679, 1697, 1703, c.1705, 1706, 1707, and 1822 are *Terrier*, 1612 are *MiscDep 332*, 1635 are *Yarb*, 1657, 1715, 1726, 1738, 1766, 1799, 1802, and 1838 are *Red*; other sources are noted.

(a) Albany (on parish boundary); Bank; Barn Fd; Beck Carr (*v.* **kjarr**), Beck Cl (3x, referring to Kingerby Beck, *v.* **bekkr**); Behind the House 1802; Bishop Bridge Cls 1838 (named from Bishop Bridge in Glentham parish LWR); Far -, Level -, Square Bottom (*v.* **botm**); Brick-Kiln Cl 1802, Brickiln Cl 1842; Bridge Cl adjoin the Ancolm 1802 (cf. foll. and Brigg Furlong Cl *infra*); Brigg end Cl 1766, - End Cl 1799 (*Brigg end close* 1657, 1726, 1738, *Briggin Close* (sic) 1715), Brigg Furlong Cl 1766 (*Briggfurlonges usque in Ancolm* 1230-50, *Brigg Furlonge* 1657, *the Brigg Furlong close* 1715, *Brigg Furlong* - 1726, *Briggfurlong* - 1738, from ON **bryggja** or the Scandinavianized form of OE **brycg**, with **furlang**); Broadgate (*v.* **brād, gata**); Brook Green (*v.* **brōc**; it adjoins Kingerby Beck, cf. Beck Cl *supra*); Broxtowe Cl 1766, Broxholme - 1799 (*great -, little Broxholme* 1612, *great -, Little Bloxholme* 1657, *Broxholme Close* 1715, 1726, 1738, perhaps *v.* **brocc** 'a badger', **holmr**); The Canons (cf. Friars Thorns· *infra*); Church Yd; The Cliff; Corn Cl 1802; Corner Cl; Court fd (*v.* **court**; it is beside Kingerby Manor *supra*); Dale Cl; Dock Carr (*v.* **docce, kjarr**); East Fd (*in campo orientali* 1284, *Eastefielde* 1635, *v.* **ēast, feld**, earlier one of the open fields of Kingerby); the Eighteen Acres 1802; The Elmhurst; Fore Cl (*v.* **fore** 'at the front'); Fotherby Fd (cf. *Fotherby Ings* 1657, presumably from the surn. *Fotherby*); The Fox Cover; Friar Thorns (adjoins The Canons *supra*); Furze Cl (*v.* **fyrs**); The Garth (*v.* **garðr**); Glebe; The Glen; Great Western; Hatfield (*v.* perhaps **hǣð, feld**, but the form is late); the Home Cls 1802, Home Cl 1842; Hopedale; Horse Pasture; Hospital Cl, - Gdn; House Cl 1842 (adjoins Redhouse Fm *supra*); Little Ings Cl 1766, 1799 (*the little Ings Close* 1715, 1726, *the Little* - 1738, *v.* **eng** 'a meadow, pasture', as elsewhere in the parish); Johnson Cl (named from the *Johnson* family, cf. William *Johnson* c.1705); Kells Cl (from a local family, cf. William *Kell* 1793 *LTR*); Level Cl; Long Acre; Middle Cl; Minster Acre (*v.* **mynster** but in what sense is uncertain, **æcer**); Morton Fd (presumably from a family name); Nine Acres; Norton Cl (*Nortons close* 1612, - *Close* 1635, named from an earlier member of the *Norton* family, cf. Robert *Norton* 1776 *BT*); Oak Tree Cl; Oatlands; The Orchard 1822; Osgodby Cl, - Lane (the reference is to Osgodby in Kirkby cum Osgodby parish *infra*); Oxlay (*v.* **oxa, lēah**); Paddock; the Pingle 1802, The - 1842 (*v.* **pingel**); Plantation; Plough Cl; Ryegrass Cl 1802; Sandbeck (*v.* **sand, bekkr**); Old Seeds 1802, Seed Cl 1842; Sheep Fd (*the sheepefielde* 1612, *Sheepfield* 1635); Shepherds Wong *v.* **vangr** 'a garden, an in-field'); Sockdike Cl (perhaps 'the drainage ditch', *v.* **soc** in the sense 'drain, drainage', **dīk**); Stackyard Cl (by a house); Swards (*y*^{e} *swares* 1577, *y*^{e} *swarres* 1679, *A Cloase called Swarchs* (sic) 1697, *a Close called the Swarrs* 1703, *Swarrs* c.1705, *y*^{e} *Swarrs* 1706,

Mr Dixon Swarrs 1707, perhaps 'greenswards, grassy land'); Sweede Cl; Taylor Cl (named from the *Taylor* family, cf. *Gylbert Tayler* 1588 *BT,* John *Taylour* 1693 *ib*); Thorofare Cl (by a road); Twelve Acres; Twenty Acres Cl 1766, twenty - 1799 (*the Twentie Acres & conteyninge by estymacion Fiftie & two Acres* 1635, *Twenty acres Close* 1657, *the Twenty Acre Close* 1715, *Twenty Acres close* 1726, 1738); Walk (probably a sheep-walk, *v.* **walk**); the Wharf 1838 (at Bishop Bridge); Westgate (*v.* **west, gata**); Woodlands (by Kingerby Wood); Yarrow; Yews.

(b) *Bowmans Yard* 1612 (probably from the surn. *Bowman*); *Church layne* 1577; *the Cottyers pasture* 1635 (i.e. Cottagers); *super Dale* 1230-50; *Esthynges* 1398 ('the eastern meadows', *v.* **ēast, eng**); *Greene close* 1577, *The Green Cloase* 1697 (perhaps an enclosure of the village green, *v.* **grēne**²); *Little Habberdings Close* 1715 (Dr John Insley suggests that *Habberdings* is a compound of the ME surn. *Habert* from *Herbert, v.* Forssner 148-9, and the pl. of **eng** 'a meadow'); *Haverdales* 1612 (*v.* **hafri** 'oats', **deill** 'an allotment, a portion of land'); *Herwichil* 1230-50 (this appears to be from **here-wīc** 'a military encampment', with **hyll**); *hestlanges* 1284 (the form in *h-* suggests that the first el. is ON **hestr** 'a horse, a stallion', the second being **lang²** 'a long strip of land' in the pl. Perhaps it denotes strips of land used for horse-racing); *holm* 1230-50 (*v.* **holmr**); *Kirkibryg* (sic) 1364 (*v.* **kirkja, bryggja,** a Scand. compound); *one layne called Kynggerbye layne* 1577 (*v.* **lane**); *Kirkeby als Kingerby Close* 1612 (*Kirkeby* is from the neighbouring Kirkby in Kirkby cum Osgodby *infra*); *Leyres* 1230-50 (perhaps 'muddy, clayey places', *v.* **leirr**); *Linholm* 1230-50 (*v.* **līn** 'flax', **holmr**); *the Milnefield* 1612, *the Myln fielde* 1635 (*v.* **myln, feld**); *in campo boriali* 1230-50 ('in the north field', an open field not separately named in later records); *ye parsonage ground* 1577, *the Parsonage Garth* 1703 (*v.* **garðr** and cf. Vicarage *supra*); *peselandes* 1230-50 (*v.* **pise, land**); *in campo australi* 1230-50, *the Southfield* 1612, *the Southfielde* 1635 (*v.* **suð, feld,** one of the open fields of the parish, cf. East Fd *supra*); *Walayscroft* 1364 (from the byname or surn. *Waleis* 'Welsh, a Welshman', but possibly in this county 'a Breton', cf. Reaney s.n. *Wallis,* with croft); *in Wellinsdowid* (sic) 1445 AASR xxix.

Kirkby cum Osgodby

KIRKBY CUM OSGODBY, 1679 *Terrier,* 1688 *TLE,* 1697 *Terrier,* 1738, 1763 *TLE,* 1789 *Td'E,* 1797 *TLE,* 1811 *Td'E,* 1824 O, *Kirkbe cum Osbarbe* 1625, 1634 *Terrier, Kirkby with Osgodby* 1657 *TLE,* - *alias Osgodby* 1785 *TLE.*

KIRKBY

Kyrchebeia 1146 RA i, *Kirchebeia* 1162 ib iv, *-bi* 1163 ib i
Kirkebi m12, l12 Dugd vi, *-by* 1209-35 LAHW, 1214 RA iv,
c.1221, 1223 Welles, 1247-48 RRG, 12254 ValNor *et passim*
to 1346 FA, 1556 *Td'E*, *-b'* 1242-43 RRG, *Cirkebi* 1200 Cur
Kyrkeb' c.1167 RA iv, *-by* 1211-12 RBE, 1272 Dugd vi, 1291
Tax, 1303 FA, 1327 *SR*, 1406-7 RRep *et passim* to 1559
Pat, *(juxta Rasyn')* 1405 RRep
Chirchebi c.1160 Dugd vi
Kyrkby 1526 Sub, 1535 VE iv, 1536-37 Dugd vi, 1539, *(iuxta
Owresby)* 1542 *Monson*, ("next" *Owresby*) 1550 Pat, *-bye*
(nere Owresbye) 1566 *Monson*, *-be* 1472 WillsPCC
Kirkbie 1562-67 LNQ v, *-bye* 1576 Saxton, 1610 Speed, *-by*
1576 LER, 1602, 1647, 1655 *Monson et passim*, - *Willoes*
1556 LNQ xiv
Kyrby 1535 VE iv, - alias *Kyrkby* next *Owresby* alias *Owersby*
1566 Pat, *Kirby* alias *Kirkby iuxta Owresby* alias *Owersby*
1566 *Monson*
Krikby 1655 *Monson*
Kerkbe 1731 *TLE*

This is a Scand. compound of **kirkja** and **bȳ**, 'the village with
a church', identical with Kirkby Green, Kirkby Laythorpe, Kirkby
Underwood Kest, and Kirkby on Bain and East Kirkby LSR, cf.
SSNEM 57-58. The generally held current view is that the
compound **kirkju-bȳ* was normally given to villages in which the
Danes found a church on their arrival and therefore that these
were existing villages taken over and renamed by the new settlers.
The place is described as near *Rasen* and *Owersby*. Only a single
example of the affix *Willoes* has been noted, *v.* **wilig** 'a willow' in
the pl.

OSGODBY

Osgotesbi (2x) 1086 DB
Osgotebi (7x) 1086 DB, c.1115 LS, 1139-41 *DuLaCh*, 1139 AC,
-b' 1185 Templar, 1232 Cl, *-by* 1214 RA iv, 1232 Ch,
1242-43 Fees, 1275 RH, 1281 QW *et passim* to 1379 AD
vi, ("by" *Kinyerbie*) 1306 Ipm, *Osgotabi* (3x) c.1115 LS,
Osgotteby 1327 *Foster*, 1344 AD vi, 1357 *Cor*, 1367, 1385

Foster, Osgotheby Hy2 (1409) Gilb (p)
Osegotabi R1 (1308) Ch, *Osegoteby* 1281 *Foster,* 1298 *Ass*
Osgotby Hy2 (e14) Selby, e14 AD ii (p), 1316 FA, 1332 *SR,*
   1374 Peace (p) *et passim* to 1438 *Foster,* 1527 Wills ii,
   Osgottby 1377 *Foster,* 1445 AASR xxix, 1447 *Foster*
Osgodby Hy2 (14) Dugd iii, 1322 Ipm, 1322 Cl, 1327 *SR,* 1392
   Foster, 1402 FA, 1428 AASR xxix, 1529, 1534 *Monson et*
   passim (*- next Kirkby*) 1588 *ib,* (*- alias Osgarby*) 1606 *Td'E,*
   -bye 1537-38 *AOMB 409,* 1576 Saxton, 1610 Speed,
   Hosgodby 1530 Wills iii
Osgodeby 1234 FF
Osegodeby 1331 Ch
Ossgardby 1445, 1446, 1449 Fine, Hy7 AASR xxix
Osgarbye 1447 *Foster,* 1552 *Monson,* 1601 *Terrier,* 1611
   Monson, -by 1535 VE iv, 1598 *Monson, -bee* 1602 *ib,*
   Osgerby 1530 Wills ii
Angoteby 1153-62 Gilb (p), 1187 (1409) ib, c.1200 RA ib
   (p), e13 *DC* (p), 1202 Ass (p), 1210 FF, 1236 Cl, *-bi* 1200
   Cur, e13 *DC,* 1206 Ass (p), 1206 P (p), 1209 ib, 1218 Ass,
   -b' e13 RA iv, *Hangotebi* 1212 Fees, *Angotheby* c.1150
   Gilb, *Angetheby* Hy2 (1409) Gilb (p)
Ansgotebi 1168, 1169 P (p)
Angotesbi 1206, 1207 Cur, 1219 Ass (p), *-by* 1208 Cur
Angotesbi 1256 FF

'Osgot's farmstead, village' *v.* **by**. The first el. is AScand
Ōsgot, -god, a regular Anglicized form of ON *Ásgautr,* ODan,
OSwed *Asgutt, -got,* in which Scand *Ás-* has been replaced by the
corresponding OE *Ōs-*. The same p.n. formation occurs as
Osgodby, now lost, in Bardney LSR and Osgodby in Lavington Kest.
Forms in *An(s)-* are due to AN influence. *v.* Feilitzen 164-6.
Osgodby is described as near to Kingerby and Kirkby. In DB, a
tenant named *Osgot* is recorded as having held land in 1066 at
South Cadeby LSR and Barlings LWR (DB f. 341b (3/48), 357a
(26/21)).

SUMMERLEDE (2x) 1086 DB. This is clearly an error for
Osgodby, and *Summerlede* is a pers.n., ON *Sumarliði,* a name which
seems to have originated in the Norse settlements of the Western
Isles, *v.* SPNLY 270-71. The mistake presumably arose from the

fact that a man called *Summerlede* held one bovate of land in Osgodby in DB (f. 371b, 68/46).

BRICK KILN FM. COTE HILL FM. THE COTTAGE. THE DALE, cf. *the Dale Close* 1711 *T'dE*, *The Dales* 1797 *TLE*, on the boundary with Market Rasen. HILL HOUSE FM. KIRKBY GLEBE FM. KIRKBY GRANGE (lost), - *Graunge* 1602 *Monson*, y^e *Mannor Ferme or Graunge of Kirkeby iuxta Kingerby* 1603 *ib*, *v.* **grange**; there is no indication which monastery it belonged to. KIRK HILL, *the* - 1797, 1798 *TLE*, cf. *Kirkhill close* 1711 *Td'E*, - *Lane* 1825 *TLE*, self-explanatory. LINCOLN LANE. MANOR FM, cf. *Kirkby Grange supra.* MANOR HO, 1830 Gre; it is in Osgodby. MILL LANE, cf. *Osgodby Mill* 1824 O, 1828 Bry, 1830 Gre, *Wind Corn-Mill* 1827 *Padley, Miln Close* 1688 *TLE, the Milne Close* 1690 *ib*; the site of the mill is marked on the map just east of Mill Lane, but is not named. It is approximately TF 078 920. NASH'S LANE, named from the family of John *Nash* 1842 White. OSGODBY GLEBE FM. OSGODBY HO, - *house* 1842 White. OSGODBY MOOR, 1817, 1820, 1822, 1823, 1824 *Monson*, 1824 O, 1828 Bry, 1830 Gre, - *moore* 1556 *Td'E*, - *more* 1557 *ib*, *the moors* 1657 *TLE*, - *Moor* 1821 *ib*, cf. *the Moor Close(s)* 1797 *ib*, *the little Moor Closes* 1818 *ib*. OSGODBY PLANTATION, 1824 O, 1828 Bry, 1830 Gre, *the Plantation on Osgodby Moor* 1817, 1822, 1824, 1825 *Monson, Plantations on the Moor* 1828 *BRA 1260, the East -, The West Plantation* 1814 *TLE*. SAND LANE. TOP ROAD FM is *Brick Yard* 1824 O, 1830 Gre. VICARAGE, *the vicaridg* (sic) *of Kirkby next Kingerby* 1601, *the Vicarage howse* 1625, - *Howse* 1638, *the vicaridge howse* 1671, - *house* 1674, y^e *vicarge House* (sic) 1679, *Vicaridge* 1697, *The Vicarage house* 1707, *The Vicarage House ... was taken down about sixty years since, and there is not any Building of any description now* 1822 all *Terrier.* WASHDYKE COTTAGES, - LANE. WILD DALES, on the boundary with Middle Rasen. THE WOODLANDS.

Field-Names

Forms dated 1162, 1163, c.1167 are RA i; e13 are *DC*; c.1280, 1280-90 are *Foster*, 1316, 14, 1586, 1647, 1711 are *Td'E*; 1327, 1332 are *SR*; 1428, 1445, Hy7 are AASR xxix; 1535-46 are *MinAcct*, 1537-38 are *AOMB 409*, 1542, 1566, 1588[1],

1602, 1603, 1611[1], 1721, 1814, 1816, 1820, 1824, and 1832 are *Monson*; 1588[2], 1611[2], 1657, 1660, 1663, 1665, 1669, 1670, 1698, e18, 1708, 1709[1], 1729, 1731, 1736, 1737, 1738, 1740, 1744, 1797, 1798, 1818, and 1825 are *TLE*; 1601, 1606, 1625, 1634, 1638, 1671, 1674, 1679, 1697, 1703, 1707, 1709[2], 1822[1] are *Terrier*; 1746 are *Dixon*; 1806 are *EnclA*; 1822[2], 1841, 1842, 1848 are *Deeds*; 1827 are *Padley*.

(a) the Barnacles 1797, 1798 (*Barnackles close* 1711; it is on the boundary with Owersby, where earlier forms are noted, s.n. Barnicle Cl in f.ns. (a)); the Beck Cl 1797 (1711), the beck - 1798 (*Beck Close* 1665, 1746), Osgodby Beck, Beck Drain 1806, 1806 *Map*, Beck Fd 1806 *Map*, the - 1806 (cf. *Osgarbye beck* 1611[2], *our Beck which soports our town* [Osgodby] *and Kerkbe* 1731, v. **bekkr**); the Bishop Bridge and Caistor Road 1806, Bishop-Bridge and - 1827, Bishop Bridge Road 1848; closes called Blackmiles 1818 (*Blackmiles close* 1711, *the Farr* -, *the Hither Black Miles* 1746, v. **blæc, mylde** 'soil, earth'; the same f.n. occurs as *Black miles* in Thornton le Moor f.ns. (a) and as *Blakmyls* in South Kelsey f.ns. (b)); Blow's Pingle 1827 (named from the *Blow* family, cf. Robert -, Thomas *Blow* 1797 *TLE*, with **pingel**); the Bottoms 1797, 1798 (v. **botm**); Burnt House Garth 1827, the - 1797, the burnt house garth 1798 (v. **garðr** 'an enclosure', as elsewhere in the parish); East -, West Chanters Closc (sic) 1841, - Chanter's Cl 1842 (*Chanterford close* 1711, perhaps alluding to a chantry, though none is found in ChancCert); the Claxby Road 1806 (self-explanatory); the Colly Cl 1797, - Colley Cl 1798, Old Colley Cl 1827, South Colley Cl 1827, 1848 (from the surn. *Colley*, cf. John *Colly* 1681 *BT*); The Cow Cl 1797, 1798 (*the Cow-close* 1711); Cowfold 1818, - Drain 1806, 1806 *Map*; the (Little -, the Great) Craycrow 1797, 1798 (*litell Cracoe* 1606, *litle* - 1638, *Cracoe* 1634, *the Cracowe* 1670, *Cracow Close* 1703, 1737, *a Close called Cracow* 1707, *Cracow* 1709[2], *Cracrow close* (sic) 1711, perhaps 'mound frequented by crows or ravens', from **haugr** 'a mound' with **kráka** 'a crow', or a derived byname); the Cream Poke 1797, 1798 (a complimentary name for fertile land or rich pasture, cf. *Creampoch Feild* in Linwood f.ns. (b) *infra*); two closes called the Dales 1825 (v. **deill**); the eight Pound Cl 1797, 1798 (*the eight pound close* 1711, possibly alluding to the rent of the land); the Farmer 1797, 1798, Farmer Fd 1806 *Map*, the Farmer(')s Fd 1806, Farmer's Fd 1848, Farmer-Field Road 1841, Farmer Fd - 1842, First -, Third Farmer Hill 1827, the third town End Cl or Farmerhill Cl 1818 (cf. *the Townsende close* 1711); the Farrow Garth 1797, 1798 (v. **garðr**); the Field Allotment 1841, 1842 (*ye feild of osgodby* 1588[1]); the Garden 1797, 1798; the Great Fd 1797, 1798; The Hall Garth 1797, 1798 (cf. *ad aulam* 1327 (p), *the Hall Croft* 1711, v. **hall, garðr, croft**); the High Cls 1797, 1798 (*High Close* 1665, *the high close* 1711, v. **hēah** 'high'); the Home Cl 1797, 1798 (1711, 1721, *Home Close* 1746); House Cl 1797, 1798; Kirkby Gate 1806 (a gate); Kirkby Rectory

1814, 1816, 1820, 1824, 1832 (*totam Rectoriam nostram* ... *de Kyrkby iuxta Ouersby* 1542, *Rectorye and parsonage of Kyrkebye nere Owersby* 1566, *the Rectory* 1647, cf. Vicarage *supra*); the Kirkby Road 1806; Lamming Close (sic) 1797 *LTR*; the Little Cl 1797, 1798; the Little Fd 1797, 1798; the Long Cls 1797, the long - 1798 (*Long Close* 1707); The Low House Bottom 1797, 1798 (*v.* **botm**); the Market Rasen and Caistor Road 1806 (*the highe way* y^t *leadeth from caister to rasen* 1556); the Market Rasen Lane 1806 ("an ancient Lane"); Methodist Chapel of Kirkby-cum-Osgodby 1822; the Middle Rasen Road 1806; Moor & Beck Fd Drain 1806, Moor Beck Fd Drain 1806 *Map* (*v.* **mōr**[1] and Beck Fd *supra*); the North Side 1797, - side 1798; the old Garth 1797, 1798 (*v.* **garðr**); the open Road Close 1797, 1798; Peterson Lane 1806 (presumably from the family name *Peterson*); the Pingle 1797, - pingle 1798 (*v.* **pingel**); the Prior Hill 1797, the prior - 1798 (*Prior close* 1711); Russell's Allotment 1827 (from a local family, whose name is variously spelt, William *Russels* 1802, - *Rousell* 1809, - *Russell* 1816 all *LTR*); the Scar Gate 1797, the Scar(e) - 1798 (*Skeithegate* e13, *Scayrgat'* (sic) c.1280, *Skeygate Close* 1711, *Scaygate Close* 1737, 1740, 1744, possibly from **skeið** 'a race(-course)' and **gata** 'a road'); the Sedcock (sic) 1797, 1798, Setup Lane (sic) 1822[1] (*Sett Copp close* 1671, *Set Coppe* 1674, *Set cop Close* e18, *Setcop Lane* 1703, *Set Cop Lane* 1707, *v.* **set-copp** 'a seat-shaped hill', probably to be identified with Sedgecopse Fm, *Set Cop* 1824 O in West Rasen parish); the third town End Cl (*v.* the Farmer *supra*); the Town Street of Osgodby 1806 (y^e *towne street* 1697, *the Towne Streete* 1698); Usselby Road 1806 (self-explanatory); the Vicars Occupation Road 1806 (i.e. the green lane providing the vicar with access to his selions in the open field); the Walesby Road 1806 (self-explanatory); West Fd 1806 *Map*, the West Fd of Kirkby-cum-Osgodby 1806, - of Kirkby cum Osgodby 1822[1], the West Fd Drain 1806 (cf. *occident' campo Osgoteby* 1360, *Westcampo de Osgoteby* 14, *Osgarbye west feilde* 1606, *the west feild of Osgarbe* 1625, 1634, 1638, *Osgarbe west feild* 1634, *the west feildes of Osgarby* 1657, *the westfeild of Osgodby* 1669, *the west field of Osgodby* 1729, *the West field of Osgodby* 1709[1] *the West Field* 1711, 1746, *the west field* 1736, *the Westfield* 1738, one of the great fields of the parish, cf. *in orientali campo* ... (b) *infra*); the West Rasen Road 1806 (self-explanatory); the (old) Yard 1797, the yard 1798.

(b) *Akerdike Close* 1611[1], *the Akadike* (sic) 1657, *Acker dyke Close* 1663, *Akerdike* 1670, *Akerdyke close* 1711, *the Akadike close* 1744, (*v.* **æcer**, **akr**, **dík**; for a discussion *v.* y^e *acredikes* in Barnetby le Wold f.ns. (b), PN L 2 13 and further the old Acre-Dike in Middle Rasen f.ns. (a)); *Blownte Howsse* Hy7 (from the surn. *Blounte*, cf. John *Blounte* 1445 AASR xxix); *Briggfoot Close* 1744 (*v.* **brycg**, **fōt**, the first el. is a Scandinavianized form); *Burtlands* 1670; *Cadby garthe* 1445 (from the surn. *Cadby* or *Cateby* with **garðr**, *Willi' fil Johis de Cateby* held land in Osgodby, Owersby and Kingerby 1281 *Foster*); *chancigate* e13; *Chaple Garth* 1537-38 (*v.* **garðr**), *Chapele Close* 1746; *Church Lane* 1709^2; *ye common* 1731; *the Cornfeild* 1731; *in orientali campo de Osgoteby* 1280-90, *the east feild of Osgodby* 1657, *the Eastfeild of* - 1669, 1709, *the East field of* - 1729, *the Eastfield* 1738, *the East field* 1711, 1736, - *Field* 1746 (one of the great fields of the parish, cf. West Fd (a) *supra*); *estiton' de Osgotby* c.1280 (p), *Estyton'* 1332 (p) ('east in the village', *v.* **ēast, ī, tūn**, cf. *Estiby* elsewhere in north L); *the farr Close* 1721; *the Firr close* 1711; *Gosedaile* e13 (*v.* **gōs, deill**); *Gosehole* 1670 (*v.* **gōs, hol**^{1}); *de la grene* c.1280, *a la grene* 1280-90, *atte Grene* 1327, 1332 all (p) (*v.* **grēne**^{2}); *Hawkes close* 1711 (from the surn. *Hawkes*, cf. William *Hawkes* 1641 LPT); *the hedplasse off Osgardby* Hy7 (*v.* **hēafod, place**); *Hill close* 1711; *Home Garth* 1746 (*v.* **garðr**); *Hopkins Garth* 1721 (from the surn. *Hopkin(s)*, cf. William *Hopkin* 1680 BT, - *Hopkins* 1682 *ib*, and **garðr**); *the Kings Pingle, Kings Yard* 1721 (*Mr King* is named in the document); *Kirkbye Nook* 1665, *Kirkby Nook* 1707, *Kirby-nook close* 1711 (*v.* **nōk**); *le kyrkegate* 1445 (*v.* **kirkja, gata**, a Scand. compound); *atte Lades* 1332 (p) (*v.* **(ge)lād** 'a watercourse'); *the Lea Closes* 1711; *Lessam plasse* Hy7 (from **place** with the surn. *Lesham*, cf. William *Lesham* 1428 AASR xxix); *super maram de Kinnerebyi* e13 (*v.* **(ge)mǣre** 'a boundary'); *marefurlang'* 14 (probably 'the boundary furlong', *v.* **(ge)mǣre, furlang**); *Martyn thing* Hy7 (from the surn. *Martyn*, cf. William *Martyn* 1540 *Inv* with **þing** 'property, premises'); *the middle Warth in Osgarbye beck* 1611^2 (*v.* **waroð** 'a shore, a flat piece of land along a stream, marsh, ground by a stream'); *Milford Close* 1708, 1729, - *close* 1791 (presumably from the surn. *Milford*); *molandino meo de Kirchebieia* 1162, *molendino de Kirchebi* 1163, - *Kyrkeb'* c.1167 ('Kirkby mill'); *Morley Close* 1721, 1746 (no doubt from a local family name); *the Neatfould* 1660, *Neat fold close* 1711 (*v.* **nēat** 'cattle', **fald**); *Netegate* 1280-90 (either 'cattle road' or (like *cowgate* etc.) 'allotment of pasture to a single head of cattle', *v.* **nēat, gata**); *New Close* 1665, *the new close* 1711, *New Close* 1746; *North syde close* 1711; *ye feild of osgodby* 1588^2, *Osgodby Field* 1707, 1709^2; *Pond Close* 1665, *the Pond close* 1711 (*v.* **ponde**); *Pound Close* 1537-38 (*v.* **pund**); *super dividam de Rasne* e13 ('on the Rasen boundary'); *the Rye-close* 1711 (*v.* **ryge**); *le Stygh* 1360 (*v.* **stīg, stīgr** 'a path'); *Suthlanges* e13 (*v.* **sūð, lang**^{2} 'a long strip of land'); *Suthtoftes* e13 (*v.* **sūð, toft** 'a messuage, a curtilage'); *thornbergsic* e13 (*v.* **þorn** 'a thorn bush', **berg** 'a mound, a hill', **sīk**); *ye Thorow-fair Close leading from Kirkeby to Owersby* e18; *three leys* 1721 (*Ley* in the pl. is a common variant of **lea** in north L in the sense 'meadow, pasture'); *Threwell'* e13 (probably 'three springs or wells', *v.* **þrēo, wella**); *vio* (sic) *de Thwergates* 1360 (*v.* **þverr** 'lying across', **gata**); *Firma ... Horrei in Kyrkbye ... voc' le tithe lathe* 1536-46 MinAcct ('the tithe barn', *v.* **hlaða** 'a barn'; *Horrei* is the gen.sg. of medieval Latin *horreum* 'a barn'); *i selionem terrae et r. p. ann. nomine Warnot* 1428 (*v.* **warnoth** and for a full discussion *Wardnoth* in Croxton f.ns. (b), PN L 2 104 and note "annual rent of 2s de Warnote" 1392 MC in Middle Rasen); *Westlanges* e13 (*v.* **west, lang**^{2} and *Suthlanges supra*); *West Osgoteby watermylne* 1445; *White house laine* 1698.

Linwood

LINWOOD

> *Lindude* (2x) 1086 DB, -*wda* c1115 (2x) LS, -*wode* p1182 Dane
> (p), a1189 RA x (p), Hy2 (1409) Gilb, 1200 Cur *et passim*
> to 1428 FA, -*wod* 1255 FF, 1281 RSu, 1323 Pat, 1353 *Cor,*
> -*uuode* 1252 Cl, -*wd* Hy2 RA iv, 1219-20 *DC* (p), -*wude*
> 1196 ChancR, 1197 P, 1225 Cur, -*woude* 1242-43 Fees
> *Lyndwod* 1213-19 RA ix (p), 1281 QW, 1340 Ipm, 1350 Pat,
> -*wode* 1261 Pat, 1261 Lib, 1297 Cl, 1310 ChancW, 1311
> Fine *et passim* to 1487 Ipm, -*wood* 1576 Saxton, 1602 *BT*
> 1610 Speed
> *Lindewode* 1187, Hy2, e13 (1409) Gilb, 1210-20 RA v,
> 1242-43 Fees, 1275 RH *et passim* to 1375 Peace, -*wod*(')
> 1219 FineR, 1249 Ipm, 1298 Ass, 1303 FA, -*wde* c.1200
> RA v (p), 1212 Fees, 1249 IpmR, 1254 ValNor, -*wud'*
> 1224 Cur
> *Lyndewode* 1260 Cl, 1285 Ipm, 1303 Cl, 1307 Pat, 1308
> Orig, 1308 Pat *et freq* to 1464 ib, - *Bayons* 1465 ib,
> *Lyndewod* 1268, 1311 Ipm, 1343 NI, 1350 Pat, 1352 Cl,
> 1430 Pat, 1431 FA, 1431 BS
> *Linwd'* 1163 RA i, l12 ib iv, -*wude* 1190, 1191 P, 1314 Ch,
> -*wod* 1234 FF, -*wode* 1251 Ch, -*wood* 1605, 1622 *BT et*
> *passim*
> *Lynwode* 1305 RA ii, 1324-25 *Extent,* 1428 FA, 1428 Fine,
> 1526 Sub, -*wod* (- *Beaumont)* 1462 Pat, 1519 DV, 1535
> *AD,* 1562 *BT,* -*wood* 1514 LP i, 1526 Sub, 1547, 1551
> Pat, 1584 *BT,* (- *Beaumounte and Bayons*) 1547 Pat,
> -*woode* 1535 VE iv, 1556 Pat
> *Lenwode Beaumond* 1467 Pat

'The lime-tree wood', *v.* **lind, wudu.** The manor was held by the *de Baiocis* family, cf. John *de Baiocis* 1242-43 Fees, *De Henrico le Bayous ... quam Willelmus de Baiocis quondam tenuit in Lindewod* 1303 FA, Robert *Bayhous* "of" *Lyndewode* 1343 Cl, John *Baiouse*

1352 ib and "Bayhous Fee" 1462 Pat. It was later held by the *Beaumont* family, cf. Alice late the wife of Henry *de Bello Monte* 1340 Cl, *Dominus de Bello Monte quondam Johannis Beyus* 1428 FA.

COLLEGE FM, *those Lands* (i.e. y^e *woodhouse Lands*) *are colledge land* 1745 *Terrier,* held by *The Masters Fellows and Scholars of Clare Hall Cambridge* 1842 *TA.* ELEANOR WOOD, 1842 ib, *Elinor Wood* 1824 O, 1830 Gre; the *Eleanor* who gave her name to the wood has not so far been identified. GLEBE FM, cf. *the Gleab Plat* c.1700, y^e *Gleabe platt* 1703, *A Plot of Ground called* y^e *Glebe* 1724 all *Terrier, v.* **plat**^{2} 'a small plot of ground'. THE HARDINGS, *the Hardinges* 1601, 1664, 1668, *-ynges* 1612, *-ings* 1671, Mr *Atkinsons Hardings* c.1700 all *Terrier, the Hardings* 1736 *Tur, - Hardens* 1674, y^e *hardinges* 1703 both *Terrier, the Great Hardinges* 1711 *LindDep* 29, *Hard Ings* 1842 *TA,* 'the meadows, pastures hard to till', *v.* **heard, eng.** LINWOOD WARREN, 1824 O, 1830 Gre, *Warren* 1828 Bry, and is marked *Moor* 1842 *TA.* LOW FM (lost, TF 115 865), *Low F.^{n}* 1828 Bry. LYNWODE WOOD, *boscum ... apud Lindewod'* 1265 Cl, y^e *Wood* 1724, *The Wood* 1745 both *Terrier, Linwood Woods* 1824 O, - *Wood* 1828 Bry, *Lin Wood* (sic) 1830 Gre. MANOR HO, 1828 Bry, 1842 White. MOOR FM, *the Moore* 1612, 1668, 1671, - *moore* 1674, y^e *Moor* 1745 all *Terrier,* 1842 *TA,* cf. *the moore groundes* 1664, Mr *Boothes Moor* 1703, Mr *Dixon's moor close* 1745, *Moor Close* 1724 all *Terrier.* NORTH WOOD, 1830 Gre, *le Northwode* c.1200 (1409) Gilb, *North W.^{d}* 1828 Bry, self-explanatory, *v.* **norð, wudu.** OLD HALL CLOSE (lost, TF 115 858), 1828 Bry, *Hall Close* 1842 *TA.* RECTORY, *the parsnedge* (sic) 1612, - *parsonage* 1668, - *house* 1679, *The Parsonage House* c.1710, 1724, *the Rectorie* 1664, - *Rectory* 1674, y^e - 1703 all *Terrier.* THE POPLARS. THE SYKE (lost), 1828 Bry; this formed part of the western boundary of the parish, approximately TF 094 862, *v.* **sīc, sík** 'a small stream', 'a ditch'. TOMFOOLS PLANTATION, apparently a derogatory nickname. TOP FM. WALK FM, *Walk House* 1842 White, cf. *the Sheep Walkes* 1711 *LindDep* 29, *The Walk* 1842 *TA, v.* **walk** 'a stretch of grass used for pasturing sheep', common in L. WARREN HO, 1842 White, *Linwood Warren Ho.* 1828 Bry, cf. Linwood Warren *supra.* WOOD HILL, 1824 O, at which date it was wooded.

Field-Names

Principal forms in (a) are 1842 *TA*. Spellings dated Hy3 and 112 are RA iv; Hy2 (1409), 1187 (1409) and c.1200 (1409) are Gilb; 1327 and 1332 are *SR*; 1334 are Ipm; 1601, 1612, 1664, 1668, 1671, 1674, 1679, 1703, c.1710, 1724, 1745, and 1822 are *Terrier*; 1711 are *LindDep 29*; 1717, 1719, and 1727 are Hungate. Other sources are noted.

(a) Ann's Cl; (Middle) Barf (*the Barfes* 1711, cf. *ye west barfe* 1601, *the west Barfe* 1612, *East Barf(e) furlonge* 1601, (*The*) *East Barf furlong(e)* 1612, *West Barf furlonge* 1601, *the West Barf* - 1612, 'the low ridge or hill', from NCy dial. *barf*, ultimately from OE **beorg**); Batchelors Moor (named from the *Batchelor* family, cf. William *Batchelor* 1810 *BT*); Bigg -, Far -, Nr Braywells, Braywells Bottom (*v.* **botm**) (perhaps 'the broad spring', *v.* **breiðr, wella** rather than from the surn. *Braywell*); Brick Cl; Causeway Cl (*v.* **caucie**); Church Cl 1822 (*the Church-close* c.1710, *ye Church close* 1724, 1745); Clacks Bottom, Upr Clacks Cl (from the surn. *Clack*, which is ultimately from the ODan pers.n. *Klak, v.* Reaney s.n. *Clack*); a close called the Clerks yard for the benefit of the Clerk 1822; Close (2x); Collar Dale (*Collodale* 1711, *Collowdale* 1745; two fields so named are in the extreme SW corner of the parish, and though at some distance apart *Collar* would seem to be identical with Collow (*Colehou* c.1250 (14) *VC*) in the neighbouring parish of Legsby LSR. This was perhaps a **deill** 'a share, a portion of land' belonging to or associated in some way with Collow); Compton Corner (*Crumton Corner* (sic) 1711, cf. *de Angulo* 1327, *in angulo* 1332 both (p), probably from the surn. *Crompton*); Coney Garth (Hill) (*the Cunny Garth* 1711, 'rabbit warren', *v.* **coni, garðr**); Cottagers Moore (*Cother More* 1702 (*sic*, perhaps for *Cotcher*, a variant of *cottager, cottier*), *Cotyers Moor* c.1710, *cottiger Moor* 1745, 'the moorland allotted to the cottagers', *v.* **cottere, mōr**); (Big -, Lt) Cow Cl, Cow House & Yd; (Far -, Nr) Dam Gates (*daynegate, dangattes* 1601, *dangates* 1612, probably from ON *Dana*, gen.pl. of *Danir* 'the Danes' and **gata**); Double Cottage; (Lr -, Upr) Fisher Cl, Fishers Gdn (named from the *Fisher* family, cf. John *Fisher* 1726 *BT*); Middle -, Upr Forty Acres; Four Acres & Far Cl (*iiii Acres* 1601, *fower Acres* 1612); First -, Upr Hill; Home Cl; the Homestead 1822, Home Stead 1842 (cf. *the Homestall* 1612, *the Homstal* c1710, *the Homestall or Onsett* 1711, the reference in Terriers is to the parsonage, cf. Rectory *supra*); North -, South Horncastle Dale (*Horne Castle dale furlonge* 1601, 1612); Kirtons Cl (named from the *Kirton* family, cf. Thomas *Kirton* 1699 *BT*); Far -, Nr Lands; Lane Cl; Laughton Cl; Lawn Cl, - Sykes (cf. *Le Launde* 1334, *Lownde* -, *ye lownde furlonge* 1601, (*The*) *Lownd* - 1612, *a close called the Lounds* c.1710, *ye Lounds* 1724, *the East* -, *the West Lound* 1711, *the lownd seyke* 1601, *the Lownd* *syke* 1612, *the Lound Flatt, the Lound Sick* 1711, *v.* **launde** 'an open space in woodland, a forest glade', with **furlang, flat** and **sik**); Linwood Cl (on the boundary with Market Rasen); Long Cl; Low Cl; Mill Hill (Bottom) (*the Mill Hill* 1711); Mitchells Cl (from the surn. *Mitchell*); Middle Moor Dam (cf. *the Moore damme plattes* 1612, *the North Moor Dammes* 1711 and Moor Fm *supra*); Nine Acres; Orchard (*the Orchard* c.1710, *The Orchard* 1745); Long Ordinary (in the SW corner of the parish) (perhaps here alluding to 'customary fare', cf. NED s.v. *ordinary* 13, and applied to a field in which refreshments were taken); Ox Cl (cf. *the Oxpasture* 1612); Paddock; Patterson Cl (from the surn. *Patterson*); Pinfold Cl (*v.* **pynd-fald**); Plots (*the Platts* 1711, *v.* **plat**^{2} 'a small plot of ground'); Lr -, Upr Rasen Gate (the road to (Market) Rasen, *v.* **gata**); Riddings, Road to Ridings (*ye Ridinges Lane* 1703, *the Ridings Lane* c.1710, *ye Ridings* - 1724, 1745, *v.* **ryding** 'a clearing')); Rogers Cl & Braize Garth (*sic*) (*the Roger Close* 1711, *ye Bass garth* 1703, *the Base Garth* c.1710, *the Bays Garths* 1711, *a close called ye Bays-garth* 1724, perhaps from **báss** 'a cowshed' and **garðr** 'an enclosure' and for the same name *v.* Baysgarth in Barton upon Humber and East Halton and Base Garth in Brocklesby f.ns. (a) PN L 2 32, 150, 68; but cf. the Bays Garth in South Kelsey f.ns. (a) *supra*); Seven Acres; Six Acres; Stack Yard; Teasdales Gdn (presumably from the surn. *Teasdale*); Ten Acres; Thorough Fare Cl (from ModE *thoroughfare* 'a road'); Upper Cl (2x); (Far -, First -, Middle -, Second) Wastings (*the wast-ings* 1601, *the wastynges* 1612, *the East -, the Great -, the Little Wastings* 1711, perhaps 'meadows in the waste-land', *v.* **wēste, eng**); West Yd; Big -, Lt Wood Bottom; Far -, Nr Woodside Cl (cf. Lynwode Wood *supra*).

(b) *An other land called an acre* 1601, *two landes called An Acre* 1601, *an Acre* 1612; *ye Acre dike, - dyke* 1601, *the Acre dyke* 1612 (*v.* **æcer, akr, dik** and *v. ye acredikes* in Barnetby le Wold f.ns. (b) PN L 2 13, and for a further note *v.* the old Acre-Dike in Middle Rasen f.ns. (b)); *Allens Close* c.1710, *Robert Allens close* 1724; *the Bays Garths* (*v.* Rogers Cl & Braize Garth in (a) *supra*); *Benrlys Garth* (sic) 1711 (possibly for *Bentlys -*, from the surn. *Bentley, v.* **garðr**); *Blanche leys* 1601, *Blaunch* - 1612 (presumably from ME **blanche** 'white', or the derived pers.n. and the pl. of **lea** 'a meadow'); *le Bondmanwode* c.1200 (1409) Gilb ('the wood of the husbandman; from ME *bondeman* 'a customary tenant, an unfree villager (villain) or farmer (husbandman)', *v.* MED s.v., and **wudu**); *Boothes groundes* 1668, *Booth Lands* 1674 (named from the *Booth* family, cf. Thomas *Booth* 1641 LPR); *Brotland* Hy2, *the brattes* 1601, *the Brattes* 1612 (*v.* **brot** 'a small piece of land'); *in campis* Hy2, *campos de Lindwode* c.1240 (14) VC (the open fields of Linwood); *Candlethrides* 1601, *Candlethredes* 1612 (presumably 'candle-wicks', though the term is not recorded in NED or MED;

the allusion may be to a plant, e.g. Candlewick Mullein or Hagtaper, *Verbascum thapsus*); *the Church Lane* c.1710, y^e *Church-Lane* 1724, *church Lane* 1745; *Corn Close* 1703, *Corn-close* c.1710; y^e *cowe bankes* 1601, *the Cowe Bankes* 1612; *the Cowfolds* 1711; *Creampoch Feild* 1711 ('cream poke', a complimentary term for rich pasture, cf. *Creame Poake Nooke* in the f.ns. (b) of Bigby, PN L 2 55); *Croftes grounds* 1671, *Croft Lays* 1674, *Croftes land* 1703 (*v.* **croft** or the surn. *Croft*); *doggedales* 1601, *Doggedales* 1612 (*v.* **deill**; the first el. may be **dogga** 'a dog', or **docce** 'a dock (plant)'); y^e *dovecote furlonge* 1601, *dove Cott* - 1612 (*v.* **dove-cot(e)**); *East landes* 1601, *Eastlandes* 1612, *the East Lands* 1711 (*v.* **land** and cf. y^e *west landes infra*); *eighty Acres* 1724; *Flidwde* Hy2 (1409), *Flidwode* 1187 (1409), *Flidwd* Hy3 (1409) (perhaps 'the disputed wood', *v.* **(ge)flit**, **wudu**); *A garinge called half an Acre* 1601, *a garinge* - 1612, *one lande ... called half an Acre* 1601, *half an Acre* 1612 (the references seem to be to two pieces of land bearing the same name (*Half an Acre*), and so needing to be distinguished, one grassland ('*A garinge*') and the other arable ('*one lande*'); *garing, gairing* is 'a triangular piece of land', *v.* EDD s.v. *gair* and recorded there only from L. It is probably a derivative of ON **geiri** 'a triangular plot of ground'. Cf. Garing Cl in Middle Rasen f.ns. (a)); *the Hall close* 1711; *Hoberges* Hy2 (1409), *boscum suum ... Houberges* 1187 (1409), *Houbergers* (sic) Hy3 (1409) (*v.* **beorg** 'a hill, a mound'; the first el. is perhaps **hōh** hence 'the hill at a spur of land'); *the hyghe meare furlonge* 1601, *The hygh meere* - 1612 (*v.* **hēah**[1], **(ge)mǣre** 'a boundary'); y^e *Ing rigges* 1601, *the Ing Rygges* 1612 (*v.* **eng, hryggr**); *the Ingthornes* 1601, 1612, *Ingthornes* 1711 (*v.* **eng, þorn**); *James close* 1668; *the Kings Lane* 1679; *the Lane Barfes* 1711 (cf. (Middle) Barf *supra*); *Lambecoteberh* 112 (*v.* **lamb, cot**, with **beorg**); *sub uia que dicitur Lata Uia* Hy2 ('below the road which is called the broad road'); *Legatts Close* 1711 (from the surn. *Legatt*); *in antiquum fossatum de Leggeshow apud septemtrionem iuxta le more de Leggesby, Liggeshow* Hy2 (1409), *extra fossatum nemoris mei de Legeshou ... et abutat versum austrum super campum de Leggesby* c.1200 (1409) (from the text of the two charters it would appear that "the ancient ditch" was in Linwood, but the situation of *Leg(g)eshou* is not certain, though it must have been close to the boundary with Legsby or just inside that boundary itself. *Leg(g)eshou* means 'the (burial) mound of *Legg*', *v.* **haugr** and Legsby 'the farmstead, village of *Legg*', the first el. of each being the OWScand pers.n. *Leggr*. There can hardly be any doubt that both the village and the (burial) mound were named from the same man. For the pers.n. *v.* Lind 731); *Lissington furlonge* 1601, *Lyssington* - 1612 (Lissington is an adjoining parish); *Lyttle furlonge* 1601, *The lyttle* - 1612; *Moor Close* 1724; *Norhberh* (sic) Hy2, *Northberth* (sic) 112 (*v.* **norð, beorg**); *The Northfeilde* 1601, *the North feylde* 1612, *the North Feild* 1711 (one of the common fields of Linwood, cf. *South feilde infra*); *North furlonge* 1601, 1612; y^e *north syke*

1601, *the North syke* 1612 (*v.* **norð, sik**); *Norwellestich* (sic) Hy2 (*v.* **norð, wella**, with **stycce** 'a bit, a piece, a small strip of land'); *parcum suum de Linduuode* 1252 Cl, *Parke lownde furlonge* 1601, *Parke Lound* - 1612, *the Park Lounds* 1711 (cf. Lawn Cl *supra* in (a)); y^e *Parsonage Orchard* 1724 (cf. Rectory *supra*); y^e *Parson's yard* 1745; *aqua de Pittes* 1225 Cur (*v.* **pytt**); *the Platche* 1601, *the platche* 1612 ('(land enclosed by) a woven or pleached fence', *v.* **pleche**); *Proctor's close* 1745 (named from the *Proctor* family, cf. Thomas *Proctor* 1662 *Inv*); *South feilde* 1601, *the South feyld* 1612, *the Southfield* 1679, *in campo australi* Hy2 (one of the common fields of Linwood, cf. *the Northfeilde supra* and *Westfeild infra*); *An other lande ... called a Stintinge* 1601, *an other land called a styntinge* 1612 (*v.* **stinting** 'an individual share of the common meadow'); *one land there called a stonge* 1601, *on land there* - (sic) (*v.* **stong** 'a pole, a stave', also used as a standard of measure 'a pole'); *the Common streete* 1664, *the Common Lane* 1668, *the townes streete* 1671; *Thorpe hedge* 1601, 1612; *iii Acres* 1601, *three Acres* 1612; *one lande called A iii stonge* 1601, *three stonges* 1612 (*v.* **stong** and cf. *a stonge supra*); *Tweene gates furlonge* 1601, 1612 ('between the roads', *v.* **betwēonan, gata**, with **furlang**); *ii Acres* 1601, *two Acres* 1612; *Waitberh* Hy2 ('wheat hill', *v.* **hveiti, beorg**); *low Wales* 1717, - *wales* 1719, *Low* - 1727; *West dail* 1314 Ch (*v.* **west, deill**); *Westfeild* 1601, *The West feylde* 1612 (one of the common fields of Linwood, cf. *the Northfeilde, South feilde supra*); y^e *west lands* c.1710, *the West Lands* 1711; *white-spottes* 1601, *White spottes* 1612 (presumably alluding to patches of chalk soil); *Willoughbys Close* 1711 (named from the *Willoughby* family, cf. Stephen *Willoughby* 1678 *BT*); y^e *woodhouse*, y^e *Woodhouse Lands* 1745 (cf. Lynwode Wood *supra*).

Toft Newton

NEWTON BY TOFT

> *Neutone* 1086 (4x) 1086 DB, *-tona* c.1155 (1409) Gilb, c.1160, Hy2 Dane, 1256 (1409) Gilb, - *iuxta West Rasen* 1279 RRGr, *-tun* p1169 Dane, 1170-75 RA iv, Hy2 Dane, e13, 1235-50 RA iv, *-tunia* Hy2 ib iv, *-ton*(') 1204 Cur, 1212 Fees, c.1220, a1224 RA iv, 1242-43 Fees, 1247, c.1250 RA iv, 1254 ValNor *et freq* to 1363 Cl, - *iuxta Tofte* Ed1 Foster, - *iuxta Westrasen* 1293 *Ass*, - "by" *Rasen* 1367 Ipm, *Neuetune* 1198 (1409) Gilb
>
> *Toft Neuton* 1324 Pat, 1331 Ch, 1397 Cl, *Toftneuton*(') 1327 Banco, 1327 *SR*, 1335 Pat, 1369 Ipm, 1364, 1290 Pat,

Tofteneuton 1397 ib, *Thoft Neuton* 1339 ib
Newetuna (4x) c.1115 LS, *-ton*(') 1202, 1211, 1272 FF, 1287
     Ipm
Niwetun 1170-75 RA iv, *-ton*' 1179, 1180, 1181, 1190, 1193,
     1194 P all (p)
Nuietun a1183 RA iv
Newenton' 1226-28 Fees, 1254 Cl
Nuton 1402 FA
Newton(') 1210-15 RA iv, 1213-23 ib ii, 1287 Ipm, 1428 FA,
     1461 Pat, 1526 Sub, 1535 VE iv, 1564 *BT,* 1576 LER,
     - *next Toft* 1601, 1611 *Terrier,* - *Iuxta Tofte* 1603 *BT,*
     - *nigh Toft* 1673 ib, - *al's Newton by Toft* 1674 *LCS*
Toftnewton(') 1335 Pat, 1335 Percy, 1412-13 RRep, c.1414
     AASR xxix, m16 *Cragg,* 1653 *ParlSur, Toft Newton alias
     Toftnewton* 1556 Pat, *Toft Newton* 1596 *BT,* 1666 VL,
     1674 *LCS,* 1756, 1762 *BT et passin, Tofte newton* 1582
     DCLB, - *Newton* 1596-97 *MinAcct, tofte Newton* 1612-14
     ib, *Tofft Newton* 1502 Ipm

'The new farmstead, village', *v.* **nīwe, tūn,** with the form
Newenton' from the dat.sg. **Nīwantūne.* It is described as near
West Rasen and Toft.

TOFT NEXT NEWTON

Tofte 1086 DB, 1198 (1409) Gilb, 1205 Cur, 1291 Cl, 1291
     Orig, 1292 Abbr, 1576 LER, - *next Newton* 1530 Wills ii,
     1577 *Terrier*
Toft c.1115 LS, p1169, Hy2 Dane, 1196-1203, e13 RA iv, 1202,
     1203 Ass, 1203 Cur, 1205 OblR, 1205 P, 1206 Cur, 1210,
     1211 FF, 1223 Cur, 1224 FF *et freq,* - *iuxta Westrasyn*
     1279 (1409) Gilb, - *juxta Westrasyn* 1424 IBL, - "by"
     Neuton 1311 Ipm, - *by Newton* 1671 *BT,* - *next Newton*
     1558 InstBen, 1602, 1612 *BT,* 1844 *TA et passim,* - *iuxta
     Newton* 1587, 1594, 1599 *BT,* 1601 *Terrier,* 1607, 1610 *BT,*
     - *nigh Newton* 1590 *Foster*
Tostes 1103 France (*s=f*), *Toftis* 1110 (c.1200) CartAnt, *Toftes*
     1275 Cl, 1281 QW
Thoft 1219 Ass, 1254 ValNor, 1265 FF, 1275 RH

'The site of a building, the curtilage, the messuage', from ODan **toft**, with four forms in the pl. It is described as near West Rasen and Newton by Toft.

AISTHORPE'S COTTAGE, named from the family of Stephen *Aisthorpe* 1770 *BT*, William *Aystrop* 1780 *ib*. CLAY LANE, CLAYLANE BRIDGE, cf. *ye Clay Pits* 1606, *the Clay pittes* 1638 both *Terrier*. DOGLAND FM, cf. *Doglands* 1671-77, 1707 *Terrier*, 1824 O, 1828 Bry, 1830 Gre, 1844 *TA, doglands close* 1664 *Terrier, Doglands or Doglanes Closes* (sic) 1670 *MiD, Doglands Close* 1690 *Foster, dogland Closes* 1706 *Terrier, Doglands Hall* 1828 Bry. On the face of it, this appears to be simply 'lands, strips of land where dogs are found' from **dogga** and the pl. of **land**, but the forms are too late and no certainty is possible. DOGLAND WOOD, *Doglands Holt* 1830 Gre, *v*. **holt**. EAST FM. GIBBET LANE, cf. *Gibbet Rath* 1828 Bry, on the south-east corner of the parish on the boundary with Faldingworth LWR, with Gibbet Hill a little further east in Buslingthorpe LWR. *Rath* is recorded in EDD s.v. as 'a circular earthwork or mound', though in England only from Y. This is, however, probably the sense here. INGS FM, cf. *Neutonenges* 1337 Cl, *The Ings Close* 1707 *Terrier, Ing Closes* 1813 *TN, - Close* 1841 *TA*, from the pl. of **eng** 'meadow, pasture'. MIDDLE FM is *Osmound Thorpe* 1828 Bry, for which no early spellings have been noted, and it may well be a late example of the use of **þorp**, common in L, in the sense of an outlying farm. The first el. appears to be the surn. *Osmond, -mund*, from ME *Osmund* (OE *Ōsmund*, Norman *Osmund*, etc.) However, it is just possible that it is to be identified with *Osberthorp(e)* in f.ns. (b) *infra*. NEWTON COVERT. NEWTON GRANGE, 1824 O, 1830 Gre, and note "granges in ... Tofte and Newton", "grange in Tofte Newton" 1537-39 LDRH, *Tofte et Newton, firma grang'* 1538-39 Dugd vi, "a grange in Tofte Newton" 1545 LP xx, where the references are to grange(s) of Sixhill Priory. PILFORD BRIDGE, 1664 *Terrier*, 1828 Bry, 1830 Gre, cf. *Pillesfordholm* e13, 1244 *MC, Pilfordeholm* 1337 Cl (*v*. **holmr**), *Pilford brig close* 1664, *- bridge close* 1671-77, 1707, *Pilford Lane* 1707 all *Terrier, Pilfrey Bridge Close* 1844 *TA*; the first el. is either OE **pīl** 'a stake, a pile' or ON **píll** 'a willow', hence 'the ford marked by a pole or where willows grow', *v*. **ford**. RECTORY, *parsonage house* 1601, *ye Parsonage* 1606, *the* *parsonage house* 1638, *the Rectory* 1706, 1707, 1864 all *Terrier* (Toft). SOUTH SKREE, on the boundary with Faldingworth LWR, presumably for *screed* 'a narrow strip of land', *v.* NED sb. I, ib, since both South Skree and West Skree, *infra*, are narrow strips of plantation. TOFT GRANGE, 1828 Bry, *v.* Newton Grange *supra*. WEST SKREE, on the boundary with East Firsby and Spridlington LWR. cf. South Skree *supra*.

Field-Names

Forms dated e13, 1321, 1323, 1324, 1350, a1374, 1448, and 1660 are *MC* (Toft); 1170-75, 1210-15, 1213-23, e14, and 1313 are RA iv; 1577, 1601^1, 1606^1, 1638, 1664^1, 1671-77, 1706, 1707^1, and 1864^1 are *Terrier* (Toft); 1601^2, 1606^2, 1611, 1664^2, 1671, 1697, 1707^2, 1822, and 1864^2 are *Terrier* (Newton); 1813 are *TN*; 1841 are *TA* 239 (Newton) and 1844 *TA* 344 (Toft).

(a) Ankholme Cloeses (sic) 1813, Ancholme Cl 1844 (named from the R. Ancholme); Arable Cl 1844; Barn Cl 1833, 1841, 1844; Beck Cls 1813, First -, Second Beck Cl 1844 (*v.* **bekkr**); Bell Cl 1841 (perhaps cf. *Bells Lane* 1707^1, from the surn *Bell*, cf. George *Bell* 1638); Blabers (Bottom) 1813, 1841 (*blabergh* e13, *Westblabergh* 1350, *the furlong called Blabere* 1577, *Blaber* 1601^1, *blaber* 1601^2, 1611, 1638, *long Blabor* 1606^1, *long blaber* 1606^2, *longe* - 1638, *Short* - 1601^2, *short* - 1606^2, 1611, *Blaberdaill* 1601^2, *blaber*- 1606^2, *-daille* 1611, *Blabar Northdale* 1606^1, *blabride dalle* (sic) 1577 (*v.* **deill** 'a share of land', as elsewhere in this parish) ('the cold hill', *v.* **blá(r)**, **be(o)rg**, perhaps a Scand. compound); Brick Cottage 1841; Brick Yard Cl 1813, 1841; Broad Acres 1813, 1844 (*Brat acres* 1577, *Bratacres* 1601^1, 1606^1, *bratacres* 1601^2, *brattacres* 1638, *brodacres* 1606^2, 1611 (*v.* **brot**, **æcer**, **brot** 'a small piece' from e17 was evidently taken to be ModEng *broad*, the 1601^2 and 1611 forms certainly referring to the same land); Calf Cl(s) 1813, Calf Cl 1841, Calf & part Home Cl 1844; Car 1813, Car(r) 1844, the Carr close 1864 (*the carre* 1577, *y^e Carre* 1601^2, 1611, - *Car* 1606^1, - *Carr* 1606^2, *the Car* 1638, *The Carr* 1664^1, *the Carre* 1671-77, *y^e Carr*, - *Town's Carr* 1707^1, - *Care dick* 1601^1, *v.* **kjarr**); Church Yd 1841, 1844 (*y^e churche close* 1606^2 (Newton), cf. *Byneyenyebirke* (*b = k*) 1276 RH, *Bynethenthekirk'* 1288 *Ass* (p) (*v.* **beneoðan** 'beneath'), *ad Ecclesiam* 1313 (p), *atte Kirke* 1332 *SR* (p), *v.* **kirkja**); Close (3x), A Close 1841; Corn Cl 1813, 1844, Corn Cl Bottom 1844 (*v.* **botm**), Corn Plat 1813, - Plot 1841; (Cl adjoin^g) Cottage & Land 1844; Cottagers Pasture 1813, Btm -, Top Cottage Pasture 1844; Cow Cl(s), - Plat(s) 1813, Cow Cl 1841, Cow Cl (Bottom) 1844 (*v.* **botm**), Cow Plot 1841 (*the Cow*

Close 1706[1], 1707[1]); the Dowgarth 1864[1] (*dowcoat garth* 1606[1], 1638, *a little Close called dourgarth or dour yard* 1664[2], *dowgarth* 1671-77, 'the dove yard, the dove-cote yard', v. **dūfe, garðr, dove-cot(e)**; in the 1664[1] form dial. *dow* 'dove' was evidently taken to be *dower*); Fen Cl 1813, 1844 (by R. Ancholme) (*ye Fen close* 1606[1], *the Fen* - 1638, v. **fenn**); Flash Cls 1813, - Cl 1844 (v. **flask**, ME **flasshe** 'a swamp, swampy grassland'); The Four-acre 1864 (*the fowor ackors* (sic) 1706[1]); Furse Pastures 1813, Furze Pasture 1841 (v. **fyrs**); Gabe Cl 1844; Adjoining Glebe Cl 1844; Grange Plats, - Cl 1813 (Newton) (v. **grange**, **plat**[2] 'a small plot of land', as elsewhere in the parish); Grass Cl 1844, - fd 1864[1], one Grass Cl 1864[2]; Hell Hole 1813, Hill Hole 1844 (the derogatory term *Hell Hole* apparently modified in later use); Hill Cl 1822 (*hillie close* 1664[2], *The Hillie close* 1671, *The Hill close* 1697, 1707[2]); Home Cl (several) 1813, 1844; The Homestall 1822 (*The Homestall* 1707[2], *the* - 1707[2], v. **hām-stall**); Homest[d] Stackyard & Little Croft 1844; Horner Cl 1841 (from the surn. *Horner*); House Cl 1813, House & Cl 1841, 1844 (v. **inntak**); Lane Cl 1844 (cf. *lane peece* 1706); Laures (sic) 1813, Lounges 1844 (perhaps 'long pieces of land;, v. **lang**[2]); Low -, Top Lay Sykes 1812, Low -, Top Ley Syke 1844 (*litel lare sick* 1601[1]. *litle laresike* 1606[1], *little Lairsike* 1638, *Larsike* 1664[1] (probably 'the muddy ditch' from **leirr** 'mud, clay' or **leira** 'a clayey place' and **sīk**; It is noteworthy that *a laier* occurs as an appellative in 1664[1]); Ley Cl 1844; Lineholmes 1813, Lincholme (sic) 1844 (*lyne-, lineholme* 1577, *Linholme* 1601[1], *ye medowe called Linholme* 1601[2], *Lynholme* 1611, *linam close* 1706, v. **līn** 'flax', **holmr**); Lt Long Bottom 1844 (v. **botm**); Long Dales 1813, 1841 (*langdailes* e14, *-dayles* 1313, *-daile* e14, *longedayles* 1577, *long dalls* 1601[1], *Lang(e)dayles* 1601[2], *Longe dales* 1606[1], *longe* - 1638, *long dailes, longdales* 1606[1], *longdayles* 1611, v. **lang**[1], **deill**); Longlands 1813, Long Lands 1844 (v. **lang**[1], **land**); Lords Platt Hill 1813, - Platt Bottom 1813, - Plat Bottom, Lords Hill 1844 (possibly from the surn. *Lord*, with **hyll**, **plat**[2] and **botm**); Lousey Lands 1813, Lousy - 1844 (v. **lūs**, **-ig**[3], probably alluding to land infested with plant-lice); Meadow 1844; Middle Plot 1841; Mill Fds 1813, First -, Second -, (etc.) ... Ninth Mill Plot 1844, Tenth -, Eleventh -, Twelfth Mill fd 1841 (cf. *ye Milne hill* 1601[2], *ye Milne hill* 1606[2], *ye Mylne hill* 1611, v. **myln**, **plot**, **hyll**); New Cl 1813, 1841; Adjoin[g] Boundary Newton, St[h] next adjoin[g] Newton, Adjoining S.W. Corner of Newton Parish 1844; Newton Lane 1864[1]; Normanton Well Bottom, - Well plot 1841, Norman Well Bottom 1844, Norman Well Plots 1813, - Well Plat 1844 (v. **wella**, **botm**, **plot**); Ox Pasture 1813, 1844 (*freq*) (*ye axe pasture* 1601[1], v. **oxa**); Adjoining Parish Boundary 1844; Parsonage Fd 1844 (cf. *ye Parsonage dale* 1606[1], *the parsonage daile* 1638 (v. **deill**), and *ye parsonage* 1606[1], *The parsonage house* 1611, *the parsonage howse* 1664[2], 1671, *The Parsonage house* 1697, in Newton, cf. The Rectory *supra*); Pasture Cl 1812, 1844, - Stand 1844; Pastures 1813; Pingle 1841 (*little pingle* 1664[1], *v.* **pingel**)' Mdw -, Pasture Pipsey 1813, 1844 (*in pipsie* 1601[2], 1606[2], 1611, *Pipsie* 1606[1], *Pipsey* 1638, obscure); Plats, Far Plat, Far -, Middle -, Nr Platt 1813, Far Plot 1841, 1844, Nr Plot 1841, Plot Bottom 1841 (*v.* **botm**), Plots 1841 (*v.* **plat**[2], **plot**); Road Fd 1813, 1844, - Plantn 1844; West Rasen Fd 1844 (alluding to the neighbouring parish of West Rasen); Sallow Cl 1813, - Holt 1844 (*v.* **salh, holt**); Sandland 1813, Sand Cl 1844 (cf. *Sandlanghole* e14, *Sand Land Close* 1660, *Sandlandes* 1671-77, *the Sand land close* 1707[1], *v.* **sand, lang**[2] 'a long strip', perhaps replaced by **land,** as is possible in *Estlang' infra*); Low -, Top Shearling Cl 1844; Sheetington Fds 1813; South Fds 1813 (Newton), - fd 1841; Stackhouse Cl 1813, 1822, Stack House Cl 1841 (*the long & little stackhouse closes* 1664[2], *The new stackhowse close* 1671, - *Stackhouse Close* 1697, - *close* 1707[2]); Standing hole 1813, 1841 (*the standinge hooles* 1577, *Standing hooles* 1601[1], *Stonyholes* (sic) 1660, *Stong holes* (sic) 1707[1], possibly 'a place where cattle may stand under shelter', *v.* NED *standing* 4 (d); it is within a nook on the boundary of the parish); Stickney Cl 1813, 1841 (*John Stickneys close* 1611, *Stickneyes Farm* 1660, and for earlier references to the family cf. Thomas *Stikeney* 1584 *BT*, William *Stickney* 1587 *ib*); the Street 1822 (*y*[e] *Street* 1707[2], *v.* **strǣt**; it is by the Parsonage); Sturton Plat (sic) 1813, Stourton Plot, Next Stourton Plat S. Side 1844; Thornhill 1813, Thorn Hill 1844 (*Thornhull'* 1210-15, *Thornhill'* e14, 1313, *Thornhillsik'* e14, *Thornhilsik'* 1313 (*v.* **sik**), *Thorn hill* 1601[2], *thornhil* 1606[2], *thornhill* 1611, *Thorahill* (sic) 1664[1], *Thornehill* 1671-77, *the Thornhill close* 1707[1], *v.* **þorn, hyll**); (Middle) Near Toft Leys 1844 (*v.* Toftley's Fm in West Rasen *infra*); Toynbee's Rd 1813 (John *Toynbee* is named in the same document); Willow Holt 1844 (*v.* **holt**, cf. *Maniwilghes* e13, 'many willows', *v.* **mānig, wilig**).

(b) *Arkelhow* 1210-15 (*v.* **haugr** 'a mound, a hill'; the first el. is the ME pers.n. *Arkel* (ON *Arnkell*, ODan *Arnketil*)); *belfra garth* 1601[1], *Bellfray Garth* 1606[1], *belfray garth* 1638 (from dial. *belfry* 'shelter-shed', with **garðr**, *v.* NED s.v. *belfry*, etymological notes, and EDD s.v. where this sense is recorded from L); *billion meare* 1577, *Billion meare, y*[e] *butes side bellion meare* 1601[1], *Bellion meare* 1606[1], *Billian Meere* 1638 (*v.* **(ge)mǣre**, the first el. being obscure); *Brigfurlanges, Brigforlang'* e14, *Brigfurlang'* 1313, (cf. *the church bryge* 1577, *Church bridg* 1601[1], *kirkbridg* 1606[2], *the Church Bridge* 1664[1], *v.* **kirkja, brycg, furlang** and cf. *ad Ecclesiam* e14, 1313, *ad ecclesiam* 1313 all (p)); *campo de Toft* 1203 FF (*v.* **feld**); *Cockthornes* 1660 (probably the same as Cockthorne Fm in West Rasen *infra*); *le Dale* 1321; *Dalleth* (') e14, 1313; *in orientali parte campi* 11170-75, *in campo orientali* e14, - *orientali campo* e14, 1313, *est fyelde* 1577, *y*[e] *East Feild* 1601[1], *y*[e] *Eastfield* 1601[2], 1611, *eastfield* 1606[2], *East feyld* 1638 (one of the open fields of Toft); *Estlang'* e14, *Estlanges* 1313, *Eastlandes* 1601[2], 1611 (*v.* **ēast, lang**[2], though it is by no means certain that the 17th century forms refer to *Estlang'*); *a meare* *furre called the Feible Meare* 1660 (*v.* **marfur** 'a boundary furrow'; the first el. may be derogatory, *feeble* in the sense 'run-down, ill-maintained'); *flyt landes* 1577, *flitlandes* 1601[1] ('strips in dispute', *v.* **(ge)flit, land**); *Froskholes, Froscholes* e14, *Froskeholes, Frosk'holes* 1313 ('frog holes', *v.* **frosc, hol**[1], the first el. being Scandinavianized); *furresby hedge* 1577, *fursbee hedge* 1601[1], *Furresbie hedge* 1601[2], *fursbie hedge* 1606[2], *furresbie hedge* 1611 (named from the adjoining East Firsby LWR); *gadmill furlonge* 1601[2], *ye furlong called godmun' gift* 1606[1], *gadmyll* 1611, *Goodburngft* (sic) 1601[2], *godburngift* 1611 (*v.* **gift** 'a dowry, a marriage portion'; the relative rarity of this el. would appear to link these otherwise dissimilar forms, but they are so late as to preclude an accurate etymology, though *Go(o)ldburngift* may contain an OE fem. pers.n. **Goldburh*); *goot furlonge* 1577, 1601[1]; *Gosbarthorpe* 1606[1], *Gosberthorpe* 1638 ('Gosbert's dependent farmstead', *v.* **þorp**, with the first el. the same pers.n. as that in Gosberton L Holland, Frankish *Gozbert, v.* Forssner 124); *le Grengat'* e14 ('the green road', *v.* **grēne**[1], **gata**, perhaps a 'green lane', or 'occupation road', giving access to strips in the common field); *Gretland* 1210-15, *great landes* 1601[2], *- Landes* 1611, *greatlandes* 1606[2], 1638 (*v.* **grēot** 'gravel'. **land**); *ye hallgarthes* 1577 (*v.* **hall, garðr**, in Toft); *the hardinge* 1577, *ye hardinge* 1601[1], *hardinges* 1611, *Newton Hardings* 1660, *the Harding* 1664[1], *the Hardings* 1671-77 (*v.* **heard** 'hard to till', **eng**, a common f.n. in north L); *ye headinges* 1606[2] (perhaps with much the same sense as **hēafod-land** denoting places at the end of a ploughed field where the plough is turned, cf. *ye headings* in Barnetby le Wold f.ns. (b), PN L 2 14, where it is noted that *headings* occurs on a number of occasions as an appellative in north L); *headland* 1601[1], *The headland close* 1671, 1697, *the -* 1606[1], *ye hedland Close* (sic) 1706, *the Headland Close* 1707[1] (*v.* **hēafod-land** and cf. prec.); *Kelgarth, le Kelegarthe* 1596-97 MinAcct (*v.* **garðr**, the first el. is a byname or nickname from the ON appellative *kjǫlr* 'a keel', cf. the Icelandic byname *Kiolfari* LindB 201); *Langaker* 1324 (*v.* **lang**[1], **æcer, akr**); *Langholm* 1210-15, *Langholme, Langtholm'* (sic) e14, *Langholm'* 1313, *longholme* 1577, 1601[1], *longe Holme* 1606[1], 1638, *Longholm* 1707[1] (*v.* **lang**[1], **holmr**, cf. *Scortholm, Southlongholme infra*); *law sleides* 1601[1]; *uie que ducit Linc'* 1210-15, *Lincolne gatte* 1577, *- gate* 1638, *Lincoln gatt* 1601[1], *- Gate* 1606[1], *- gate* 1606[2], *Lincolne -* 1601[2], 1611 ('the road to Lincoln', *v.* **gata**); *Litteldale* e14, 1313, *lyteledaill* 1577, *liteldalle* 1601[1], *litle dale* 1606[1], *little daile* 1638 (*v.* **lytel, deill**); *Lollesties* e14, 1313, *Lollestyes* e14 (*v.* **stīg**, (it is just possible that *Lolle* is a nickname for *Lawrence*); *ye long close* 1671, *ye Long Close* 1697; *Lyng* 1606[1] (*v.* **lyng** 'heather'); *market dale* (*v. South dale infra*); *bosci mei apud swynelund q apellatur menelund* eHy2 (Ed1) Barl ('of my wood in the swine grove which is called the common grove' *v.* **swīn, (ge)mǣne, lundr**); *merfurlandes* 1601[2], *merfurland* 1606[2], *merrfurlonge* 1611 ('boundary furlong', *v.* **(ge)mǣre, furlang** *furland* is common in north L for *furlang, furlong); le*

Mikkylgate 1323 (*v.* **mikill, gata**); *the minster garthe* 1562 *DCLB* (*v.* **mynster, garðr,** held by the Dean and Chapter of Lincoln Minster); *molendino de Toft* 112 Dane (*v.* **myin**); *Naylergarth* 1601[2], *Mathew Nayler his close* 1601[2], *Nayler yate* 1606[2], *Naylers close* 1611, *Nayler gate* (sic) 1611 (*v.* **garðr,** the 1601[2] and 1611 forms referring to the same name; the first el. is the surn. *Nayler*, cf. *mathew nealer* 1601[2], *Mathew Nayler* 1606); *in campis eiusden ville* 1170-75, *in campo de Neuton*[a] Ed1 *HarlCh* (*v.* **feld**); *the new close* 1664[2]; *Neyerscortholm'* e14, 1313 (-*y*- = -*þ*-), *Netherstertholm* e14 (*st*- = *sc*) (*v.* **neoðera, sceort, holmr,** cf. *scortholm infra*); *North feyld* 1638 (of Toft); *Osbernthorp(e)* e14, *Osbernthorp'* 1313 (*v.* **þorp;** the first el. is the ME pers.n. or surn. *Osbern, Osborn,* from AScand *Ōsbe(o)rn,* from ON *Asbiǫrn* or Norman *Osbern;* cf. Middle Fm *supra*); *Paitefinfurlang',* *Paytefinforlang'* e14. *Paitefinfurlang'* 1313, *Nether paytfield* 1601[2], - *paytefield* 1611, *Payte feild* 1606[1], *Netherpatfield* 1606[2] (the early forms are from the AFr pers.n. or surn. *Peitevyn* (*v.* Reaney s.v. *Poidevin*) and **furlang;** this occurs as a surn. in William *Paytefyne alias Paytfyn* of Normanby le Wold 1395 Pat. Whether the 17th century spellings represent a reduced form of the pers.n. is, of course, uncertain, but the similarity does appear too great to be coincidental); *Potterstygh* 1448 (*v.* **stīg** 'a path', **pottere** 'a potter', or its derived surn.); *Pratting* 1596-97 *MinAcct* (*v.* **eng;** the first el. is the surn. *Pratt* for which *v.* Reaney s.n. *Pratt*); *Prestgat'* e14, *Prestgate* e14, 1313, *priestgates* 1601[2], -*gayte* 1606[2], *preystgates* 1611 (*v.* **prēost, gata**); *Rasynge gatte* 1577, *Rasen gatt* 1601[1] ('the road (from Toft) to (West) Rasen', *v.* **gata**); *the Rectory* 1664[2] (Newton); *Sandholm* 1210-15, (*v.* **sand, holmr**); *scharthedge* 1577, *y*[e] *scarhedge* 1601[1], *y*[e] *short hedge* (sic) 1606[1], *the* - 1638, (*v.* **skarð** 'an opening, a gap', **ecg,** cf, *Skarth infra); Scortholm* 1210-15 (*v.* **sc(e)ort, holmr**); *Scortwang'* e14, 1313 (*v.* **sc(e)ort, vangr** 'a garden, an in-field'); *Scotgate* 1601[2], 1611, *scotgate* 1606[2] (*v.* **gata;** the first el. may be the surn. *Scot,* cf. Walter *Scot* 1235-40 RA iv); *A Close of six Acres* 1664[1]; *Sixil Dawnter* 1601[2], 1611 (the first el. is perhaps a surn. from Sixhill LSR, but *Dawnter* is obscure); *Skarth* a1374 (apparently from **skarð** 'an opening, a gap', though it is described in the text as 'a watercourse'); *a dale of meadow called South dale or market dale* 1660 (*v.* **deill**); *Southlongholme* 1601[2], 1606[2], *southlongholme* 1611 (*v.* **lang**[1], **holmr** with **sūð,** cf. *Langholm supra*); *Squar close* (sic) 1664[1]; *Stone Gate* 1606[1], *stone gate* 1638 ('the stoney road', *v.* **stān, gata**); *Sudwell* 1210-15 (*v.* **sūð, wella**); *Swerdlandes* 1601[2], 1611, *short* - 1611, *Short Swerlandes* (sic) 1601[2], *Swerdlandes* 1606[2] (probably 'the narrow strips of land with the shape of a sword', *v.* **land,** the first el. being ON *sverð*); *Taselbrough* 1577, *Taselbrough* 1601[1] (*v.* **tæsel** 'teasel', **beorg,** but both forms appear to be corrupt); *Thornacr'* e14. *Thornacre* e14, 1313 (*v.* **þorn, æcer**); *in campo de Toft* 1203 FF, *Toft townes end* 1601[2], *Toftes towne end* 1611, *y*[e] *towne end* 1606[1], *the* - 1638 (named from Toft); *the Towne bridge* 1671-77; *Wadelandes* e14, 1313, *Wadlanides* (sic) e14 (*v.* **land,** the first el. being the OE pers.n. *Wada* rather than **wād** 'woad'); *Warlottes* e14, 1313 (*v.* **warlot**, discussed under Waterhill Wood in Brocklesby, PN L 2 67); *Whaitebergh* 1210-15, 1313, *Waytberk'* (sic) e14, *-berg* 1313, *Whaitberg'* e14, 1313, *Whaitbergh'* e14, *Whaiteberg'* 1313, *Whatborowe* 1601^2, 1611, *Whaytborowe* 1606^2, 1611 ('wheat hill', *v.* **hveiti, berg,** a Scand. compound); *the furlonge called Wheytefurres* 1577, *Whyetfurs* 1601^1, *wetfurs* (sic) 1606^1, *west furrs* (sic) 1638, *waite feyld* (sic) 1638 ('wheat furrows', 'wheat field', *v.* **hveiti, furh, feld**); *in occidentali* (sic) 1170-75, *in campo occidentali* e14, *in occidentali campo* 1313, *the west fyelde* 1577, *West feild* 1601^1, *ye Westfield* 1601^2, 1611, *westfield* 1606^2 (self-explanatory); *ye furlonge called Whitewell furres* 1601^2, *whitwell furres* 1606^2 (*v.* **furh**); *Wranglandes* e14, 1313, 1321, 1601^2, 1611, *Wrang landes* 1606^2 ('the crooked strips of land', *v.* **vrangr, land**); *in the Wro* 1213-23 (p) (*v.* **vrá**); *le Wyndemylne* 1323.

Normanby le Wold

NORMANBY LE WOLD

    Normanesbi (2) 1086 DB, *Normannesby* 1208 Cur
    Normanebi 1086 DB, c.1200 RA iv, *-by* 1190 (1301) Dugd
        vi, 1223 Cur, 1242-43 Fees, 1338 Pat, *Normanabi* 1150-60
        Dane, *Normannebi* 1150-60 ib, R1 (c.1331) *Spald i*, c.1200
        RA iv, 1204 P, 1205, 1206 Cur, 1206 Ass (p), 1207 Cur,
        1218 Ass, Hy3 *HarlCh*, *-by* 1199 (1330) Ch, John *HarlCh*,
        1202 Ass, 1210 FF, 1230-40 RA iv, *-b'* c.1200 ib iv
    Nordmanabi (4x) LS
    Northmannebi c.1200 RA iv
    Normanb' l12 RA iv, *-bi* 1200 (c.1330) *Spald i*, e13 *HarlCh*,
        e13 RA iv, 1209 P, 1200 RA iv, 1212 Fees, *-bia* 1212
        ib, *-by* 1192-1205, e13 RA iv, 1205 Abbr, 1231 FF,
        1242-43 Fees, 1250 FF, 1254 ValNor, 1267 AD ii, 1269
        FF, 1275, 1276 RH, 1282 Ipm, 1288 *Ass*, 1291 Tax, 1303
        FA *et freq*, *-bye* 1576 Saxton, 1610 Speed, *-bie* 1576 LER,
        Normanby iuxta Claxby 1329 *Cor*, 1375, 1383 Peace, 1597
        Foster, *- juxta Claxby* 1506-7 Lanc, 1697 *Terrier*,
        Normanbie iuxta Claxbie 1626 *Yarb*, *Normanby next Claxby*
        1620 ib, *Claxbynormanby* 1369 *FF*, *Normanby juxta*
        Castrum 1343 NI
    Normanbey on the Hyll 1563 *BT*, *Normanby on ye hill* 1664,
        1674, 1724 *Terrier*, *- upon the Hill* 1665 *Dixon*, 1688
        Foster, *Normanbie sup' montem* 1601 *Terrier*, *Normanbye*

super montem 1588 *BT,* Normanby *sup montem* 1636
Foster, - *super montem* 1666 VL, - *Montem* 1762 *Terrier,*
Normanby *super Waldam* 1294 RSu, - *on the Wolds* 1824
O, 1828 Bry, 1832 Gre, 1852 *Yarb*

Three of the early spellings with medial -*es*- suggest that the
first el. is the OE pers.n. *Norðmann,* well-evidenced in DB, Feilitzen
331-32, but all the rest indicate that the name is from the gen.pl.
and means 'the farmstead, village of the Northmen or Norwegians',
v. **Norðmann, bý,** as suggested by Ekwall, DEPN s.n. and
Fellows-Jensen, SSNEM 60-61. The latter points out that the name
indicates an isolated settlement of Norwegians. The same name
occurs three times in LWR. Normanby is described as near Claxby,
Caistor, on the Hill and on the Wolds, *v.* **wald.**

ACRE HO, *Acreholes* 1150-60, 1160-66 Dane, Hy3 *HarlCh,* 1327 *SR*
(p), *Akerholes* c.1160 (Ed1) *Newh,* (*Bercarie de*) John *HarlCh,*
(*grangia de*) e13 *CottCh,* eHy3 *HarlCh,* (*Grangie ... de*) Hy3 *ib,* Hy3
CottCh, (*Grangiam de*) c.1250 *HarlCh,* 1332 *SR* (p), 1365 Bodl,
(*dominium siue grangiam uulgaritur nuncupatur*) 1527 *HarlCh,*
Akyrholes 1401-2 FA, *Akeroles* c.1200 RA iv, (*grangie ... de*) e13
HarlCh, (*grangie ... de*) *Hakerholes* m13 *ib,* (*grangie ... de*) *Akerhol'*
Hy3 (Ed1) *Newh, Akerhols* l13 *HarlCh, Acreholm* 1155-66 *ib,*
Akerholme alias Akerhowse 1539 LP xiv, - *als Akerhouse* 1609 *Foster,*
Acre howse 1569 *Yarb, acer howse* 1577 *Terrier,* - *house* c.1580 *ib,*
Acrehowse 1571, 1628 *ib,* - *alias Acrus* 1629 *Yarb, Acrehows* 1587 *ib,*
Acrehouse 1602 YD iv, 1628 *Yarb,* (*in the parish of Claxbie*) 1648
Inv, 1652, 1663 *Yarb, Acre House* 1700 PR (St Martins, Lincoln),
1779 *Yarb,* 1824 O, 1828 Bry, and cf. *Akyrholesbusckes in campo de*
Normanby 1330 *CottCh* (*v.* **busc** in a Scandinavianized form), *acre*
hous Cliffe 1601 *Terrier, the acrehouse land* 1644, 1662 *ib, acre*
house closses 1717 *LPE, Acrehouse close* 1768 *Yarb.* The earlier
name is derived from OE **æcer** 'a field, a plot of arable or
cultivated land' and the pl. of **hol**[1] 'a hollow'. There are numerous
deep hollows here, as well as modern chalk diggings. It was a
grange of Newsham Abbey in Brocklesby, PN L 2 64. There are
three forms in which the second el. is **holmr** 'an island of land, a
water-meadow', which seems singularly inappropriate here. By the
16th century -*holes* had been replaced by *house.* The grange itself

was situated in Claxby parish, the buildings being demolished during the last century and the present farm built to the east in Normanby parish, *v.* further DB xlvii.

CLAXBY WOOD, 1824, 1828 Bry, 1830 Gre, from the neighbouring parish of Claxby. LLOYD'S FM. NORMANBY CLUMP, 1828 Bry. NORMANBY DALES. NORMANBY GRANGE, apparently a late use of **grange**, for which cf. The Grange in Claxby *supra.* NORMANBY LODGE is *White Ho* 1828 Bry. NORMANBY MILL, 1824 O, 1830 Gre, *Mill* 1848 *TAMap*, cf. *Milne Hill* e17 *AddRoll, the miln hill* 1671, *ye Miln hill* 1664 both *Terrier* (Claxby), *ye mill hill* 1674 *ib, Mill Close* 1717 *LPE.* THE PARSONAGE, *the Parsonage House* 1601, *- house* 1822, *the Rectory* 1674, 1700, 1762 all *Terrier.* ROOKERY CLUMP (lost), 1828 Bry.

Field-Names

Asterisked forms in (a) are found in both 1848 *TA* and 1852 *Yarb*; spellings occurring in only one of these are dated. Forms dated l12 (13) and e13^1 (13) are *Alv*; c.1200, e13^2, a1205, 1276 are RA iv; e13^3, Hy3^1, m13, c.1250, l13 are *HarlCh*; Hy3^2, 1330 are *CottCh*; 1566, 1601, 1662, 1664, 1671, 1674, 1677, 1697, 1703, 1707, 1724, 1762, and 1822 are *Terrier*; 1587, 1713, 1734, 1739, 1747, and 1768 are *Yarb*; 1590, 1597, 1631, and 1636 are *Foster*; e17^1 are *AddR 37691*; e17^2 are *AddR 39927*; 1717 are *LPE.*

(a) Lt -, Gt Barf 1768, Barf Cl 1768, 1848, 1852 (*Barfes* 1674, *Barfe* 1677, 1717, *v.* **beorg** 'a hill, a mound' and note *barf* is common in L dial.); Blayfield, - Cl 1768 (probably from ME *blā, blō* (ON **blá(r)**) 'bluish grey, lead- or ash-coloured, dark', *v.* MED s.v. and **feld** and cf. *Blafen* e13^2, *Blayfin* 1601, *v.* **fenn**, seemingly rare in north L); Broken hill Cl 1768, 1852, - Hill Cl 1848 (*Broken hills* 1566, - *Hills* 1601, *Brockenhills* 1631, *v.* **brocen** 'broken, broken up, uneven', **hyll**; *Bracken Hills* 1717 is perhaps an error for *Brocken -*); Cow pasture Cl 1768; Croft 1768, 1848, 1852 (*v.* **croft**); Crow Wood Cl*; Far -, Nr Deepdale Cliff 1768, Deep Dale 1848, - dale 1852 (*ad Depedale* e13^3, *two deep dales* 1713, *Deepdale* 1717, *the Deep Dales* 1734, *The Deepdales* 1739, *Two -* 1747, cf. *deepe dale gate* 1597, *v.* **dēop, dalr**, it was on the boundary with Claxby, *v.* (Lt) Deepdale Cl in Claxby f.ns. (a)); Eastdale -, East Dale Cl 1768, East-dale Cl 1848, East dale Cl 1852 (*v.* **ēast, dalr**); the East Fd 1762, - fd 1822 (*ye East* *feild* 1703, - *field* 1724 (one of the open fields of Normanby, but apparently later in origin than North Fd *infra* or *ye southe fild* (b) *infra*, cf. also the West Field *infra*); Folly, Melton (sic) 1848, Folly Melton 1852, (*v.* **folie**, cf. Melton Cl *infra*); Fox Cover Pce 1848, - pce 1852; Lt Graft Cl 1768 (*le Graft, dimidiam acram terre que dicitur Graft, terram que dicitur grafte* e13^3, *terram ... que vocatur grafte* Hy3^2, *super graftum* Hy3^1, *Le Grafte* c.1250, *graft litle close* e17^2, *Grafts* 1717, *v.* **graft** 'a ditch', found also as an appellative *unum graftum* in Hy3 *HarlCh*); Grass 1848; The great Cl 1768, Great Cl*; The Home Cl 1768, Home Cl* (cf. Parson's Cliff *infra*); Hornseys Cottage 1768 (from the surn. *Hornsey*, cf. Mark *Hornsey* 1768 *BT*); The Lane Cl 1768; Little Chapel 1852; Lt Cliffe* (*y*^{e} *litle Cliffe* 1662, 1671, *y*^{e} *Litle cliffe* 1664, *y*^{e} *Litle cliffe* 1674, *v.* **clif**, North Cliff and the South Cliff *infra*); Little Walk* (*v.* **walk** 'a sheep-walk', as elsewhere in this parish); Meadow Cliff 1768 (*v.* **clif**); Melton Cl 1768 (*Meltons Close* 1717, named from the *Melton* family, a member of which is named in this document without a forename); Nr Middle Walk 1852 (*v.* **walk**); (Far -, Nr -, Nether -, upr) Middle Would, Middle Would Cl 1768, (Far) Middle Wold*, Nr - 1848 (*v.* **wald**); Middle Wood 1768; Mill Cl 1768, 1848, 1852, Mill Dam (Fd) 1852 ('Water Mill' is shown and named on *Plan* 1852); The New Cl 1768; North Cliff 1768 (*y*^{e} *northe Clyffe* 1590, *le north Cliffe* 1597, *the north Cliffe* e17^2, *v.* **norð**, **clif**, and cf. North Cliffe in Claxby f.ns. (a)); North Fd*, Nr - 1848 (*in aquilonali campo de Normanbi* e13^3, m13, *y*^{e} *Northfild* 1566, *in boriali campo de Normanby* 1597, *Northe fielde, the northe fielde* 1601, *y*^{e} *Northfield* 1662, 1664, 1674, *Northfield* 1671, *the north hie feild of Normanbie* e17^2; one of the open fields of Normanby, of which at first there were evidently only two, cf. the East Fd *supra* and *le South fild* (b) *infra*); (The far -, The Near) North Would 1768, North Wold 1848 (*v.* **wald**); Old Walk* (*v.* **walk**); Old Wives Nook 1768, - Nooking*; The out Walk 1768 (*the out walk* 1717, *v.* **ūt**, **walk**); Ox Pasture 1852 (cf. *oxe pece* 1713, *The Oxpeice* 1734, - *Oxpiece* 1739, *The Oxe piece* 1747, *v.* **oxa**); Paddock* (several); Parson's Cliff 1852 (*v.* **clif**; this land is identical with two of the fields named Home Cl 1848); Pudding hole 1768, - Hole* (a frequent fanciful name for boggy land); the sheldow 1768 (obscure); Skeltonwood 1768 (presumably *Skelton* is a surn. here); Sleight Debden 1768 (*v.* **sléttr** 'smooth, level'; *Debden* may be a surn.); Smiths Cottage 1768 (from a local family, cf. William *Smyth* 1590 *BT*); Gt Snape Cl 1768, Gt Snape-Dale 1848, Gt Snape dale 1852 (cf. *i Close called Snapes* 1566, *Snaypes* 1601, *Snapps Close* e17^2, *Snapes* 1674, 1677, *v.* **snæp** 'a boggy piece of land' and cf. the same name and feature in Far -, Nr Snapedale Cl in the f.ns. (a) of Claxby *supra*); the South Cliff 1762, - cliff 1822, South Cliffe*, South Cliff next Skeltonwood 1768, South Cliff Cl 1768 (*the Southcliffe* 1674, *y*^{e} *South Cliff* 1707, *Sooth Cliffe* 1717, *the South Cliffe* 1724, *v.* **sūð**, **clif** and cf. *North Clife* in f.ns. (b) *infra* and South Cliff

Pingle in Claxby f.ns. (a)); Southwell Cls 1762, - Cl 1768, - cls 1822 (*Southe well* 1662, *Southwell* 1671, 1674, 1677, *A close by ye Name of Southwell* 1664, *Southwell Closes* 1707, *farr -, neare Southwell Close* 1717, v. **sūð, wella**); The Far -, Near South Would 1768, South Wold* (v. **sūð, wald**); Gt Stonebridge Cl 1768, (Gt) Stone Bridge Cl* (*Stonebridge Close* 1717, cf. Stonebridge Lane in Claxby f.ns. (a)); Templedale 1768, Temple Dale 1848, - dale 1852 (*Temple dale* 1717, commemorating the former holdings here of the Knights Templars, cf. *De priore milice Templi tenente iiij. partem j.f. in Normanby* 1303 FA, v. **temple, deill** 'a share of land', as elsewhere in the parish); The upper Cliff 1768, Upper Cliff 1848 (v. **clif**); Walkwell Spring (Cliff) 1768, Well Walk Spring (sic) 1848, Walk Well Spring 1852 (cf. *Walk well Close* 1717, v. **wella**; *Walk* here may be from **walca** 'fulling, the dressing of cloth', which was a streamside operation); the West Fd 1762, 1822 (y^e *West feild* 1703, *the west* - 1707, *the Westfield* 1724, cf. the East Fd *supra*); Wold Cl* (cf. *the Would fields* 1717, v. **wald**); The far -, The near Wood Steer 1768, Wood Steers* (cf. *Wood close* 1717; *steer* may perhaps be the rare *steer* NED s.v. 4, ? 'a pile of wood', not recorded there before 1837).

(b) *Acer dale* 1601 (v. **æcer, deill**); *toftum* (*predicte*) *Aldusie* c.1200, *toftum Alduse* e13^2, a1205 ('the toft of Aldus', v. **toft**; for the ME fem. pers.n. *Aldus, v.* also *selionem Alduse* in the f.ns. (b) of the adjoining parish of Claxby *supra*); *Asuir medowe* 1587 (the reading is uncertain); *Auethlande* a1205, e13^2, *Auethlonde* a1205 (perhaps 'Awi's grazing land', from the ODan pers.n. *Awi* as in Avethorpe Kest, and OE *eteland*, v. **ete**); *Barton yatt* 1566, *Barton yate* 1601 (this appears from the text to refer to a way, v. **gata**, though derivation from **geat** 'a gate' seems more likely in view ·of the forms themselves); *betwixt bankes* e17^2 (v. **banke**); *One pece of medow Called Becke dale* 1587 (v. **bekkr, deill**, cf. *East dale Becke infra*); *Bistalls* e17^2 (perhaps v. **bī** 'by, beside', **stall**); *bladewell' heades* (v. *Crosse hookes infra*); *Blasin Prick Holes* 1717; *Bonde graft* l12 (13), e13 (v. **bondi, graft**, cf. Lt Graft Cl in (a) *supra*); *bowacer* 1566, *Bowe aker* 1601 (v. **boga, æcer**, the same name occurs in f.ns. (b) in the neighbouring parish of Claxby *supra*); *ad bra iuxta prat' Rob'ti Lecurtais* Hy3^2; *brendrid Daill* 1566, *Branderye dale* (the reading is uncertain) 1601 (the same name, no doubt referring to the same piece of land, occurs in Claxby f.ns. (b) *supra*, where it is suggested that it is derived from the surn. *Brandreth*, with **deill**); *Calke holes* l12 (13), *Calkeholes* e13 (13) (cf. *Calck Pitt* 1601, v. **calc** 'chalk', **hol**[1], **pytt**); *Capewellclose* 1717; *vie de caster* m13 ('the road from Caistor'); y^e *Church close* 1674; *Clife Scrub Wood* 1717 (cf. Lt Cliffe *supra*); *ye comon Lane* 1601; *le comon Mere* 1597 (v. **(ge)mǣre** 'a boundary, land on a boundary'); *Le Cougate* 1330 (v. **cow(e)-gate** '(right of) pasture for a cow'); *Crosse hookes & bladewell' heades* e17^2, *Cross* *hooks* (sic) 1707, 1724, *Crow hooks* 1664, - *hookes* 1671, *Crowhooks* 1674 (*v.* hōc. *Cross* and *Crow* seem to have become confused; the sense may therefore be either 'angles of land lying across others', *v.* **cros**, or 'angles of land on which crows were seen', *v.* **crāwe**; the first el. of *bladewell'* is obscure); *derhau* Hy3^1, *direhau* Hy3 (Ed1) *Newh, Derhou* m13, *Derowe, derow* e17^1 (*v.* **haugr**, the first el. is possibly **djúr, dýr** 'an animal, a beast'); *dockedaile* e13^3 (*v.* **docce, deill**); *dock hill closes* 1713, *Dockhill Close* 1717, 1739, *Dockhill hill close* 1734, *Dockhill Closes* 1747 (*v.* **docce, hyll**); *Dovehill' Furlonge* e17^2 (*v.* **dūfe, hyll**); *East dale Becke* 1597, *Eastdailebecke* e17^2 (*v.* **bekkr** and Eastdale Cl in (a) *supra*); *Eastwell Closes* 1717; *Edie gaite, - gate* e17^2 (probably from the surn. *Edie* with **gata**); *Estiby* 1288 *Ass* (p) ('east in the village', *v.* **ēast, í, bý**); *the farr close* 1717; *in campo de Normanbi* c.1200, *in campis de Normanby* e13 (13), 113^3, Hy3^2, *inter campum de Normanby* 1276 RH, *in campo de Normanby* 1330 (*v.* **feld**); *ye fildes meare* 1566. *ye fieldes mear, - meere* 1601 (*v.* **feld, (ge)mǣre**, cf. *le comon Mere supra*); *flintilandes* Hy3^2 ('the strips with flinty soil', *v.* **flint, -ig, land**); *frontehoweeudland'* 112 (13), *front hov* (sic) e13^3 (*v.* **haugr**, with **hofuð, land**); *ad hou iuxta prat' Rob'ti Lecurtais* Hy3 (*v.* **haugr**); *lez furlonges, le furlonge* 1597 (*v.* **furlang**); *Garly hill, garly gat* 1566, *garlie gate* e17^2, *Gallye hill* 1601 (perhaps from **galga** 'a gallows' and **hyll**, with **gata**); *gat a furlonge* 1566, *gate a furlonge* 1601 (perhaps comparable with *Lingadaile infra* with a reduced form of **haugr**, the first el. being doubtful, unless it is associated with *Gaithou* in Claxby f.ns. (b)); *graindegates* Hy3^2 (perhaps alluding to a forked road-junction, *v.* **grein** 'a branch, a fork (of a river), etc.', **gata**); *Grenlokehouue* e13^2, *Grenloke Houue* a1205 (*v.* **haugr**, more early forms are needed to explain *Grenloke*); *Hallibread dale* e17^2 (Mr John Field suggests that this appears to be a share of land (*v.* **deill**) subject to a payment of *holy bread silver* (*v.* NED s.v. *holy bread* (c)), an impost used to purchase *holy bread*, the bread provided for the Eucharist. 16th and 17th century spellings in *Halli-* are recorded in NED s.v. *holy bread*, the term is being used, therefore, in a similar way to *Smoke Close* etc., alluding to *smoke silver*); *a piece of medow called halling* 1587 (*v.* **hall, eng** 'a meadow', as elsewhere in the parish, cf. *a Farme called Normanby Hall infra*); *Harber slacke Furlonge* e17^1 (*v.* **here-beorg** 'a shelter', with **slakki, furlang**); *le headland* 1597, *ye headland* 1601 (*v.* **hēafod-land**); *Herberde Hille the laune* e13^2, a1205 (*Herberde* is perhaps from **here-beorg**, cf. *Harber slacke Furlonge supra*, which would then presumably refer to the same feature; alternatively, it may well be the ME pers.n. or derived surn. *Herberd, Herbert* (*v.* Forssner 148); this is an early example of the addition of an explanatory phrase to a name); *the high or low cloase or feilde* 1587; *sub colle de Normanby* 113, *super montem in campo de Normanbi* 113, *hillclif* 1566 (*v.* **hyll, clif**); *hodg grene* 1566 (*v.* **grēne**^{2}; the first el. is the pers.n. or surn. *Hodge*); *Holliwinge Sneps* (sic) 1717; *Howfar hill* 1717;

hupwarpe 112 (13), *Vpwarp* e13³ (*v.* **up, wearp** 'something thrown up', referring to spoil from trench digging and the like); *Jon headland* e17² (*v.* **hĕafod-land**; *Jon* is the pers.n. *John*); *Karewelbec* Hy3², *Carewell' furlonge* e17², *Carwell gaite,* - *garth,* - *gutter* e17² (this is probably 'Kari's spring', from the ON pers.n. *Kari* and **wella,** to which is added **bekkr** 'a stream', **furlang, gata, garðr** and ME **goter** 'a gutter'; cf. *semita de Carewell* in the neighbouring parish of Claxby, f.ns. (b), which must refer to the same spring); *Lady mill* 1597, *Lady milne Furlonge* e17² (*v.* **hlǣfdige, myln,** with **furlang,** though the significance of *Lady* is uncertain); *lambe daile* e17¹ (*v.* **lamb, deill**); *lauedistayes* (sic) e13¹ (the first el. is **hlǣfdige** 'a lady', cf. *Lady mill supra*), the second is obscure); *Lindale Close* 1636 (*v.* **lin** 'flax', **deill**); *Lingadaile* e17¹ (perhaps '(a share of land on) the heather-covered mound', from **lyng, haugr** with **deill,** and for a possible comparison, *v. Lingam furlonge* in Binbrook f.ns. (b)); *Little Close* 1717; *Lowe would* e17² (*v.* **wald**); *Mardaile Furlonge* e17² (*v.* **(ge)mǣre, deill**); *minster headland* e17¹, *mynster headland* 1601 (*v.* **mynster, hĕafod-land**; the Church of Lincoln held land in Normanby, *v.* RA iv, 189-201); *yᵉ moore* 1674, *Moore Cliffe* 1713, 1739, *More Clife* 1717, *Moorcliffe* 1734, *Moor Cliff* 1747 (*v.* **mor¹, clif**); *nettelcoate feild* 1587 (*v.* **netel, cot,** alluding to the vegetation surrounding the cottage or hut); *Nettelton gate* 1566, 1597, e17², - *gaite* e17², *netleton gat* 1601, *viam ducent' ad Nettleton* 1597 ('the road to Nettleton', *v.* **gata**); *Nettleton meare* 1566, - *Mere* 1597, - *mere* e17² ('the boundary with Nettleton (a neighbouring parish)', *v.* **(ge)mǣre**); *Normanby Field* 1717; *a Farme called Normanby Hall* 1713, *a farm called* - 1734, *a Farm called* - 1739, 1747; *the north coattes dick* 1587 (*v.* **norð, cot, dik**); *Northinge* 1566, 1601, *the Northings* 1717 (*v.* **norð, eng;** the same name occurs in Claxby f.ns. (a) and may well refer to the same piece of meadow); *ye north stong* 1566 (*v.* **norð, stong,** cf. *tenstang infra*); *the owerwhart dyche* 1601 ('ditch or dike running across', *v.* **ofer³, þvert, dic**); *land called Peale* e17²; *Redde daill* 1566, *Redd daile* 1601 (*v.* **deill;** the first el. may be **rĕad** 'red' or **hrĕod** 'reed'); *Redegate* Hy3¹, m13 (*v.* **gata** and cf. the prec.); *Rose garthes* 1601, e17² (*v.* **garðr**); *inter semitam de Rowelle & divisam de Nettletun* e13¹ (13) ('between the path from Rothwell (LSR) and the Nettleton boundary'); *scharth* 112 (13), *ad Scarth* e13¹ (13) (*v.* **skarð** 'an opening'); *Short furlonge* 1566, *le short furlonges* 1597, *Shorte furlonge* 1601, *Short Furlonges* e17² (*v.* **sceort, furlang**); *Slawell* m13 (*v.* **slāh** 'a sloe', **wella**); *yᵉ southe fild* 1566, *the south hie feild of Normanbie* e17², *the -, ye South fielde* 1601, *yᵉ South field* 1662, *yᵉ Southfield* 1664, 1671, 1674, *super australem camp ... de Normanby* 1597 (one of the open fields of Normanby); *stretfurlonges* e13³, *street Furlonge* e17¹, *yᵉ street* 1724 (*v.* **strǣt;** the reference is presumably to High Street, which forms in part the eastern boundary of the parish, *v.* Margary 240); *tenstang* e13¹ (13) (*v.* **stong,,** no doubt as a measure of land 'a pole'); *þirsheteles dailes* (sic) e13¹ (13) (apparently

'Thirketil's portion of land', *v.* **deill**; the first el. being the Anglo-Scandinavian pers.n. *Þurketel* from ON *Þorketil* etc., which, as Dr John Insley points out, has here been influenced by ODan *Thyrkil* with secondary palatal mutation, and would, therefore, be indicative of late Scand. influence in the area); *thornetree* e17[2] (self-explanatory, *v.* **þorn, trēow**); *thorsway gate* 1566, *Thoreswaie gate* e17[1], *Thorseway gate* 1664 ('the road to Thoresway (an adjacent village)', *v.* **gata**); *the Towne end Close* 1717; *Thrawbankes* 1566, *Thrawbanckes* 1601, *the Throw Banks* 1713, - *banks* 1747, *the throw* - 1734, *The Throw Canks* (sic) 1739 (perhaps from *v.* **þrūh** 'a conduit' and **banke**); *Toft becke* l13 (*v.* **toft** 'a curtilage', **bekkr**; the same name occurs eight times in Claxby f.ns. (b) and must refer to the same stream); *y*[e] *Walk* 1674, *An enclosed Sheepwalk* 1697, *an enclosed Sheep walk* 1703, *y*[e] *greatwalk* 1724 (*v.* **walk**; the explanation is to be found in the 1697 and 1703 forms, the references being to the same feature); *Wascuuell'* e13[2], *Wascuuelle* a1205 (*v.* **wæsce** 'a washing place (for sheep etc.)', **wella**); *Watter Close* 1717; *toftum ... quod fuit Beatricis ad fontem* e13[2] (*v.* **wella**); *Wickdaile* e17[2]; *the windinges* e17[2] (for an explanation of this name, *v. two wyndinges* in Binbrook f.ns. (b) *supra*); *Wrang Landes* m13 (*v.* **vrangr** 'crooked', **land**); *a mare de Wycham* l12 (13), *Wichammare* e13[3] (*v.* **(ge)mære**), *a uiam de Wycham* e13[1] (13), *wickham Cloase* 1587 (all named from *Wykeham* (in Nettleton parish) PN L 2 238); *yadewordehou* l12 (13), *Jadewurthehou* e13[3] (*v.* **haugr** 'a hill, a mound, a burial mound', the first el. being the ON pers.n. *Játvarðr*, ODan *Iadwarth* from OE *Éadweard*).

Owersby

NORTH OWERSBY

Aresbi 1086 DB (3x)
Auresbeia 1189 (c.1200) CartAnt, -*by* e13 RA iv, 1210 FF, 1238-39 RRG, 1254 ValNor, -*bi* 1203 FF
Oresbi 1086 DB (4x), 1182 P, 1193, 1194, 1197, 1198, 1199, 1200, 1201, 1202, 1203 ib all (p), *Oresbi Josleni* 1190, 1191, 1193 ib, -*by* 1517 *Monson*, 1551, 1557 Pat, 1586 *Monson*, 1629 *TLE*, 1651 *Red*, - als *Longoresby* 1586 *Monson*, *Horesbi* 1203 Cur
Oresbi hundred 1086 DB
Orresbi 1191 P (p), 1202 ChancR (p), 1233 WellesW, 1233 RA ii
Ouresbi c.1115 LS (5x), 1139-41 *DuLaCh*, m12 Dugd vi, c.1160 RA iv, 1166 RBE (p), Hy2 LN (p), 1196-1203 RA

iv, 1203 Cur, 1203, 1219 Ass, *-bia* c.1162 RA ii, *-by*
c.1200 ib iv, 1203 Cur, 1203 Abbr, 1206 Cur, 1234 FF,
1236 Cl, 1239-49 *Foster*, 1242-43 Fees, 1245, 1262 FF,
1265 Cl, 1267 AD ii, 1274 Ipm, 1275 RH, 1281 *Foster*,
1281 QW, 1287 RA iv *et freq* to 1528-29 *AD*, *-bie* 1549
Pat, *Ourisby* 1280 RSu, *Houresbi* 1161 RA i, a1168 ib iv,
1205 P, 1219 Ass, *-bia* 1212 Fees, *-by* Hy2 (1409) Gilb,
1205 OblR, 1275 Peace, a1291 RA ix, *Hourisby* 1300 *Stix*
Owresbi 1198 FF, *-by*, 1209-35 LAHW, c.1212 Welles, 1412
AD i, 1445 AASR xxix, 1447, 1460 *Foster*, 1498 *Monson*,
1525-26, 1528-29 *AD*, 1529 Wills ii *et passim* to 1634
Monson, - *Stiddolfe* 1684 *ib*, - *otherwise longe Owresby*
1563, 1625 *ib*, - *als Long Owresby* 1607 *ib*, - *als Longe*
Oresby 1647 ib, - *als Oresby* 1655, 1684 *ib*, *-bye* 1540 *ib*,
1610 Speed, *-bie* 1552 Pat, *-bey* 1585 *Monson*, *Owrysby*
Hy7 AASR xxix, 1509 Monson
Oureby (sic) 1206 Cur, *-bi* 1221 FineR
Oueresby 1230 Cur, 1373 Peace, *Ouersby* 1598 *LCS*, *Owereby*
1395 Pat
Owersby 1436 Fine, 1529 Wills ii, 1655 *Monson et passim*,
- *als Owresby* 1604 *ib*, - *als Oresby* 1627, 1655 *ib*, *-be*
1472 WillsPC, 1625 *Terrier*, *-bye* 1552, 1609 *Monson*,
-bie 1558 InstBen, 1576 *LER*, 1607, 1610 *Monson*, 1611
Holywell, - *otherwise called longe Owersbie* 1608 *Monson*
Owrsbie 1589 *Monson*, *-by* 1605 *ib*, - *called Owrsby Stidolfe*
alias Owrsby Southall 1612 *Td'E*, - *otherwise called Long*
Owrsby 1612 *Monson*, - *als Oresby* 1655 *ib*
Owasby (sic) 1790 *Monson*
Orsby 1630 *Monson*
Longe Ouresbi 1219 Ass, *Langehouresby* 1272 *Ass* (p),
Langouresby ll3 *Foster*, ll3 Percy, 1308 *Monson*, 1318
FF, 1328 Banco, 1329 *Foster*, 1384 *FF*, *-houresby* 1349
Monson, *lang Ouersby* 1383, 1392 *ib*, *long Ouersby* 1612
ib, - *Owresby* 1510 *DCFabRents*, *longe Owresby* 1604,
1609 *Monson*, *Longowresbye* 1595 *ib*, *longe Owersby*
1596 *ib*, *Longe* - 1597 *Inv*, - *Oresbye* 1604 *Monson*,
Long Owersbe 1624 *Terrier*, - *Oursby* 1664 *ib*, - *Owersby*
1693 *ib*, 1728 *MiscDep 16*, *Longe Owersby als Oresby*
1684 *Monson*
North Owersby 1824 O, 1828 Bry, 1830 Gre, *Owersby North*

End 1822 *LTR*

Fellows-Jensen, SSNEM 63, points out that Ekwall's suggestion, DEPN s.n., that the first el. is the Scand. pers.n. *Ávarr*, has been shown by *Bower*, s.n., to be phonologically unsatisfactory. The latter proposed that the first el. might be the OWScand pers.n. **Aurr*, from *aurr* 'gravel, mud, mire' and this might be phonologically acceptable, but it presents semantic difficulties with regard to its significance as a byname. Fellows-Jensen then suggests alternatively that the first el. might be the appellative *aurr* "perhaps indicating a stretch of gravelly land". However, both the occurrence of the gen.sg. and local topography point to a pers.n. as first el. here. Dr John Insley points out that the phonological development of "ON *Ávarr*, *OD*an (runic) *Āve*R, OSwed *Āver* follows the sequence **Anu-*( )*aiRaR*> **AnwœiRR*> *ĀvœaRR*, *v.* J. Brøndum-Nielsen, *Gammeldansk Grammatik*, I, 2. andrede udgave, København, 1950, p. 194. As runic *œuaiR*, the name occurs in the Runic inscriptions of Helnæs and Flemløse (both on Fyn), which are generally dated to the ninth century (*v.* E. Moltke, *Runes and their Origin: Denmark and Elsewhere*, Copenhagen, 1985, p. 156, and for a rather narrow dating to 800-825, L.F.A. Wimmer, *De danske runemindesmærker*, Haandudgave ved Lis Jacobsen, København-Kristiania, pp. 100-102)". Dr Insley goes on "Runic *œuaiR* retains a nasalized vowel. In English, this would have been rendered initially as *An-, On-* ... Later loans, from c1000 on, have already lost the nasalized vowel, *v.* J. Insley, *Studia Anthroponymica Scandinavica*, 3, 1985, 23-24, 45". He concludes "Given that the p.ns. in -bȳ represent a secondary phase of Scand. colonization, it is quite possible that the first el. of Owersby is the pers.n. *Ávarr*, borrowed at a date subsequent to the loss of the nasal quality of the initial vowel, i.e. at the end of the tenth or the beginning of the eleventh century". No certainty is possible but '*Ávarr*'s farmstead, village', *v.* bȳ, seems a likely solution of the name.

The affix *Long* presumably refers to the straggling village, eventually distinguished as North Owersby and South Owersby. The *Joselin* found in forms dated 1190, 1191, 1193 is derived from *Joselinus de Barewe* 1193 P, while *Stiddolf* is from the Surrey family which held the manor in 1571, cf. Thomas *Stydolff* 1571 *Monson* and later *Sigismund Stydolfe* 1683 *ib.* This is referred to in *manerium de Owersby alias Owersby Stidolfe alias Owersby Stydolff* 1571 *Monson*, *the Mannor of Owersbie Stidolfe als Owersbie Southall*

1575 *ib, the Mannor or Lordshipp of Owersbie als Owersbie Stydolfe als Owersbie Southall als Southall Garthe* 1683 *ib* (*v.* **garðr** 'an enclosure or small plot of ground, especially one near a house'). The latter has been first noted in *manerio de Southall'* 1384 *FF* and later as *the mannor of Owersbie als Owresbye Southall als Southalle garthe* 1566 *Monson, the mannor of Owersbe Southall als Southall garthes* 1578 *ib* and *the Mannor of Owersby als Owersby Southall als Southall garth* 1684 *ib. South Hall* has been noted independently in *atte Southall' de Langouresby* 1329 *Foster* (p), - *de Ouresby* 1333 *ib* (p), *Southalle* 1365 *FF, the South Hall* 1638 *Terrier* (Kirkby cum Osgodby), and cf. *the South hall dayle* 1608 *Monson* (*v.* **deill**). There is no real indication of the site of the *South Hall*, but it is tempting to associate it with South Owersby, a name not so far noted before 1822, though there is, however, no substantial evidence to support this hypothesis. Cf. Hall Fm *infra* and *maner ... de Newhall* in f.ns. (b) *infra.* Note also *ad finem borealem de Owresby* 1445 *AASR* xxix ('at the north end of Owersby'), which presumably refers to North Owersby.

SOUTH OWERSBY, 1824 O, 1828 Bry, 1830 Gre, *South Owersby otherwise Owersby South End* 1822 *Terrier, Owersby South End* 1822 *LTR.*

THE BELT. BLACKHALL is *The Bottoms* 1824 O, 1830 Gre. BLACKHALL WOOD. CADES TOP FM is *Sheepfield House* 1824 O, 1830 Bry, cf. *Sheepfield* 1721 *Monson, Sheep Field* 1805 *ib.* The later name is presumably from the surn. and note *Cadeheuedland* 1330 *Foster* (*v.* **hēafod-land**). CATER LANE is mainly in Thornton le Moor parish and most references occur in Thornton documents; *v. infra* for forms and discussion. COCKED HAT PLANTATION (lost, TF 090 959), 1828 Bry, *Cock'd Hat Plantation* 1824 O, 1830 Gre, named from its shape. COLEM PLANTATION, probably cf. *Cowlam Close* 1805 *Monson.* DARK PLANTATION. THE GRANGE, apparently a late use of **grange,** cf. The Grange in Claxby *supra.* NORTH GULHAM, *Gullham* 1824 O, 1830 Gre, *High Gullum* 1828 Bry, *Gulham* 1837 *BT* (Owersby) cf. *Goulholm, Goulholmendes* 1280-85 *Foster, Gulholm* 1358 *Monson, Gullholme* 15 (c.1570) *ib, Gulholme* 1707 *Terrier,* -*holm* 1709 *ib, Gullholme bancke* 1611 *TLE, Gullam* 1608 *Monson,*

1700, 1706 *Terrier*, - *als Gulholme* 1625 *Monson, Gullam ground*
1664 *Terrier, gullam Cloase* 1693 *ib*, upper *Gullam* 1721 *Monson,*
Gullum 1646 *Inv, Gulham groundes* 1674 *Terrier, Gullham Grounds*
1679 *ib*. This appears to be a compound of ME **goule** 'a ditch, a
stream, a channel' (probably derived from OE **golu* denoting some
form of watercourse) and ON **holmr** 'an island of land, higher
ground in marshy land', the latter showing developments to *-um,*
-am, -ham, common in north L. The place lies on the slope above
the R. Ancholme, though personal examination suggests that there
is now no obvious watercourse to which *goule* referred. Just to the
west is Peaseholm *infra*, another name in **holmr**. SOUTH
GULHAM, *low Gullam* 1721 *Monson, Low Gullum* 1828 Bry and is
Low Farm 1824 O, *Lower F.ᵐ* 1830 Gre. HALL FM is *Manor*
House 1824 O, 1830 Gre, - *Ho*. 1828 Bry, but cf. *at Hall* 1428
AASR xxix (p), *Hall Close* 1805 *Monson, hall Garths* 1821 *ib*.
HAPPYLANDS FM is apparently a modern name, presumably of
the complimentary nickname type. HIGHFIELD FM. This
appears to be a comparatively modern name for a farm which is
situated in the highest part of the parish on Owersby Moor.
However, it may be reminiscent of the former open fields of
Owersby, recorded in *campo de Ouresby* 1287 RA iv, 1428 AASR
xxix, *campum de Ouresby* 1320-27 *Foster, campis de Ouresby* 1330
ib, 1339 *Monson*, 1352 *Foster*, 1390 *AD*, 1404 AASR xxix. HOOK'S
FM, cf. *the howke* 1601 *Terrier*, (*the*) *Hook* 1684, 1721 *Monson,*
1767 *Stubbs*, 1790, 1805 *Monson*. The forms are late, but the
earliest spelling seems to suggest that it is comparable to Hook, PN
YW 2 20, and is derived from OE **hūc* (*v*. **hōc** 'a hook, a bend'),
referring to its situation lying low, but with a bend in the Old R.
Ancholme to the west. The later spellings certainly suggest
association with *hook* from **hōc**. INGS MEADOW, (lost,
approximately TF 030 090), 1805 *Monson*, 1824 O, - *Meadows* 1830
Gre, *The Ings* 1828 Bry, *les Inges* 1616 *Monson, the Inngs* 1629
TLE, the Inges 1630 *ib*, (*the*) *Ings* 1684, 1705 *Monson, Inges* 1767
Stubbs, cf. *north Ings* 1601 *Terrier, the South Inges* 1611 *Monson, the*
West Inges of Owersby 1603 *ib, the Ing-fure* 1608 *ib* (*v*. **furh**), *the*
ynge furre, - dyke 1611 *TLE*, 'the meadows, the pastures', from the
pl. of **eng**. They lay in the south-west corner of the parish by the
R. Ancholme. LOW CLOSE (lost, approximately TF 026 952),
1805 *Monson*, 1828 Bry, 1839 *TA*. MANOR FM. MILL FM,
Owersby Mill 1824 O, 1830 Gre, *Mill* 1828 Gre, cf. *molendinum*

domini abbati Thornton' 1280-85 *Foster, unum modendinum ventriticum in Ouresby ... quod vocat le Northmiln'* 1339 *Monson, Molendinum Ventriticum* 1531 *ib, molendini ventrit' voc' mancknolle myln* 1558 *ib, molend' ventrit'* 15 (c.1570) *ib, Wind Mill standing in one platt of arrable ground called Brownedaile hill* 1609 *ib, the Mill* 1684 *ib* (a windmill), cf. *milne enge* 1320-27 *Foster, milnheng'* 1345 *ib* (*v.* **eng**), *Myll-hill, Millfeild* 1608 *Monson, Mill Plat, - Lane* 1805 *ib.* It is uncertain whether all these references are to the same mill. For the family named in the 1558 form, cf. *Robert Manknowle* 1583 *BT,* Robert *Mancknall* 1607 *Monson, the heirs of William Mancknele* 1634 *ib* and for *Browne-daile hill,* s.n. Browns Farm in f.ns. (a) *infra.* NEW FM. NEW PLANTATION. OWERSBY BECK (lost), *Oresbie Beck* 1601 *Terrier* (Thornton le Moor), *Owresby beck* 1622 *Monson, north Owersby beck* 1723 *LPC, Owersby Beck* 1755 *ib,* 1777 *Red, atte Beck'* 1327 *SR* (p) *atte Bek* 1358 *Monson* (p), *le Bek* 1360 *ib, le beke* 15 (c.1570) *ib,* cf. *Bekfurlanges* 1360 *ib, Beck(e)furlanges* 1345 *Foster, Beckefurlonges* 1327 *ib, beckfurlanges* 15 (c.1570) *Monson, le bek furlong* 1536 *ib, the becke furlonge* 1634 *ib, Bekgate* 1453 *ib, the Becke Leyes* 1607 *ib, Beck-leas* 1608 *ib, Beckclose* 1684 *ib, beck Close* 1721 *ib, Far -, Near Beck Close* 1839 *TA, v.* **bekkr** 'a stream', with **furlang, gata** and **clos(e)**. Some of the references are from Thornton le Moor documents, suggesting that the *Beck* formed part of the boundary between the two parishes. On the other hand, *le becke* 1552 *Monson* was certainly between Osgodby and Owersby as the text makes clear. OWERSBY BRIDGE is on the site of *Kirkby Ford* 1828 *Bry,* from Kirkby in Kirkby cum Osgodby parish *supra.* OWERSBY DRAIN (lost), 1824 *O;* now unnamed on the map, it formed the northern boundary of the parish with Thornton le Moor and may be a later name for *Owersby Beck supra.* OWERSBY MOOR, 1683 *MiD,* 1790 *Monson,* 1824 *O,* 1828 *Bry,* 1829 *Td'E,* 1830 *Gre, Owresby More* Hy7 *Lanc, The More of Owersby* 1608 *Monson, Owerby More* 1669 *MiD, owersby moore* 1679 *ib, the Moor* 1721, 1805 *Monson,* cf. *Moor Lane* 1805 *ib, Moor Lane Close* 1805 *ib, Aldemore* 1338 *ib, Aldmore hedland* 15 (c.1570) *ib* (*v.* **ald** 'old' in the sense long used or formerly used), *Moredayle* 1339 *ib, the more dayle furres cloose* 1611 *TLE* (*v.* **hēafod-land, deill, furh**), *Moor Close, - Land* 1889 *TA.* MOOR FM, 1824 *O,* 1830 *Gre.* OWERSBY TOP. TOP FM. PEASEHOLME, *peseholme* 15 (c.1570) *Monson, Pease holme* 1608 *ib, peaseholme* 1684 *ib, Pease* *holm* 1721 *ib, Pease Holmes* 1805 *ib,* cf. *peasholme cloose, - headland* 1611 *TLE, Pease holme close* 1634 *Terrier, upper Peaseholme Closes* 1822 *ib,* 'the island of land, the higher ground in marshy land where pease grow', *v.* **pise, holmr**. ROUND PLANTATION. SOUTH OWERSBY HO is *Kirkby Old Hall* 1824 O, 1830 Gre, - *old Hall* 1842 White, *Kirkby Hall* 1828 Bry, from Kirkby in Kirkby cum Osgodby parish. STEPPING STONES, over Kingerby Beck. THE VICARAGE, *the vicarage of ye Church of Owresbey* 1585 *Monson, the vicaridge house of Owersby* 1601 *Terrier, the vicaredge of Owresby als Oresbye* 1618 *Monson, the vicarage of Owersbe* 1625 *Terrier, the viccaridge of Owresby als Oresby* 1626 *Monson, ye Vicaridge House* 1664 *Terrier, the viccaredge* 1684 *Monson, ye Vicarage house* 1697 *Terrier, ye Vicarage* 1700 *ib, Vicarage House Garden and Paddock* 1829 *TA,* cf. *Vicarage close* 1721 *Monson.*

Field-Names

Forms dated c.1160, c.1200, e13, 1287 are RA iv; those dated 1280-85, 1284, 1299, l13, 1320-27, 1326, 1327^1, 1330, 1345, 1348 are *Foster;* 1327^2 are *SR;* 1343 are NI; 1370 are *FF;* 1377 are Fine; 1381 are Peace; 1393 are Works; 1395, 1549 are Pat; 1404, 1428, 1445, Hy7 are AASR xxix; 1529 are Wills ii; m16 are *Cragg,* 1562 are *Inv;* a1567 are LNQ v; 1601, 1606, 1625^2, 1633, 1634^2, 1638, 1664, 1674, 1679, 1693, 1697, 1700, 1706, 1707, 1709, and 1822 are *Terrier;* 1611^2, 1629, 1630, 1665 are *TLE;* 1723 are SDL; 1767 are *Stubbs;* 1839 are *TA 248.* All other forms are *Monson.*

(a) Answer Dike 1790 (*v.* **dík**); Barnicle Cl 1805 (*East -, west Barnackles* 1629, *East -, West Barnakles* 1630; it is on the boundary with Kirkby, cf. the Barnacles in the f.ns. (a) of Kirkby cum Osgodby *supra;* earlier forms are needed to suggest a convincing etymology); Birch Cl 1805; Bottom 1805 (*v.* **botm**); Branbury Btm 1751, Bramberry - 1805 (possibly 'broom-covered hill', *v.* **brōm, beorg** with **botm**, though it may be from a local surn., cf. William *Brainbrow* (sic) 1805); Brickyard Pces 1839 (cf. *lee* ('the') *bricke kylne close* 1588, *Brickkill Close* 1665); A Farm called Browns Farm 1839 (cf. *Broundayle* 1339, *browne daille hill* 1588, named from the *Brown(e)* family, cf. John *Broune latt of owersbye* 1571 *Inv,* Thomas *Browne* 1597 *ib,* and many references in *BT,* though so far no medieval references to the family have been noted, *v.* **deill**); Bull Garth 1805 (*v.* **garðr**); A Farm called Busts Farm 1839 (named from the *Bust* family, cf. Joseph *Bust* 1839 *BT*); Butcher Cl 1805, 1839 (possibly from the surn. *Butcher*); Calf Cl 1805 (*The Calfe close* 1608); Car (Head), Top Car 1805 (*the Karre* 1601, *the Cars* 1608, *North Carr close* 1609, *lower -, upper Car* 1721, cf. *prato de Kergate* c.1160, *Kergate* 1299, 1330, *Kergat'* 1327[1], 1330, 1345, *Carregate* 15 (c.1570), *kerlandes* 1280-85, *Kerfurlanges* 1320-27, from **kjarr** 'marshland', with **gata, land, furlang**); Cats Cl 1805; Clay Pits 1805, the Clay Pits 1839, Clay Pit Fd, - Plat 1805 (*v.* **clæg, pytt**, with **feld** and **plat**[2] 'a small piece of ground', as elsewhere in this parish); Far -, Middle -, Nr Close 1805; Clover Cl, Low Clover Plat 1805 (*v.* **plat**[2]); Common Cl 1805 (1721); Corn Cl 1805 (1721); Nort[h] Cotchers Cl 1751, Cottager Carr 1767, 1790, Cottage Car (sic) 1805 (*Cotcher Car* 1721, *v.* **kjarr**; *cotcher* is a freq. dial. form of *cottager*); (North -, South) Cottage Pasture 1805; A Farm called Cotterills Farm 1839 (from the surn. *Cotterill*, cf. John *Cotterill* 1821 *LTR*); Coulman plat 1751 (named from the *Coulman* family, cf. George -, James *Coulman* 1751, with **plat**[2]); Cow Cl, - Marsh 1805 (*the cowe cloose* 1611[2], (*far or South*) *Cow Close* 1721); Crabtree Marsh 1805 (probably from the surn. *Crabtree*); Cream Poke Cl 1805 (*Cream poak Close, - lane, Creampoke* 1721, a complimentary term for rich pasture, cf. *Creame Poake Nooke* in the f.ns. (b) of Bigby, PN L 2 55); Cringle 1805 (*Cringlands* 1721, *v.* **kringla** 'a circle', **land**); Dale Cl 1805, the Dale 1822 (*y*[e] *dale Close* 1674, *dale Close* 1679, *Dale Cloase* 1693, *v.* **dalr**); A Farm called Darleys Farm 1839 (from the surn. *Darley*); A Farm Called Davy's Farm 1839 (named from the *Davy* family, cf. William *Davy* 1607 *Inv*); A Farm called Dixon's Farm 1839 (named from the *Dixon* family, cf. Robert *Dixon* 1593 *Inv*); Dole and Garth Cls 1839; the East fd 1751 (*in orientali campo de Ouresby* 1284, *in oriental' campo de Ouresby* 1299, 1327[1], *in Orientali campo de Owersby* 1330, *in orient' campo* 1339, 1344, - *de Ouresby* 1339, 1360, *in orientali campo de Ouresby* 1345, *campus orientalis* 1404, *in orient' campo de Owersbye* 1552, - *de Owresby* 1588, *the est Feyld* 1561, - *Est feilde* 1612, *the east feild* 1601, 1604, - *fielde* 1611[1], - *feilde* 1611[2], *the East feild* 1603, 1629, - *Feild* 1630, *The East Feild* 1608, - *Eastfeild* 1609, *Eastfield* 1721, *the East feildes of Owrsby* (sic) 1605, *the East feildes* 1609, 1634, one of the open fields of Owersby, cf. West Field *infra, v.* **east, feld**); A Farm called Easts Farm 1839 (named from the *East* family, well represented in the parish, cf. Thomas *Est* 1585 *Inv*); Eight Pound Cl 1805 (doubtless alluding to a rent); Fat Beast Cl 1805 (*fat Beast('s) Garth* 1721); A Farm called Fatchetts Farm 1839 (named from the *Fatchet* family, cf. John *Fatchet* 1804 *BT*); Feeding Cl (Btm) 1805 (*feeding Close* 1721); Flamber Cl 1805 (*Flaumber* 1609, *Flambrughe hill cloose, Flambrughe hedge* 1611[2], *Flamber* 1721); Foley Cl 1805 (presumably from the surn. *Foley*); Foot Ball Cl 1805, - ball Cl 1822, Football cl 1839 (*football Close* 1721); Forty Acres Cl, Forty Acres Btm 1805 (*v.* **botm**); Furnish Cl 1805 (named from the *Furnish* family, cf. William *Furnish* 1707 *BT*); Garth 1805 (*v.* **garðr** 'an enclosure

or small plot of ground', as elsewhere in this parish); Gilliat Cl 1805 (named
from the *Gilliatt* family; *John Gilliatt* is mentioned in the document); the Glebe
Cls 1822; Goose Garth 1805, 1839 (cf. *Goose Close* 1721, *v.* **gōs, garðr**); Great -,
Little Cl 1805, 1839; Green Cl 1805 (1721) (*The Green close* 1608, *v.* **grēne**[1], cf.
foll. n.); Greens Cl 1805, 1839 (probably named from the family of Andrew
Green 1705 *Monson*, the surn. going back at least to Thomas *de Grene* 1393,
John *de Grene* 1404. Note also *Grenesparting* (probable reading) 15 (c.1570), the
source being a 16th century copy of a charter probably to be dated 15th
century. The first el. is the surn. *Grene*, but the second el. is difficult. Mr
John Field draws attention to the occurrence of *parting* as the first el. of f.ns.
with *Close, Mead*, and *Acre* (PN O 460), where it is suggested that "the
reference could be either to land which can be divided, or to land which forms
a boundary", with a preference for the latter. In these instances *parting* is an
adj., whereas in the present example it is clearly a noun. MED s.v. *partinge*
gives, among others, two meanings which may be pertinent here - 'a boundary,
barrier' and 'a share, portion'. Each is clearly possible and a meaning 'Green's
boundary' or 'Green's share, portion of land' may be suggested. The use of
parting as a noun does not seem to have been noted previously in the Survey);
A Farm called Greenwoods Farm 1839 (named from the *Greenwood* family, cf.
George *Greenwood* 1751 *Monson*); Ground Cl 1805; Hazard Wells Cl 1805
(*hazardwel* 1721, perhaps to be identified with *hatherswelhyl* in (b) *infra*); (North
-, South) Hill Cl, East -, West Hills 1805, Far Hill Cls 1822, Hilly Cl 1839
((*East -, West*) *Hill Close, West hill Close, - hill Close*(*s*) 1721); Low -, Middle -,
Top Hollow 1805 (*west hollow* 1552, *a fullong* (sic) *called Short Hollowe* 1611[2],
Hollowses (sic), *East -, Low -, Middle -, West Hollows* 1721, *v.* **hol**[1]); Home Cl
1839, - Garth 1805, 1839 (*v.* **garðr**); Hornsey Cl 1805 (from the surn. *Hornsey*, cf.
William *Horncy* 1721); House Cl 1805, South of the House 1839; Hunger Hill
1805 (1721, a common name for 'infertile land' or 'poor land', *v.* **hungor**);
Husbandman Carr 1767, Husband(man's) Car 1805 (*v.* **kjarr**, *Husbandman* in
contrast to *Cottager*, cf. North Cotchers Cl *supra*); Inhams plat 1751, Inams Cl
1805 (*v.* **innām, plat**[2]); Intack 1805 (1721, *v.* **inntak**); Kirk Cl 1805, 1822, - Btm
1805; Kirkhill 1751, (Rush) Kirkhill Cl 1805 (*Kerchyll* 1280-85, *kirkhill*(*e*) 15
(c.1570), *kirke hill, Kirkehill furlong* 1601, *Kyrke hill* 1611[2], *East Kirkhills* 1609,
Kirkhill's (sic), (*Great -, Little*) *Kirkhill* 1721, *v.* **kirkja, hyll**); Lairsykes Btm 1805
(*layrsikes* 1280-85, *lairsykes* 15 (c.1570), *lawsickes* (sic) 1601, *laresikes lands* 1608,
Lare Sykes Lane 1721, from **leirr** 'clay' and **sīk** 'a ditch', with **botm**); Land Cl
1805, 1839 (1721); Lane Cl 1805 (1721); Long Cl 1805, 1822 (1721, *the Long
close* 1608); Long Flg 1751 (*longfurlong* 1536, *the lonnge furlongs* 1601, *longe
furlonge* 1634[1], *v.* **lang**[1], **furlang**); Long Lands 1751 (*cultur' voc' longe lands* 1588,
West Longlands, east-Long lands 1608, *West Longlands cloose* 1611[2], 'the long

strips', *v.* **lang**[1], **land**); Lovedale (Cl) 1805, - Cl 1822, East End of Top Lovedale Cls 1832 (*Lufdayl* 1404, *lovedaille* 1588, *Love dayle hole,* - *hill* 1608, *Lovedale* 1684, 1700, 1706, 1707, 1709, *the upper Love dayle cloose* 1611[2], *Loudall ground* 1664, *A Cloase Cauld Louedale* 1693, *Lovedales* 1721, from either the OE feminine personal name *Lufu* or the ME surn. *Loue* (Reaney s.n. *Love*) and **deill** 'a share, a portion of land', apparently identical with Love-dale in South Kelsey f.ns. (a) *supra*); the low Dam 1790; High -, Low Marsh 1805, 1839 (*y*[e] *mershe* 1404, *ye mershe of Owresby* 1445, *Owresby Marsh* 1601, - *marche* 1606, *the Marsh* 1608, 1638, (*bottom* -, *high* & *low*) *Marsh* 1721, *lee* ('*the*') *southe marshe* 1588, *the south marshe cloose* 1611[2], *South Marshe* 1625[2], *Marsh-end* 1608, *the marsh end cloose* 1611[2], *the Marshe in the furlonges* 1607, *le Marshe in the Furlonges* (sic) 1610); Mellfield 1805 (*Melfeild* 1607, 1610, *Melfield* 1721, perhaps from **mylen** 'a mill' and **feld**, but note *Millfeild* 1608 under Mill Fm *supra*); Mickledale 1767, Mickle Dale 1790, 1839, Mickledale Cl 1805 (*Mickeldayel* 1705, *Michaeldale* 1721, 'the big share of land', *v.* **mikill**, **deill**, a Scand. compound); Moor Cl, - Land 1839; Morris Garth 1805 (from the surn. *Morris* with **garðr**); Munk Moor Pitt 1790; New Cl 1805 (1665, 1721, *The New Close* 1608, *the new close* 1610); Obkin Cl 1805 (named from the *Hobkin* family, cf. Robert *Hobkin* 1711[2]); A Farm called Ogg's Farm 1839 (named from the *Ogg* family, cf. William *Ogg* 1831 *BT*); Ogsley Cl (sic) 1805 (cf. *Hoggesly-hill* 1608, *Hogsley* 1721; it is not clear whether this is to be associated with the prec.); Old Orchard 1751 (1721), Old Orchard Plat 1805 (*v.* **plat**[2]); One Cl 1839; Osgodby Lane (Cl) 1805 (from the road leading to the neighbouring township of Osgodby); Osgodby's Fm 1839 (from the family name, cf. Mathew *Osgodby* 1601 and cf. *Osgarby Wray* in (b) *infra*); Owersby Lane End 1839; Paddock 1805; Paradise 1805 (*Paradise, Paradise Close* 1721, probably a complimentary name, but ME **paradis(e)** also denoted 'a garden, an orchard, a pleasure garden'; this name is fairly common in north L); Parsonage Cl 1805, Two closes each called - 1839 (cf. *y*[e] *parsons pingle* 1679, *y*[e] *Vicaridge Pingle* 1664); Passley Btm 1805 (if not from a surn. possibly alluding to corn parsley (*Petroselinum segetum*) found on grasslands in the Midlands); Pinfold Garth 1805 (*v.* **pynd-fald**, **garðr**); Pingle 1805 (1665, 1721), the - 1822 (*y*[e] *pingle Close* 1679, *one Little Cloase Cauled y*[e] *Pingle* 1693, *y*[e] *pingle* 1697, *y*[e] *Pingle* 1700, *v.* **pingel** 'a small enclosure'); Plantation 1805, - Garden 1839; Plat Bottom, - Top, Bottom -, East -, Low -, Top -, West Plat, Middle -, Top Plat Cl 1805 (*Platt, Plats, South plat(s)* 1721, *v.* **plat**[2]); Plot Cl 1839; Plough Cl, - Plat 1805 (*v.* **plat**[2]); Prior Cl 1839 (*Pryor close* 1608, cf. *terram Prioris de Croy roys* 1299, *terr' prioris de Roiston* 15 (c.1570) ('the land of the Prior of Royston, Hrt'), *Pryor hill* 1611[2], *Prier hill* 1721, *Prior hill Close* 1629, - *close* 1630); Pry Hill 1805; Ray Cl 1805 (cf. *East* -, *West Ray Close, Ray Nookin* 1721); Rush Cl 1839; Rye Car 1805 (*v.* **ryge**, **kjarr**); Sand Sykes 1805 (1721, *sande sicke*

1601, *Sand syke* 1611[2], 1721, *Sandsyke* 1721, *v.* **sand, sik**); Sash Gate 1822
(*Saxgate* 1280-85, 1327[2] (p), 15 (c.1570), 1562, 1608, - *de Ouresby* 1361 (p),
Saxgate syk' 1339, *Saxgaitt* 1536, *Saxgaite* 15 (c.1570), *a Streete Called Saxgate in
Owersbie* 1607, *Saxgate syke, Saxegate* 1611[2], *Sasgate, Sasgate Road, Sasgate syke
Close* 1721; formally this could be 'Saxi's way, road' from the ODan pers.n. *Saxi*
and **gata**, a Scand. compound; this pers.n. occurs in Saxby All Saints, PN L 2
254 and Saxby LWR. Alternatively, we might think of 'the road of the Saxons'
Saxons here denoting Anglo-Saxons. The first el. would then be an uninflected
OE *Seax-*, ON *Sax-, v.* **Seaxe**. Dr J. Insley compares this with ON *Sax-land*
'Germany' (literally 'Saxon-land')); Scornills Cl 1805, Scornhills Cl 1839 (*Bywest
Schornhilles* 1339, *scornhill* 1601, *Scornils* 1721, *v.* **hyll**, the first el. is uncertain,
but may be ON *skarn* 'dung'); Middle -, Top Seeds, Seed(s) Cl 1805, the Top
Seeds 1822; Severals Cl 1805 (*Severall close* 1608, *v.* **several** 'land in private
ownership'); Short Flg 1751; Short Lands 1751 (*Short lands* 1721, *v.* **sceort, land**);
Six Acres Cl 1805; South Beck 1792, 1799, 1806, 1807 (1608, cf. *the North Beck*
1608, *the northbeck cloose* 1611[2], and Beck Cl *supra, v.* **bekkr**); South Cl 1805
(1721, *the south Close* 1684); Stack Yard 1805; Low -, Top Stewills 1805 (*Stiuel*
1280-85, *Steuele, Steuel pyttes* 15 (c.1570), *the stole* (sic) 1601, *Stewell closes* 1625[1],
Great -, Little Stewels 1721, obscure); Stray Pingle 1805 (1721), - Cl 1839
(perhaps from ME **stray** 'a piece of unenclosed common land' with **pingel** and
clos(e)); Stripe Moor 1790 (perhaps from **strip** 'a narrow tract of land' and
mōr[1]); Thoroughfare Btm 1805, - Cl, Two closes called Thoroughfare Closes
1839 (*v.* **thoroughfare, botm**); Townend Plat 1805 ('the plot of land at the end
of the village', *v.* **tūn, ende**[1], **plat**[2]); Gt -, Lt Towler Cl 1839 (probably from
the surn. *Towler*); Land North of the Triangle 1839; Usleby Cl 1805 (*Usleby
Close* 1721; it was on the boundary of the neighbouring parish of Usselby);
Upper Grounds 1805 (*Upper Ground* 1721); Urns Cl 1805; Walk Btm 1805, -
Top 1805 (*Walk* 1721, *v.* **walk**); Weetfurs 1751, Wetfurrow Cl 1805 (cf. *short weet
furrows* 1601, *Weetfurrs* 1721, *v.* **furh**); Wells Cl 1805 (1721, possibly from the
surn. *Well(s)*, cf. *Jon a well* Hy7); the West Fd 1751, West Field Cl 1805 (*in
occident' campo* 1339, ·1344, - *de Owresby* 1588, *in occident' campo de Ouresby*
1339, 1358, *in campis occidentali campo de Ouresby* (sic) 1404, *in campo
occidentali de Owresby* 1536, *in occidentall' campo de Owersbye* 1560, *in occident'
campo de Owresby* 1588, *Owersby west field* 1601, *the west feild off Owresby* 1606,
- *of Owresby* 1625[2], 1638, *the west feilds of Owersbe* 1634[2], *the West Feild* 1608,
The West feilde 1611[2], *the Westfeild* 1612, *the west field* 1607, 1611[1], 1625[2], *West
Field* 1700, 1706, *Westfield* 1707, *the West field* 1721, one of the open fields of
Owersby, cf. the East fd *supra*); West Field Cl 1805 (*y*[e] *west field Close* 1664,
West feild Close 1679, *Westfeild Cloase* 1693, *Westfield Close* 1721); the West plat
1751 (*v.* **plat**[2]); Wheat Cl 1805; Wickers Btm, Wickers Cl 1805, 1822 (*Wythcaste*

(sic) 1280-85, *wycaste, foreram voc' Wicaste hedlande* 15 (c.1570), *The close ... called Wycaste* 1608, *Wicas Close* 1674, *Close Cald Wickass* 1679, *Wykas* 1707, *east Wycaste close, the West Wycast Cloose* 1611^2, *Lower -, upper Wikas* 1721, obscure); *Wood* 1767, 1790, *Wood Cl* 1805.

(b) *abbot clos* m16, *Abbottes Garth* 1549 (Thornton Abbey owned land here, v. **clos(e)**, **garðr**); *Agatztoft'* 1330 (from **toft** perhaps with ME *agas* (OFr *agace*) 'a magpie', probably used here as a byname); *Aldolgate* c.1200, 1299, *Adolgate, -gaite* 15 (c.1570) (from the OE pers.n. *Eadwulf* and **gata** 'a way, a road'); *Allingtons lande* 1607 (from **land** and the surn. *Allington*); *Atherby thyngge* 1503, *atherby thynge* 1509 (from **þing** with the surn. *Atherby*, Catherine *Atherby* held land in the adjoining parish of Usselby 1445 AASR xxix); *Bartons Plat* 1721 (Robert *Barton* is named in the document, v. **plat**^{2}); *Beack-, Beckmontofte* 1610, *Beakman Tofts* (sic) 1611^2 (from **toft** 'a curtilage, a messuage' with the surn. *Beckman* or *Beakman*; *Beakman* may be from ME *Becheman*, v. Fransson 203, with Scand. *-k-* for *-ch-*); *Beech Close* 1721; *le beregate* 1503, 1509 (for a discussion v. the same name in South Kelsey f.ns. (b)); *bihill, byhill slak* 15 (c.1570) ('(place) by the hill', v. **bī**, **hyll**, with **slakki** 'a shallow valley'); *Black myles* 1608 ('black earth', v. **blæc**, **mylde**); *Bradegate* 1280-85, *bradegat'* 1299, *Braythegate* 1326, 1344, *Braythegat'* 1330, *braygatte* 1404 (where it is noted that it *gose owre ye lande*, i.e. 'goes over the lands'), *Braigate, Braygate* 15 (c.1570) ('the broad road', v. **breiðr**, **gata**, a Scand. compound, though the earliest forms are from **brād** and **gata**); *Brayzemerhill'* 1299, *brathmerhille* 15 (c.1570) (perhaps 'Breiðr's boundary, land on a boundary', from the ON pers.n. *Breiðr* and **(ge)mǣre**, with **hyll**); *brigg Close* 1721 (v. **brycg**, **clos(e)**, with the first el. in a Scandinavianized form); *Brottes* 1385, *brattes* 1588, *the furlonge called the Brattes, Bratte steighe* 1611^2 (v. **brot** 'a small piece of land'); *Broxholme garres* 1666, *Brockson -, Brocksholm Close, Brom. Hotham's Brokson Garth alias Stockwell* (sic) 1721 (named from the family of Robert *Broxholme* 1445 with **gara** 'a gore' and **garðr**; *Brom. Hotham* is for *Broomhead Hotham*, named in the document; for the alternative name v. *stockwell infra*); *Burwell Close* 1608 (named from the *Burwell, Burrill* family, cf. Henry *Burrill* 1596; the name is common in the parish into the 17th century); *Busedeile* (var. *Buledeile*) c.1160 (the latter is probably correct, v. **bula**, **deill**); *Butt Lea* 1721 ('the meadow, pasture near a mound or tree-stump', v. **butt**^{2}, **butt**^{1} and **lea** (OE **lēah**)); *Castergate* 1280-85 ('the road to Caistor', v. **gata**); *Cawath leas* 1552 (perhaps 'the ford frequented by jackdaws', v. **cā**, **vað** with **lea** in the pl.); *Chapmans dale* 1629, *- Dale* 1630 (from the surn. *Chapman*, cf. Francis *Chapman* 1608 with **deill** 'an allotment, a portion of land'); *Cherygarthe, Chery garthe* 1607 ('the cherry orchard', v. **chiri**, **garðr**); *Clawman hill* (*cloose*) 1611^2 (presumably *Clawman* is a surn.); *Combreworth Landis* 1510 MC ('lately belonging to Sir Thomas Combreworth'); *Conyholme* 15 (c.1570) (v. **coni** 'a rabbit', **holmr**); *Cotcher Mill Close, The Mill Cotcher Close* (sic) 1721 (cf. North Cotchers Cl in (a) *supra*); *Coterigges furlanges* 1339 (*v.* **cot, hryggr** with **furlang**); *Cotes* 1280-85 (*v.* **cot** 'a cottage, a shed'); *Couperlane* 1453, *Couper Close* 1721 (named from the *Couper* family, cf. Robert *Couper* 1453, Bartholomew *Cowper* 1581); *Crase crofte* 1445, *toftum vocat' Crasse Croft* 1453 (from the surn. *Crasse* (*v.* Reaney s.n. *Crass*) with **croft**); *crossefurlanges* 15 (c.1570) (*v.* **cros** 'a cross', **furlang**); *Crosse hill* 1609, *Crosse hill cloose* 1611² (*v.* **cros, hyll**); *Dannot Thyng* 1529 (from **þing** 'premises, property', with the surn. *Dannot,* cf. *Robert Dannot,* named in the 1529 will, and John *Daynot* 1404); *Darsy Dayll* 1529 (from **deill** 'an allotment, a portion of land' with the surn. *Darcy,* Dame Agnes Tournay, daughter of Philip Lord Darcy of Knaith, held estates in Owersby, *v.* AASR xxix 13, 23; *v.* also *Turneys thinge infra*); *Dicksons close* 1608 (named from the *Dickson* family, cf. Edward *Dickson* 1608 (same document), Patrick *Dixson* 1574 *BT*); *le Dowthehall* (sic) 1445; *Dry leas* 1609, *dry leas, Dryeleas* ("being furry", i.e. overgrown with furze) 1721 (the second el. is from **lea** (itself from **lēah**) in pl. in its later sense 'meadow, pasture', as elsewhere in this parish); *East Close* 1721 (*v.* **ēast, clos(e)**); *Easter gates* 1601; *Eight Lands* 1721 (*v.* **land** 'a strip of arable land in the common field'); *Elshame land* 1588 (*v.* **land;** Elsham Priory held land here); *Emerson Garth* 1721 (from the surn. *Emerson* with **garðr**); *les Estfurlanges* 1358, *Est furlang* 1404, *Esturlang* (sic) 1445 (*v.* **ēast, furlang**); *Estlanges* 1280-85, *hestlanges* 1284 (*v.* **ēast, lang²** 'a long strip of land'); *lez ferthinges* 15 (c.1570) (*v.* **feorðung** 'a quarter', freq. 'a quarter of a virgate'); *a Street commonly called ye finckle Street* 1664 (probably 'a lovers' lane', a frequent st.n., cf. Finkle St (Lincoln), PN L 1 65-6 and Finkelgate (Norwich), PN Nf 1 105); *forty Acres* (area: 60.0.10) 1721; *Fosters Car, - folly, - Greens, - Wyhasses* 1721 (*Foster* is mentioned in the document, without forename); *Fouldayel hole* 1608 (*v.* **fūl** 'foul', **deill** with **hol¹**); *fourteen lands* 1721 (*v.* **land** and cf. *Eight Lands supra*); *fower Acres and three Swathes* 1607, *Three Swathes* 1607 (*v.* **æcer, swæð** in its later sense 'a strip of grassland'; the references are to the same piece of land); *the Free School at Owersby* 1744; *frosckholles* 1601 ('the frog holes', *v.* **frosc, hol¹**); *furlanges* 1299, *les furlanges* 1330, *les Furlanges* 1344, 1348, *the furlonges* 1612 (*v.* **furlang**); *furlanghend* 1280-85, *the furlong end* 1611² (*v.* **furlang, ende¹**); *the grippe* 1611² (*v.* **grype** 'a ditch, a drain'); *hatherswelhyl, hathereswelsyk* 1280-95, *hatherswell hall* (sic) 1601, *Hadderswell goote* 1608, *- gote* 1611² (*v.* **hyll, sīk, hall, gotu,** the first el. being possibly the OE pers.n. *Heaðurǣd;* this perhaps became Hazard Well in (a) *supra*); *Mr Headons Walk* 1721 (*v.* **walk**); *hedlandaile* 15 (c.1570), *unam selionem pasture voc' a headeland* 1588 (*v.* **hēafod-land, deill**); *Holdernesse thyngge* 1503, *holdnesse thynge* 1509, *Howdernes Thyng* 1529 (from **þing** 'property, premises' with the surn. *Holderness,* cf. Hugh *de Holdrenesse* 1320-27); *holowgate* 1320-27,

Holowgat' 1345, *hollow gate* 1552, *Holowsik'* 1320-27, *Holosyk'* 1345, *Holousyk'* 1360, *hallow closse* (sic) 1561, *Hollowe hill* (*cloose*) 1611^2 (*v.* **hol**^{1} 'a hollow', **gata**, with **sik**, **clos(e)**, **hyll** and cf. Low -, Middle -, Top Hollow in (a) *supra*); *Houstlanges* 1339 (possibly 'eastern long strips', *v.* **austr**, **lang**^{2}); *in a Howe place there* 1634^1 (*v.* perhaps **haugr** 'a hill, a mound'); *Hychehome* (sic) 1503, *hychehome* (sic) 1509, *hetch holme* 1608, *Hetcheholme cloose* 1611^2 (perhaps from **hiche** 'an enclosure for penning sheep' and **holmr**, though the ME surn. *Hiche* from a short form of *Richard*, for which *v.* Reaney s.n. *Hitch*, might also be considered as first el.); *Hynghow* 1280-85, 1299, *Hynghowsyke* 1299, *hynghowe syke* 15 (c.1570), *Hyngebergh* 1330, *Hyngbergh*, *Hynberghsyk'* 1339 (*v.* **haugr**, **sik**, and **berg**. The first el. may be the ON pers.n. *Ingi* masc. or *Inga* fem.); *Ingoll hill, - Sike* 1609, *Ingoll syke cloose* 1611^2, *Ingoul, Hurns* (sic) 1721 (*v.* **hyll**, **sik**, **hyrne** 'a corner of land'; *Ingoll* is the ME pers.n. *Ingolf* from ON *Ingolfr* or ME *Ingald* from ODan *Ingæld*); *Croftum Isolde* 1344 ('Isolda's croft', *v.* **croft**, with the OFr pers.n. *Iseut, Isout*, in its normalized Latin form); *la Kyrkgate* 1344 (*v.* **kirkja**, **gata**, a Scand. compound; no doubt identical is the later *church lane* 1601, *Church Lane* 1706); *Kyrkby nowke* 1561, *Kirkby nooke* 1611^2, 1630, *close called - * 1629, *Kirkby Nook* 1665 (alluding to the neighbouring parish of Kirkby, *v.* **nōk**); *Lamb-Garth* 1608 (*v.* **lamb**, **garðr**); *Widdowe lammynge melfeilde cloose* 1611^2 (*v.* Mellfield in (a) *supra*); *Langedeila* c.1160, *Langgedayl* 1299 (*v.* **lang**^{1}, **deill**); *Langfers* (sic) 1358, *Langfurse* 1445, *the long furrowes, south long-furres* 1601 (*v.* **lang**^{1}, **furh**); *lawnders land* Hy7 (named from the *Lawnder* family, cf. John *Lawnder* 1404, Richard *Lawnder* 1445); *lyne landes* 15 (c.1570) ('the strips on which flax was grown' *v.* **lin**, **land**); *Mancknols Close* 1608 (named from the *Mancknol* family, for which *v.* Mill Fm *supra*); *mariot furres* 15 (c.1570) (from **furh** 'a furrow', in the pl., with the surn. *Mariot*, cf. Richard *Mariot de Ouresby* 1348); *melton pytte* 15 (c.1570) (named from the family of *John melton, Richard de Melton* mentioned in the same document and note earlier Richard *de Melton* 1330, with **pytt**); *medelfen* 1320-27, *methelphen* (sic) 1327, *methelfen* 1345 (*v.* **meðal** 'middle', **fenn**); *Middle Close* 1721; *the middle Warth* 1611^2 (*v.* **waroð** 'a flat piece of land by a stream, etc.'); *molle wray* 15 (c.1570) (from the surn. *Molle*, cf. *terram Will'i molle* 1280-85, and **vrá** 'a corner of land'); *Moredalefurrs* 1609, *the moredayle furres cloose* 1611^2 (*v.* **mōr**^{1}, **deill**, with the pl. of **furh** 'a furrow'); *Mownson gronde* Hy7 (named from the *Mondeson* family, cf. John *Mundeson* 1396, John *Mondeson* 1453); *Nettle bush plaine* 1608, *Nettlebushe playne, the Upper Nettle bushe plaine cloose, the Lowe nettle bushe playne cloose* 1611^2 ('the nettle-patch, nettle-bed', *v.* **netle-bush** and cf. *nettle-busk* in Binbrook f.ns. *supra*; *v.* also **plain** 'a piece of flat meadowland'); *Neudayl* 1320-27, *Neudayle* 1345 ('the new allotment or portion of land', *v.* **nīwe**, **deill**); *maner de Newhall* (*in Oresby*) 1586, 1591 (*v.* **nīwe**, **hall**); *Nichall head land* 1608, *Nicholl Headlande*

Close 1612[1] (from **hēafod-land** with the ME pers.n. or surn. *Nicol*); *Nine Lands* 1721 (*v.* **land** and cf. *Eight Lands supra*); *Nordlanges* 1299, *Northe langes* 15 (c.1570) (*v.* **norð, lang**[2] and cf. *Eastlanges supra*); *Northal* 1343 (*v.* **norð, hall** and cf. *Southall* under Owersby *supra, Newhall supra, Westhall infra*); *the North feild* 1629, *the north feild* 1630 (probably one of the open fields of Owersby, cf. the East fd, the West Fd, in (a) *supra*); *the (short) north furlonges* 1601 (*v.* **norð, furlang**); *North'holm* c.1200, *Norpholm'* e13, *Northolme* 15 (c.1570) (*v.* **norð, holmr** 'an island of land, higher land in marsh'); *Northyby* 1280-85 (p) ('north in the village', *v.* **norð, í, bý**); *oat close* 1608, *Oat -, Oats field* 1721; *one Land* 1721 (*v.* **land** and cf. *Eight Lands supra*); *Open Car* 1721 (*v.* **kjarr**); *Osgarby Wraye* 1608, *William Osgarbye platte* 1611[2] (*v.* **vrá, platt**[2]; the surn. *Osgarby* is from Osgodby (Kirkby cum Osgodby) *supra* for which *Osgarby* occurs as a spelling in the 15th-17th centuries); *Outlanges* 15 (c.1570) (*v.* **ūt. lang**[2]); *Owersbieend* a1567, *Owersby-end* 1723; *Pademor* c.1160, *Pademoreholm* 1280-85, *padmore holme* 15 (c.1570) (*v.* **padde** 'a toad', **mōr**[1] with **holmr**); *Pasty-hill, Pasty dayle* 1608, *pastye hill cloose* 1611[2], *Pasty Dale, Pasty hill* 1721 (obscure); *Potterstygh* 1358, *potterstigh hedland* 15 (c.1570), *Potters -, potter steighe* 1611[2], *Potterwathe* 1428, 1445, *Potter Wath* 1453 (from the surn. *Potter* with **stīg, stigr** and **vað** 'a ford'); *Pottos Closes* 1721 (named from the *Potto* family, cf. William *Potto* 1709 *BT,* 1821); *in pratis de Ouresby* 1299, 1320-27, 1327, 1330, 15 (c.1570), *prat' de Owersbie* 1607 ('in the meadow(s) of Owersby'); *prestholme, pristholme* 15 (c.1570), *Priest-holme close* 1608, *priestholme cloose, - headland* 1611[2], *y[e] parson Priest holme* 1679, *Priestholm* 1700, 1705, 1706, 1709, *Lower Priestholme* 1664, *low prest holme* 1674, *low priest holme* 1679, *A Cloase Cauld Low prist holme* 1693, *One Cloase Cauled upper Pristholme* 1693, *Uper* (sic) *priestholme* 1697, *upper Priestholme* 1707 (*v.* **prēost, holmr**); *Pygot Thyng* 1529 (*v.* **þing** 'property, premises', with the surn. *Pygot,* cf. Robert *Pygot* 1404); *pyme hedland, - well* 15 (c.1570) (named from the *Pyme* family, cf. Walter *pyme* 15 (c.1570)); *maner ... de Rayleston garth* 1591 (probably from a surn. with **garðr**); *Rayson head land* 1611 (from the surn. *Rayson* etc., from Rasen *infra,* and **hēafod-land**, cf. Thomas *Ryeson* 1562 *BT,* William *Reason* 1565 *ib*); *Risadeila* c.1160, *Rysdayle* 1280-85, *Risedayle* 1330, *Rissedaile* 15 (c.1570) (*v.* **hrīs** 'brushwood', **deill** 'an allotment, a portion of land'); *riven-daile close* 1608, *Rivendale* 1721; *Rodesmare* 1280-85 (perhaps literally 'the boundary of the clearing', *v.* **rōd**[1], **-es**[2], **(ge)mǣre**); *Salter Garth* 1721 (named from the *Salter* family, cf. John *Salter of long Owersby* 1624 *Inv,* with **garðr**); *sandome ponde furlonge* 1588 (*v.* **sand, holmr**); *apud Sandwath* 1428 (*v.* **sand, vað** 'a ford'); *ye Sondy lane* 1674 (i.e. *Sandy Lane*); *Schaldail* 1327[1], *Schaldayle* 1345 (*v.* **deill**); *Schawthorn* 1428, *Schawthorne* 1445, *Shawthornes* 15 (c.1570), 1609, *cultur' voc' shawthorns* 1588, *Shaw thornes* 1601, *Shawthorns* 1609 (*v.* **sceaga** 'a small wood, a copse', **þorn**); *Schotgate* 1299 (perhaps 'the road with a steep slope', *v.* **sceot**[3],

gata); *del See de Ouresby* 1370, *atte See de Ouersby* 1377, 1383, *at See de Ouresby* 1381, *att See de lang Ouresby* 1392, *othe See* "of" *Oweresby* 1395, *de ly See* 1406, *othe See* 1412, *del See* 1428, 1478 all (p) (*v.* **sǽ** perhaps in the sense of 'marsh', since 'lake' is not topographically appropriate); *Slegtheng* 1280-85 (probably 'the smooth, level meadow', *v.* **sléttr, eng,** a Scand. compound, cf. *Sleightings* in East Halton f.ns. (a), PN L 2 154); *South Furlongs* 1608 (*v.* **furlang**); *Sowthebutt* 1404 (p) (*v.* **sūð, butte**); *Stamps Plat* 1721 (named from the *Stamp* family, cf. *Mathias Stamp*, named in the document, with **plat**^{2}); *staneholmlandes* 1280-85 (*v.* **stān, holmr** with **land**); *Stanny lands* 1608, *y*^{e} *Stonie Landes* 1664 (it is impossible to tell whether these are associated with prec.); *Stocke pyttes* 1612 (*v.* **stocc, pytt**); *stokwel* 1280-85, *Stockewell'* 1299, *Stockwell* 1606, *Stocwelforlang'* 1299, *Stockwell gate* 1344, *Stokwellegrippe* 15 (c.1570), *Stockwell platt* 1609 ('the spring or well by a post or tree-stump', *v.* **stocc, wella,** with **furlang, grype** 'a ditch, a drain' and **plat**^{2}, *v.* also *Brocksom Close* under *Broxholme garres supra*); *prati in Stonge* (sic) 1428, *vno le stong* 1536 (*v.* **stong** 'a pole', a measure of land); *studdalle land* 1588 (*v.* **stōd** i.e. stud, 'a herd of horses', **deill** with **land**); *Stute nooke* 1608 (*v.* **nōk;** the first el. is obscure, though it might possibly be ON **stútr** 'a bull'); *Stykenaytoft'* 1329 (from **toft** with the surn. *Stickney*); *le Swaithe* 1536 (*v.* **swæð,** and cf. *fower Acres and three Swathes supra*); *Longe Swynstye Dykes* 1607, 1610 (*v.* **swin**^{1}, **stigu**); *Taler Robinsons lane* (*Close*) ("*in tenor of William Robinson Taler*" (sic) 1721, *Robinson lane Close* 1721 (he is also referred to as *Taler Robinson* in the same document); *Thakdayl* 1330 ('the portion of land where thatch was got', *v.* **þakkr, deill,** a Scand. compound; this might have been a reed-bed or a piece of land on which wheat was specially grown for the straw used in thatching); *thingehowe* (*syke*) 15 (c.1570) ('the assembly hill or mound', *v.* **þing, haugr,** with **sík**); *pratas Elisabethe Halley vocat'* xiij *stonges* 1536 (*v.* **stong** and cf. *Stonge supra*); *Three Swathes* 1607 (*v.* **swæð** in the late sense 'a strip of grassland, etc.'; *swathe* is common in north L Terriers as an appellative, cf. *fower Acres supra*); *Toyn Garth* 1721 (named from the *Toyn* family, cf. John *Toyn* 1701 BT, with **garðr**); *Turneys thinge* 1611^1 (*v.* **þing** 'property, premises', cf. *Rentale dominæ Agnetis Tournay de terris … in villis et campis de Ouresby* 1414 and for a long discussion of the history of the family, AASR xxix 1-42); *twelve Lands* 1721, *twenty Seven Lands* 1721, *two Lands* 1721 (*v.* **land** and for the three names cf. *Eight Lands supra*); *The Close on the South side of the tyled Hall, The Tyled Hall close* 1608; *The Walnut Close* 1608; *the Wathe cloose* 1611^2 (*v.* **vað, clos(e)**); *venellam que vocatam Warynlaupym* (sic) 1348 (the reading seems certain and this may be 'Warin Laupym's lane', though the surn. *Laupym* is obscure); *the Waterie Laine* 1666; *Wengland* 113 (*v.* **land;** the first el. is perhaps ME *weng* from ON *vǣngr* 'a wing' probably used as a byname); *a pasture called Westhall* 1611^1; *lee White howse als wright thinge* 1581, *the White howse* 1602, *The*

White-House 1608, *the White house or Wrightes thing* 1609 (a *White House* was often one built of stone; the alternative name is from **þing** 'property, premises', with the name of the family of Richard *Wright* 1498); *Wilson Garth* 1721 (named from the *Wilson* family, cf. *Dawud Wylson* Hy7, John *Wilson* 1701 *BT*, with **garðr**); *Wrangbek* 1393 ('the crooked stream', v. **vrangr, bekkr**); *Wrawby hous* 1684, *The Wrawbys Pingle* 1721 (named from the *Wrawby* family, cf. William *Wrawby* 1684, Thomas *Wrawby* 1721 (each named in the relevant document), v. **pingel**); *Wrenwode* 1445 (presumably self-explanatory, v. **wrenna, wudu**); *terram prioris de Wynghall'* 1498, *Wyngall Howse* 1583, *Wingall garthes, - House* 1602, - *house, - landes* 1609, - *playne* 1611[2], *Wingal-pits, Wyngall garths* 1608 (named from Wingale Priory in South Kelsey *supra* with **garðr, hūs, land, plain, pytt**).

Market Rasen

MARKET RASEN

> *æt ræsnan* 973 (13) ECEE (S 792) (checked from MS)
> *Resne* 1086 DB
> *Rase* (8x) 1086 DB, 1203 Cur, 1203 FF, 1206, 1207 P both
>   (p), 1233 Cl
> *Rasa* 1086 (2x) DB, 1090-1100 (1402), 1100-08 (17) YCh vi,
>   c.1115 (2x) LS, 1147-53 YCh vi, 1150-54 *AddCh*,
>   1166-79 (1402) YCh vi, 1152-67 France, 1166, 1167 P
>   (p), 1180-1215 *HarlCh*, 1220 Cur, *Rasan* 1100-08 (1356)
>   YCh vi
> *Rasne* 1135-40 (1464) Pat, 1185, 1186, 1187, 1188 P, lHy2
>   Dane, 1201, 1202, 1203, 1204 P all (p), 1219 Cur, 1240
>   FF, 1255 Cl, 1279 Fine, *Rasna* 1152-67 France, 1199,
>   1200 P (p), *Rasn'* 1210 FF, 1234, 1242, 1245 Cl, *Rasum*
>   1235 Dugd vi
> *Rasen(')* 1202 Ass (p), 1210 Cur (p), 1275 RH, 1325, 1332
>   Cl, 1331 Fine, 1349 Cl, 1386 Pat, *Rasene* 1202
>   SelectPleas (p), 1230 Cur, 1252, 1278 Cl, 1327 Pat
> *Rasyn* 1228-32 (1409) Gilb, 1165 Cl, *Rasin* 1397 Cl, *Razin*
>   1620 Polyolbion
> *Reson* 1536 LP xi, *Reason* 1576 LER

All the simplex forms for Market, Middle and West Rasen have been collected together for convenience.

Parua Rasa c.1115 LS, *Parva Rasen'* 1254 ValNor
Estrase 1193, 1194 P
Estrasne 1153-62, 1187 (1409) Gilb, 1209-35 LAHW, 1250 FF,
1250 *HarlCh*, 1303 Ipm, *Est Rasne* l12 RA ii, c.1221
Welles, 1242-43 Fees
Estrasen l12 (1409) Gilb, 1254 ValNor, 1261 RRGr, 1268
Ipm, 1272 *Ass*, 1276 RH, 1303 FA, 1318 YearBk, 1332
SR, 1340 Ipm *et passim* to 1545 LP xx, *Estrase'* 1275
RH, *-rasene* 1325 *Extent, Est Rasen* 1241, 1251 RRG,
1275 RH, 1295 RSu, 1301 Misc *et passim* to 1544 LP
xix, *Est Rasen* 1610 Speed, 1615 *Amc*, 1689 VisitN,
1709 *Terrier*, 1738, 1748 *BT*
Estrasyn Hy2 (1409) Gilb, 1267 Pat, 1346 FA, 1395 Pat, 1402
FA, 1422 *Anc et passim* to 1531 Wills iii, *Hestrasyn*
Hy3 (1409) Gilb, *Est Rasyn* 1291 Tax, 1327 *SR*, 1428,
1431 FA, *Estrasyn* 1539 LP xiv, *East Rasin* 1665
Terrier, - als Market Rasin 1601 *ib, Estrason'* 1374,
1375 Peace, *-rason* 1387 ib, 1495 Pat
Estraysen 1535 VE iv, *Este Raysen'* 1535-46 *MinAcct, East
Raysen* 1628 *Td'E, - als Markett Raysen* 1630 *Amc*
East Raisen 1558-79 ChancP, *Est Raisen* 1592 *Amc*
Estreyson 1551 Pat, *Eastrayson* 1608-09 *LRMB 256, East
Rayson* 1651 WillsPCC, *Est Raison* 1625 *Terrier,
East Raison* 1651 WillsPCC
Estreason a1567 LNQ v
East Raysin 1650 *Cragg*, 1695 *Td'E, - als Markett Raysin*
1715 *ib, East Raisin al's Market Raisin* 1722, 1723
Td'E, 1723 *Deeds, East Raisin* 1773 *BT*
Marketrasyn 1358 *Cor, Market Rasyn* 1427 Pat
Market Rasyng 1418 *FF, - Rasynge* 1435 Cl, *Mercate Rasing*
1577 Harrison
Markyt Reson 1509-10 LP i, *Market Reason* 1536 ib xi
Markyt Rasen 1529 Wills ii, *Marketrasen* 1610 Speed, *Market
Rasen* 1619 *LCS*, 1707 *Terrier*, 1824 O, 1830 Gre *et
passim, Markett Rasen* 1621 *LCS*
Raysun market 1536-39 Leland, *Market Raison* 1575 SP i,
Markett Rayson 1634 *Foster*, 1658 *Td'E, Market Raizon*
1688 *Deeds*
Marketreyson 1547 Pat, *Merketrasin* 1576 Saxton, *Market Rasin*
1697 *Terrier, - Rasine alias East Rasine* 1612 *Amc*

Markett Raisin 1594 *Amc, Market* - 1734 *Deeds,* 1745 *Td'E,*
1779 *MiscDep 118, Markett Raysin* 1638 *LCS, Market* -
1717 *Td'E,* - *Raisin* 1651 WillsPCC

According to Ekwall, DEPN s.n., and Smith, EPN 2, 79,
Rasen is derived from OE **ræsn** 'a plank', perhaps in the sense 'a
plank-bridge'. The form dated 973 has been tentatively identified
with Market Rasen by Hart, ECCE 178, and indeed there does not
seem to be any other surviving English p.n. with which it could
otherwise be associated. It should be noted that the charter in
which the name occurs is a Confirmation of Estates and Privileges
of Thorney Abbey and there is apparently no other evidence linking
Thorney and Rasen. Nonetheless, the identification on philological
grounds is reasonable enough. The 973 form does not seem to
have been known to either Ekwall or Smith.

   The likely explanation of *æt ræsnan* is that it is a late OE
form of *æt ræsnum.* Spellings with *-an, -on* for the dat.pl. ending
-um are frequent enough in late AS charters and in texts like *The
Battle of Maldon,* while ME forms in *-e* are similarly common in
OE p.ns. derived from the dat.pl. *-um, v.* further Nils Wrander,
English Place-Names in the Dative Plural, Lund 1983, Chapters 1
and 3 and John Insley, *Archiv,* 225 (1988), 367. It may, therefore,
be suggested that Rasen is derived from OE *æt Ræsnum* 'at the
planks', perhaps indeed with reference to a plank-bridge, for the R.
Rase is certainly narrow enough here to be crossed by such a
bridge.

   Market Rasen is apparently distinguished from Middle and
West Rasen *infra* first as *Little* (cf. West Rasen), then as *East* and
later still as *Market,* for which *v.* Market Place *infra.*

R. RASE, 1824 O, *ad aquam current' que vocat' Ras'* 1326 *MC,*
and is referred to as *the river* 1601 *Terrier;* the r.n. is a
back-formation from Rasen, *v.* Ekwall, DEPN s.n. Rasen and RN
334.

CRANE BRIDGE. GRAMMAR SCHOOL (lost), *the* - 1734, 1762,
1796 *Deeds* and is *the Schoolhouse* 1726, 1775 *Td'E,* - *School House*
1747, 1761, 1770 *ib,* 1783 *MiscDon 140.* HOSPITAL, *the Hospitall*
1726 *Td'E,* - *Hospital* 1748, 1788 *ib,* cf. *Hospital Close,* - *Pingle* 1781
Encl4, the Hospital Close, Hospital Yard 1783 *MiscDon 140,*

Hospital Close 1864 *Terrier,* the last document records a gift of £500 to build *an hospital* in 1612. JAMESON BRIDGE, 1781 *EnclA, pons voc Jamesbrig* (sic) 1422 *Anc,* and cf. *Janson brigge close* (sic) 1608-09 *LRMB 256, firma* ... *messuagii voc' Jamson' thinges* 1535-46 *MinAcct* (*v.* þing 'property, possessions') taking its name from the family of William *Jameson'* 1440, 1443 *Anc.* MORLEY PARK (local), named from the family of John *Morley* 1851 *MiscDon 140.* RACE COURSE, 1828 Bry. SANDHILLS, 1842 White. VICARAGE, *the vicarige* 1601, *on* (i.e. one) *vickeredg house* 1625, *viccarage house* 1665, *the Vicarage house* 1707, 1822 all *Terrier.* WATERLOO BRIDGE, presumably commemorating the Battle of Waterloo, 1815.

STREETS, ROADS AND LANES

BACK LANE, 1781 *EnclA, Backlane Road* 1782 *Deeds, Back-lane Road* 1783 *MiscDon 140.* BELLAMY'S RD, 1782 *Deeds,* named from the family of Nicholas *Bellamy* 1792 *ib.* BLIND LANE, 1704, 1714, 1750, 1755, 1779 all *Td'E,* 1807 *Deeds, the blind lane* 1755 *Td'E, a laine called Blind* 1695 *ib,* a common name for a cul-de-sac. BRIDGE ST., - *street* 1753 *Td'E,* 1826, 1842 White. CAISTOR RD, 1781 *EnclA,* 1783 *MiscDon 140,* - *road* 1842 White, leading to Caistor. CHAPEL ST., - *street* 1842 White, leading to the Methodist Chapel. CHURCH ST., cf. *super viam eccl'ie* 1433 *Anc, Church Meadow* 1781 *EnclA.* CUTTHROAT LANE (lost), 1781 *EnclA,* 1783 *MiscDon 140, Cut-throat Lane* 1784 *Td'E.* DEAR ST., 1861 *Padley,* - *street* 1842 White. GEORGE ST., 1826, 1842 ib. JAMESON BRIDGE ST., 1781 *EnclA,* 1783 *MiscDon 140,* - *Brigg Street* 1745, 1747, 1764, 1779, 1781 all *Td'E, Jameson-brigg Street* 1777 *ib,* leading to Jameson Bridge *supra;* the Scandinavianized forms in *brigg* are noteworthy and are paralleled by the 1422 and 1608-09 forms for Jameson Bridge. JOHN ST., 1842 White. KILNWELL RD, *Kiln Well Road* 1781 *EnclA,* 1861 *Padley,* cf. *Kiln Yard* 1781 *EnclA, Kilns* 1828 Bry, self-explanatory, *v.* cyln, wella. KING ST., *King's Street* 1781 *EnclA, King street* 1826, 1842 White. LAMMAS LEAS RD, 1787, 1801 *Deeds,* cf. *Lammas leas* 1781 *EnclA,* 1857 *Deeds, the Lammas Leas* 1783 *MiscDon 140,* 'the meadow lands used for grazing after August 1st'. Field 121 comments "The hay harvest occupied the time between 24 June and

Lammas, when the fences were removed and the reapers turned
their attention to the corn. The cattle were meanwhile allowed to
graze on the aftermath. Loaves made from the new wheat were
taken to the church at this time for a blessing and a thanksgiving -
hence the name *hlāf-mæsse*, 'loaf festival'". Leas is probably the pl.
of **lea** (OE **lēah**) 'a meadow', apparently a common form in L.
LEGSBY RD, 1781 *EnclA*, 1783 *MiscDon 140*, leading to Legsby
LSR. LINWOOD RD, 1781 *EnclA*, 1842 White, leading to Linwood
supra. MARKET PLACE, 1727 *MiscDep 238*, the - 1689 VisitN,
1732, 1796 *Deeds*, the *Markett place* 1594 *Anc*, 1634 *Td'E*, the
market place 1601 *Terrier*, *Markett place* 1634 *ib*, the *markett place*
1689 *Maddison*, *Market place* 1697 *Terrier*, 1826, 1842 White and cf.
mercatum de Rasne 1255 *Cl*, *forum de Rasen* 1275 RH, *mercato de*
Rasen 1329 *Ass*, *foro de Rasene* 1374 Peace, - *de Estrasyn* 1456
Anc. MERCER ROW (lost), *Mersurrawe* 1454 *Anc*, self-explanatory;
it was in the market; cf. *Mercers Row* in Chester, PN Ch 5 21.
MIDDLE RASEN RD is perhaps *Rasen Lane* 1828 *Deeds*. MILL
LANE, 1851 *MiscDon 140*, 1861 *Padley*, the *Mill Road* 1781 *EnclA*,
1783 *MiscDon 140* and cf. *ante molend'* 1433 *Anc*, *Este Rasen, firma*
molend' 1538-39 Dugd vi. MILL ST., - *street* 1842 White.
OXFORD ST., - *street* 1826, 1842 White. PASTURE LANE, 1781
EnclA, 1880 *Padley*, cf. the *Pastures* 1783 *Deeds*, self-explanatory.
PINFOLD ST. (lost), 1726, 1736, 1741 *et freq* to 1775 all *Td'E*, 1781
EnclA, 1783 *MiscDon 140*, leading to the village pinfold, which is
the *Pinfold* 1775 *Td'E*, v. **pynd-fald**. PROSPECT PLACE, 1828 Bry,
- *place* 1826, 1842 White, such names being sometimes given
because of the view. QUEEN ST., 1726, 1727, 1732, 1762 *et freq* to
1796 all *Deeds*, 1826 White, perhaps with reference to Queen Anne.
SADLER LANE (lost), *sadler lane* 1594 *Anc*, *Sadlers Lane* 1781
EnclA, *Saddlers lane* 1826 White, self-explanatory. SERPENTINE
ST., - *street* 1842 ib, - *Road* 1838 *Deeds*, described in the text as *a*
new road. UNION ST., 1838 *Deeds*, - *street* 1842 White.
WALESBY RD, - *road* 1842 ib, leading to Walesby *infra*.
WATERLOO ST., *a new road called Waterloo Road* 1834 *Deeds*, cf.
Waterloo Bridge *supra*. WILLINGHAM RD, - *street* 1826, 1842
White, leading to North Willingham *infra*.

INNS, PUBLIC HOUSES & HOTELS

ASTON ARMS, *The* - 1864 *Terrier*. BUTCHER'S ARMS (lost), 1842 White. DOLPHIN INN (lost), *the* - 1775 *Td'E, Dolphin* 1826 White. YE OLDE GEORGE INN, *the George Inn* 1865 *Terrier*. GORDON ARMS, 1842 White, *The* - 1864 *Terrier, Gordon Arms Inn* 1872 *Padley*. GREYHOUND INN, 1781 *EnclA, Greyhound* 1826, 1842 White, *where a certain Messuage Tenement or Inn called the Greyhound and Buildings thereto formerly stood* 1830 *Td'E*. KING'S HEAD, 1842 White, *The* - 1864 *Terrier*. RAILWAY HOTEL, *The Railway Inn* 1864 ib, *the Railway Tavern* 1872 *Padley*. RED LION INN, *The Red Lion* 1822 *Terrier, Red Lion* 1842 White. SWAN HOTEL, *Swan* 1842 ib, *The White Swan* 1864 *Terrier*, WHITE HART HOTEL, 1828 *Nethorpe*, - *Inn* 1842 White, *The White Hart Inn* 1846 *NW*. WHITE LION, 1842 White, *The White Lion* 1864 *Terrier*.

Field-Names

Forms dated 1271-2 are *Ass*; 1327, 1332 are *SR*; 1422, 1433, 1438, 1440, 1450, 1453, 1454, 1456, 1457, 1459, 1461, and 1463 are *Anc*; 1528-9 are *AddR*; 1576, 1587, 1594, 1615, 1618, and 1630 are *Amc*; 1601, 1625, 1665, 1822, 1864[1] are *Terrier*; 1608-9 are *LRMB 256*; 1628, 1634, 1658, 1695, 1704, 1705, 1707, 1712, 1713, 1714, 1715, 1724, 1750, 1753, 1755, 1756, 1758, 1775, 1776, 1779, 1781[1], 1784, 1788, 1830, and 1837 are *Td'E*; 1781[2] are *EnclA*; 1782, 1789, 1796, 1807, 1819, 1824, 1834, 1838, 1846, and 1864[2] are *Deeds*; 1747 and 1783 are *MiscDon 140*; 1828[1] are Bry; 1828[2] are *BRA 647*; 1846 are *NW*; 1852, 1861, 1864[3], and 1872 are *Padley*.

(a) Back Cl 1852, - Fd 1864[3]; Barn Cl 1852; Beck Cl 1781[2], Far -, Near Beck Cl 1852 (*now called Beck Closes but formerly called Coney Close* 1704, *Beck Closes* 1705, 1707, *the Beck Closes formerly Called Coney Close* 1714, *the Beck Closes* 1715, 1724, cf. *the beck* 1601, *the Beck* 1665 v. **bekkr** and Coney Cl *infra*); Bellamys Road 1781[2] (Nicholas *Bellamy* is named in the document); Bottom Cl 1864[3]; High -, Low Breckon Waites (sic) 1828[2] (perhaps from **brakni** 'bracken' with **þveit** 'a clearing'); Brewster's Cl 1781[2] (the Brewsters were apparently a Roman Catholic family in Market Rasen, cf. William *Brewster* 1781 *BT*, described as *Popish Infant*); Brickkiln Cl 1781[2], 1783, Brick-kiln Cls 1796, Brick Kiln Cl 1852; Burnt Plantation 1852; the Butt Cl 1775, 1846, Butt - 1781[2], 1838; Caistor

Common 1781[2], 1782, Caistor Common Pingle 1784 (*v.* **pingel** and presumably cf.
Caistor Rd *supra*); the Calf Cl 1781[2]; Chancey Moor (sic) 1750, Chancery - 1755,
1779, 1807 (cf. the Moor Cl *infra*); the Church Bridge 1781[2], - Mdw 1864[1]; M[r]
Clark's Stile Cl 1775; Claxby Road 1781[2], the - 1783 (the road to Claxby *supra*;
the same name occurs in Middle Rasen f.ns. (a)); Clifton Cl (*v.* Coney Cl *infra*);
the Common 1775 (*the coman* 1576, *the Commons* 1618, - *comons* 1628); North -,
South -, West Common House Cl 1852; Coney Cl 1750, 1755, 1779 (1705, 1707,
1714), - but now Clifton Cl 1807 (*the Cunnery Close* 1628, *Cunny Close* 1695,
1704, *Coney Close* 1704, 1705, 1707, *v.* **coni** 'a rabbit', **coninger** 'a rabbit warren';
William *Clifton* is named in 1807 *Deeds*; *v.* also Beck Cl *supra*); East -, North -,
South-east -, South-west -, West Cottage Pasture 1852; Cow Cl 1781[2]; Cowdike
Pingle 1781[2] (*v.* **cū, dīk**, with **pingel**); Cropper's Cl 1864[3] (from the surn.
Cropper, cf. Joseph *Cropper* 1814 *BT*); Cuckoo Cl 1781[2]; Far -, Middle -, Nr
Dale 1852; Dean's Plantn 1852 (named from the *Dean* family, cf. John *Dean*
1851 Kelly and Dean's Paddock in Middle Rasen f.ns. (a) *infra*); Dovecote Cl
1830, 1837 (*v.* **douve-cote**); the East Fd 1776, 1781[2], 1783 (*in campo or'* 1463, *in
Orient Campo de East Raisen, the East fieldes* 1587, *the Eastfeilde* 1608-9, *v.* **ēast**;
one of the open fields of the parish, cf. the West Fd *infra*); Far -, Upper End
Cl 1852; Fauldingworth Footway 1781[2] (leading to Faldingworth LWR); Field Cl
1781[2]; Fifteen Acres 1852; Flats 1828[2] (*v.* **flat**); Flegg Cl 1781[2] (*v.* **flegge** 'a
marsh plant such as the iris'); the Folly Pingle 1781[2] (*v.* **folie, pingel**); Middle -,
North -, South Fox Covert 1852; Garth flats 1828[2] (*v.* **garðr** 'an enclosure, a
small plot of ground'); Gibbets 1852; Great Cl 1852; Hall Cl 1852 (*formerly
called or knowne by the Name of the Hall Closes but now called or knowne by
the Name of the Reedpond close & the Hall Close* 1715, *formerly called or knowne
by the Name of the Hall Closes but now called or knowne by the Name of the
Hall Close and the Reed Pond Close* 1724; cf. Reed Pond *infra*); the High Fd
1822, High Fd 1828[2] (cf. Low Fd *infra*); the Home Paddock 1830, 1837; Horses
or Home Cl 1834, Home Cl 1852; Ings Cl, - Pingle 1781[2], the Ings Cl 1789 (*v.*
eng 'meadow, pasture'); Lane Cl 1852; Limekiln Leys 1784; Lincoln Lane Cl 1852
(cf. Lincoln Lane in Middle Rasen *infra*); Lissington Footway 1781[2] (leading to
Lissington LSR); East Little Cl 1788, South little cl 1788 (*v.* the Moor Cl *infra*);
Little Ings 1781[2] (*v.* **eng**); Little Paddock 1781[2]; Long lea Cl 1781[2]; Long Mdw
1852; Low Fd 1828[2] (cf. the High Fd *supra*); Low Plantation 1852; High -, low
Marrs 1828[2] (*v.* probably **(ge)mǣre** 'a boundary'); Middle Carrs 1781[2], (the)
Middle Carrs 1783, the Middle Carr 1822 (*v.* **kjarr**); Midsummer Leas 1781[1],
1781[2], - Leys 1781[2] (*v.* **lea** 'a meadow' in the pl.); the Mill Cl 1781[2]; Middle
Moor 1846, 1852, Top - 1846, East -, North -, South -, West - 1852 (*le mor*
lHy2 (1409) Gilb, *le more* 1463, *v.* **mōr**[1]); the Moor Cl and formerly called or
known by the name of Chancey Moor (sic) 1758, that Close of Meadow or

Pasture Ground formerly in two Closes ... or (sic) one of them formerly called or known by the Names of the Chancery Moor closes afterwards the Moor Closes and now the South little close and Middle Close 1788 (*Chansey more* 1658, *Chauntry Moore* 1695, *Chansey Moore* (sic) 1704, *Chantry Moor* 1705, - *Moore* 1707, *the Moor Closes* ... *formerly Called or known by the Name of Chancey Moore* (sic) 1714, *Moore Closes but formerly Chantrey Moore* 1715; in spite of the variations in form, this is no doubt the same as *Chantree Moore* in Middle Rasen f.ns. (b), *v.* **chaunterie**); the New Beck or Main Drain 1781^2 (*v.* **bekkr**); Nocton's Pingle 1781^2 (*v.* **pingel**; *Nocton* is presumably a surn., but it has not so far been noted in the sources searched); North Ings 1781^2 (*the North Inggs* 1608-9, *v.* **eng** and cf. South Ings *infra*); Oak Tree Cl 1781^2; the outfall Cl 1864^2 (an *outfall* is 'the outlet of a river drain &c.', cf. *an Outgoate* used appellatively in the context of *the Beck Closes* 1715 Td'E); Paddock 1854^3; Palas Garden 1781^2; the Peck Mill Common 1781^2, Peck Mill Paddock 1872 (cf. "water mills called" *Peck Mylne* 1558-79 ChancP; presumably named from the *Peck* family, cf. Nicholas *del Peck* 1327, - *de Pek*' 1332); the pepper ley 1819; Peppermill Hall 1828^1; the Pingle 1781^2 (*v.* **pingel**); Poor House Cl 1824; Reed Pond 1834, Reedpond Cl 1781^2 (*Reed Pond Close* 1747, cf. also the early forms of Hall Cl *supra*); Rollinsons Lea 1756, 1758 (*Rollinsons Lea* 1658, *Rollinson* - 1714; probably from the surn. *Rawlinson*, common in *BT*s, cf. William *rawlinson* 1561 *BT*, Christopher *Rawlingsonne* 1575 *ib*, with **lea** (OE **lēah**) 'a meadow, pasture', as elsewhere in the parish); Round Plantation Pce 1864^3; Sampson's Garden 1781^2 (named from the *Sampson* family, cf. Michael *Sampson* 1641 *BT*); the Sand-pit allotment 1784; where two Houses commonly called or known by the Name of Scolding-Meg houses formerly stood 1784; Seed Cl 1852; Seg Cl 1846 (*v.* **secg**^{1} 'a reed, a rush', in a Scandinavianized form); Skinner's Lane Cl, Skinner's Plantn 1852 (from the surn. *Skinner*, not so far noted in local records, and cf. Skinner's Lane in Middle Rasen *infra*); South Ings 1781^2, 1783 (*The South Ings, the South Inggs* 1608-9, *v.* **eng**, cf. North Ings *supra*); Spring fd 1828^2 (*v.* **spring**); Stackyard Cl 1852; Stile -, Style Cl 1781^2, the Stile Cl 1819; Sturton Cl 1781^2 (presumably from the surn. *Sturton* which has not so far been noted in the sources searched); Teathering Leys 1775 (*v.* **lea** in the pl.); Tee Cl 1781^2; Top Fd 1864^3; the Town's end Cl 1775, Townend Cl, - Pingle 1781^2, Town end Cl 1783 (cf. *ad finem orientale ville* 1422, *Townende close* 1608-9); the Town Street 1781^2; the Tree Cl 1830, 1837; Turnpike Cl 1852; the Turnpike Rd 1781^2, 1783; Twelve Row 1872; the Water Mill Common 1781^2; West Fd 1781^2, 1783, the - 1782 (*the westfeilde* 1608-9, *the west feildes of East Rason* 1618, *v.* **west**; one of the open fields of the parish, cf. the East Fd *supra*); Wharton Cl 1776 (named from the *Wharton* family, cf. John *Wharton* 1664 *BT*); East -, West Whipit Lease 1852; Widow's Cl 1781^2; Willow Mires 1775, 1781^2, the - 1819 (*v.*

mýrr); Witch Spots 1781², the Witch Spots 1783; Middle -, North -, South Wood 1852; East -, West Wood Cl 1852; Woodhill Cl 1781².

(b) *Acrids House* 1651 WillsPCC (perhaps an error for *Acrils*. The Acril family is well evidenced in *BT*s in 116th and 17th centuries, cf. William *Acril* 1590 *BT*); *Brygmyln* lHy2 (1409) Gilb (*v.* **brycg, myln**, the first el. being Scandinavianized); *Burrell Lands* 1712, 1713 (named from the *Burrell* or *Burrill* family, Mary *Burrell* being named in 1713 and cf. Robert *Burrel* 1541 *BT*, Robert *Burrill* 1679 *BT*); *del Chaumbre* 1422 (p) (*v.* **chaumbre** 'a dwelling'); *meadow called comonly by the name or Church of meadow* (sic) 1625; *super coem via v'sus or'* 1456 ('on the common way towards the east'); *Davies House* 1651 WillsPCC (named from the *Davie* family, cf. Ann *Davie* 1594 *BT*, William *Davie* 1615 *ib*, John *Davy* 1662 *ib*); *le Engebek* 1457 (*v.* **eng, bekkr**); *situm unius grangie* Hy3 Gilb (*v.* **grange** (i.e. of. Sixhills Priory)); *atte Grene* 1332 (p) (*v.* **grēne**² 'a village green'); *Hamond place* 1528-9 (*v.* **place**; *Hamond* is presumably a surn.); *Attehil* 1271-2, *oth Hille* 1453, *del Hille* 1459, 1463, *del Hill'* 1461 all (p) (*v.* **hyll**); *hockinge close* 1608-9; *Hollme leas* 1587, *Holme -* 1615, *Holmeley close* 1608-9 (*v.* **holmr, lea**, in the pl. in the earlier forms); *the key close* 1634; *kyrkstyle* 1433, *Southe kyrkstyle* 1438, *le kyrkstyle* 1440 (*v.* **kirkja, stīgel** 'a stile' or 'a steep ascent'); *George Parker's Close* 1747; *the Rye Close* 1615 (*v.* **ryge**); *Sandy Close* 1658; *Miln or Kiln ... for the Bolting of Oats comonly called the Shooting Mill* 1747 (*v.* **myln**; the process referrred to is the sifting of the oatmeal in order to separate it from the bran; the term *Shooting* in this sense does not appear in NED); *the windemyll close* 1608-9 (cf. Mill Lane *supra*).

Middle Rasen

MIDDLE RASEN
    Media Rasa c.1115 LS, 1130-39 (1311) YCh vi, - *Rasum* 1212
       Fees, - *Rasum* 1235 Dugd vi, *Media Rasnea* 1202 FF,
       - *Rasne* 1209-35 LAHW, 1221 Welles, 1242-43 Fees,
       1271 RRGr, *Media Rasen* 1259 ib, 1272 *Ass*, Hy3 (1409)
       Gilb *et passim* to 1332 *SR*, 1535 VE iv, *medie* - c.1218
       (1409) Gilb, *media* - 1267 (13) *Alv, Med'* - 1282 QW,
       Rasen Media 1244 RRGr, *Media Rasin* 1281 QW,
       - *Reason* 1535 *AOMB 100*
    Middelrasen(') 1201 FF, 1276 RH, 1303, 1325 Pat, 1325 Fine,
       1325 Ipm, 1326 Orig *et passim* to 1374 Peace, *Middel* -
       1206 Ass, 1275 RH, 1327, 1355 Pat, 1536-7 Dugd vi,
       Middelrason 1375 Peace, *-rasyn* 1375 ib, - *Rasyn* 1402

FA, *Middelrasyn* 1502 Fine, - "or" *Middelrayson* 1502
Ipm, *Middelrasne* 1227 Cur, 1303 Ipm, *Middilrasne*
1263 FF, - *Rasyn* 1495 Ipm, *Myddelrasen* 1367 Ipm, 1552
Pat, -*reyson* 1553 ib, *Myddyl Rasyn* 1531 Wills i
Midelrasene 1201 FF, *Midel Rasen* 1311 Ipm, 1346 FA, -*rasen*
1337 Selby, 1349 Orig, 1359 Ipm, -*rasyn* 1400 Pap,
- *Rasyn* 1428 FA, *Midell Rasyn* 1402 ib, *Mydell Raysen'*
1535-46 *MinAcct, Mydelrasyn* 1539 LP xiv, *Midilrasen*
1349 Fine, 1446 Pat, *Midyl Rasyn* 1428 FA
Medelrasen 1327 *SR, Medilrasen* 1349 Ipm, *Medylrasyn* 1550
VisitN, *Medle Rasyn* 1431 FA
Mid Rasenn' 1206 Ass, *Midrasen* 1272 *Ass,* 1373 Peace
Middlerasne 1256 FF, - *Rasne* 1272 ib, - *Rasene* 1268 Ipm,
1305 Abbr, *Middlerasen* 1256 *Alv,* - *Rasen* 1282 Ipm,
1283 ChancW, 1388 Pat, 1582 Admin *et passim, Middle
Reason* 1547 Pat, 1576 LER, - *Reasonne* 1577 Admin,
middle Rasing 1577 Harrison, *Rasen Middle* 1558-79
ChancP, *Middle Raison* 1634 VisitN, - *Raisin* 1775,
1825 *Td'E, Myddle Reason* 1556 Pat, -*rasen* 1576
Saxton, *Rasen Midle* 1326 Inqaqd, *Midle Raysin*
1536-37 Dugd vi, - *Rasyng* 1537-38 *AOMB 409,* -*rayson*
1552 Pat, *Midlerasen* 1610 Speed, *Mydle Reyson* 1551
Pat
Rasen Drax 1343 NI, 1519 DV i, *Reasonne* - 1577 Admin,
1585 SC, *Reason* - 1585 Admin, *Raysen* - 1601 *Terrier,
Raison* - 1723 SDL, *Drack Middle Raisin* 1723 *Td'E*
Middel Rasen Drax 1331 Pat
Middle Rasyn Drax 1409 ChancCert. - *Rasen Drax* 1586,
1590, 1591 Admin, 1809 *Deeds,* 1824 O, 1828 Bry,
1830 Gre, - *Rayson draxe* 1574 *BT,* - *Raysen Drax*
1595 ib, - *Rasin Drax* 1606 *Terrier,* - *Rasine Drax* 1606,
1616, 1626 ib, - *Rasin Drax* 1691 ib, - *Rason Drax* 1767
ib, *Myddle Raysen Drax* 1584 ib
Myddyll Rasen Drax 1529 Wills ii, *Middylrasyndraxe* 1563 *BT*
Midlereasonne Drax a1567 LNQ v, - *Rasine drax'* 1567
InstBen, - *drakes* 1612 Terrier, *Midleraisin Draxe* 1640
BT, - *Rason Drax* 1664 ib, *midle Rasin drakes* 1662
Terrier
Media Rasen' Thuph' 1254 ValNor, *media Rasyn Tupholm*
1291 Tax

Rasen Tupholm 1343 NI, *Rasyntupholm'* 1345 *FF, Rasen
Tupholm* 1525 Sub, *Rasine Tupholme* 1597 Admin
Medylrasyn Thoophom 1563 *BT
Middil Rasyn Tupholme* 1428 FA, *Mydlereason Tupholme* 1573
VisitN
Middle Raysen tuphollm 1579 *BT,* - *Rasen Tupholme* 1582,
1585, 1587 Admin, 1809 *Deeds,* 1824 O, 1828 Bry, 1830
Gre, - *Raisen Tupholme* 1595, 1602 *BT,* - *Raison
Tupholme* 1674 *Terrier, Middleraisonne Tupholme* 1597
Admin, - *Rasin Tupholme* 1653 WillsPCC, *middle Rasin
Tupholme* 1664 *Terrier, Mid-Raisen Tupholm* 1723 SDL

For the etymology of Rasen, *v.* Market Rasen *supra.* The
affixes *Drax* and *Tupholme* are from Drax Priory, YW, and
Tupholme Abbey, LWR, both of which had extensive holdings in
Middle Rasen. It is *Middle* in contrast to *Market* and *West* Rasen.

BELL'S COTTAGE, no doubt named from a local family, cf. Dixon
Bell 1774 *EnclA.* THE BRIDGE (local), *pontem de Middel Rasen*
1332 *SR* (p), *the brige* 1634, 1638 *Terrier, the Bridge in Middle Rasin*
1725, 1728, - *Raysin* 1742, - *Rasen* 1776, *the Bridge* 1781 all *Deeds,
High Bridge Road* 1880 *Padley.* BRIDGE ST. (lost), 1848 *ib.*
BRIMMER BECK, *Braimer* - 1772 *ib, a Publick Drain called
Braimer Drain,* - *Bramer Drain* 1774 *EnclA, Bramer Drain* 1787
Deeds, a Public Drain called - 1802 *ib, a certain Drain or Beck
called Barmer Drain* (sic) 1824 *MiscDon 140, Brimer Drain* 1824,
1830 Gre, cf. *Bramer Close* 1724, 1726 *Td'E*; it is perhaps referred
to as *the beck(e)* 1612, 1638, 1662, - *Beck* 1634 all *Terrier* and as
Rasen Beck 1806, 1813, 1819, 1826 all *Deeds, Middle Raisin Beck*
1774 *EnclA.* The forms are late, so that it is impossible to suggest
a certain etymology. It would appear that *Brimmer* is originally a
p.n. and if the *-aim-* and *-ame-* spellings represent the earliest form,
this might be a compound of **breiðr** 'broad' and **(ge)mǣre** 'a
boundary, land on a boundary', but it is difficult to reconcile this
with the topography, since it does not seem to form a known or
suspected boundary. CAISTOR RD, *the* - 1774 *EnclA,* 1792, 1807
Td'E, Caister-Road 1784 *Deeds* and cf. *Caistor Gate furlong* 1772
Deeds, self-explanatory. CHURCH ST. LOW CHURCH RD, 1880
Padley. DALE FM, cf. *in vallem* l13, a1290 *MC, the Dale* 1724, *th'
dale nooke* 1601, *the Dail Nook* 1698, *yᵉ Dale Nook* 1707, *the Dayle*

Corner 1606, - *dale corner* 1612, 1634, - *dale goate* 1638, 1662 (*v.* **gotu** 'a water-channel') all *Terrier, Great -, Little Dale Bottom* 1806 *Deeds*, self-explanatory. DEAR FIELD COTTAGE, cf. Dear St. in Market Rasen *supra*. DUNGEON LANE, cf. *a Place called the Dungeon* 1774 *EnclA*, 1776 *Deeds*, (*the*) *Dungeon Close* 1774 *EnclA*, 1708, 1715, 1742, 1795, 1827 *Deeds, a place called the Dungeon Close* 1781 *Deeds, Dungeon Close or Plumbtree Close* 1728, 1742, 1776, 1781, 1792 all *Deeds*, - *Plumtree Close* 1725 *ib*; the significance of *Dungeon* is not apparent and for Plumbtree Close, *v.* f.ns. (a) *infra*. GIBBET HILL, on the boundary with Buslingthorpe LWR. THE GRANGE, *iuxta grang'* m13 *Drax, ad grangiam* 1332 *SR, ther grange in Middle Raysen* 1519 *MC, firma Graungie de Middell Raysen'* 1535-46 *MinAcct, Grangie de Mydell Rasen* 1538 *MC, the Graunge* 1575, 1584, 1602, 1603, 1653 *ib*, - *grange* 1601 *Terrier*, 1628 *Td'E*, 1662 *Terrier*, - *Grange* 1707, 1724 *ib, the Grainge* 1675 *MC, Grange Farm* 1880 *Padley*, cf. *Grange headland* 1606, 1612, 1634 *Terrier*, it is not clear in all cases from the sources searched to which Abbey the **grange** belonged, though the m13th century form clearly refers to Drax Priory and those in *MC* apparently to Tupholme Abbey. GRAPHA FM. GREEN LANE, 1828 Bry, 1863 *TSJ*, a *green* lane and not one leading to the village green. GREEN LEYS (lost TF 093 905), 1830 Gre, *Grene Leas* 1575, (*Close called*) *Greene Leas* 1575, 1603, *greene lees* 1575, *the Grene Lease* 1584, - *greene leas* 1675 all *MC*, cf. *Green leys Close* 1774 *EnclA*; the 1584 form suggests that this is 'the green pasture, meadow-land' *v.* **grēne**[1], **lǣs**, but spellings in *leas, leys* are frequent as appellatives in north L Terriers for the pl. of **lea** 'meadows, pastures'. GULLYMORE LANE (locally GALLYMORE LANE), *gallow mor lane* 1619 *LCS, Gallamore Lane* 1679 *Deeds*, 1783 *MiscDon 140*, - *moore Lane* 1679 *Deeds, Gallymore Lane* 1774 *EnclA, Gallomore Lane* 1797 *Deeds, Gallimore Lane* 1861 *ib*, cf. *Gallamore Close* 1608-09 *LRMB 256, Gallamore* 1651 WillsPCC, 'the moor where a gallows stands', *v.* **galga, mōr**[1]. HAMILTON VILLA, *that Close ... called or known by the Name of Hambleton* 1724 *Td'E, Hamilton Closes* 1774 *EnclA*, for Hamilton *v.* Hamilton Hill in Tealby *infra*. HIGH ST. is presumably *the Town Street* 1774 *EnclA*. HIGH HARBOUR FM. LINCOLN LANE, 1839 *TSJ*, 1806 *Deeds*, 1828 Bry, (*the*) *Lincoln Road* 1774 *EnclA*, 1776, 1817, 1825, 1838, 1845, 1854 *Deeds, the Lincoln Road or Lane* 1829 *ib*, self-explanatory. LOW GRANGE. LOW LANE. MANOR FM, cf. *Manor House, Old Manor House* (*now in two cottages*) 1880 *Padley.* MARSH LANE, *the Marsh Road* 1774 *EnclA,* cf. *Marshes* 1622 *Amc, the Marsh* 1628 *Td'E,* 1772, 1829, *the North -, the South Marsh* 1772 all *Deeds, y*^{e} *Marsh Banck* 1696 *Terrier, the Marsh banck* 1724 *ib,* selfexplanatory. MILL LANE, 1880 *Padley, v.* Windmill *infra.* MOOR FM, *Rasen More* 1537 LP xii, *y*^{e} *more of midlerasen* 1556, 1557 *Td'E, the mores of Middle Rayson* 1575 *MC, the Moores of Middle Raisen* 1584 *ib, midleraisin moore* 1579 *Terrier, the moores of Mydle Raisen* 1583 NCWills i, - *Raysine more* 1601, - *Raisen moore* 1638, - *Raisine Moor* 1664 all *Terrier, the Moor* 1775, 1792, 1801, 1809, 1812 *Td'E,* - *of Middle Rasen* 1822 *Terrier, the comon moor* 1612 *ib,* - *Common Moor* 1774 *EnclA,* and *the Moor Close and now of late called Green Leys Close* 1759 *Deeds, heretofore ... Moor Close but now called Green Leys Close* 1802 *ib;* self-explanatory and for the comparative name *v. Green Leys supra.* MOUNT PLEASANT, usually a complimentary nickname, but sometimes used ironically in towns. NAG'S HEAD P.H., *Nag's Head* 1842 White, - *Inn* 1852 *Padley.* NORTH ST., 1852, 1880 *Padley, Kirkings or North Lane* 1828 Bry, cf. *Kirking Lane Close* 1880 *Padley and Kirkin Bank* 1772 *Deeds;* the alternative name means 'the meadow(s) near or belonging to the church', *v.* **kirkja, eng.** NOVA SCOTIA BRIDGE, on the boundary with Walesby, an obvious nickname of remoteness. OSGODBY RD, leading to Osgodby in Kirkby cum Osgodby parish *supra.* PARK HO. PROSPECT FM, often referring to the view. PYEWIPE FM, cf. *Pywipe* 1829 *TA, East -, West Pywipe* 1852 *Padley;* this is comparable with Pyepipe (Inn), PN L 1 34, where it is pointed out that it alludes to the lapwing, *Vanellus vanellus,* by the dial. term not uncommonly used as a farm name. SKINNER'S LANE, 1874 *Padley,* named from a local family. SLATE HO. SOUTH LANE. STOCKMOOR FM, cf. *Stockmoor* 1857 *Padley,* - *Furlong* 1769 *Td'E.* THREE LEYS (lost, TF 091 912), 1824 O, 1830 Gre, cf. *Green Leys supra,* the use of the numeral clearly suggests that we have to do with the pl. of **lea (OE lēah),** hence 'the three meadows'. VICARAGE, *vicarage house* 1606 *Terrier, There is no Vicarage house* 1707 *ib.* These refer to the vicarage of St Peter's. In addition the following forms have been noted in Terriers of Middle Rasen Drax -- *the vicarag house* (sic) 1606, - *Vicarage house* 1612 and - *vicarage house* 1634. WALESBY ROAD PLANTATION, cf. *Walesby Lane North Close* 1848 *Deeds,* the road leading to Walesby *infra*. WATER MILL, (*the*) - 1747 *MiscDon 140*, 1774 *EnclA*, 1779 *Deeds*, *Water Corn Mill* 1719 *Deeds*, 1747 *MiscDon 140*, 1754 *et freq* in *Deeds* to 1816, 1817 *Dixon*, the references are to the same mill. WICKENTREE FM. No early forms have been noted for this name, but *wicken* is a dial. variant of *quicken* 'the rowan, the mountain-ash' or a similar tree, cf. PN Db 723, s.v. **cwicen**. WILD DALES, 1824 O, 1830 Gre, cf. *Wiles Dale Close* 1774 *EnclA*; it is on the boundary with Kirkby cum Osgodby plarish *supra*. WINDMILL, "a windmill" 1337 Cl, *the miln* 1601, - *Mill 1698 both Terrier*, cf. *the ould Mill hill* 1606, - *miln hill* 1612, - *ould miln hill* 1634, *ould Milne Hill* 1638 all *Terrier, le milnecroft* m13 (m14) *Drax, the Mill Close* 1698, 1724 *Terrier*, 1863 *TSJ*, y^e *Mill Close* 1707, 1781 *Terrier*, and *Sylcock Mill* 1606, *Silcok miln* 1612 both *ib*; it is not certain that all these forms refer to the same mill, though it seems at least likely; *Silcock* is a family name found frequently in the parish, cf. Christopher *Sylcocke* 1588 *BT*.

Field-Names

Forms dated 1235 are Dugd vi, ii; those dated m13 (Ed1) are *Barl*; 1259, c.1260, a1290, l13, 1320, 1332[1], 1364, 1369, 1392, 1399, c.1407, 1492, 1495, 1519, 1538, a1575, 1575, 1584, 1602, 1603, 1610, 1653, 1675, are *MC*; 1327 and 1332[2] are *SR*; 1370 and 1495 are Ipm; 1375 are Peace; 1537-8 are *AOMB*; 1558-79 are ChancP; 1577, 1601, 1606, 1612 (Kirkby cum Osgodby), 1634[1], 1638, 1662, 1664, 1671, 1674, 1697, 1698, 1703, 1707, 1709 (Kirkby cum Osgodby), 1724[1], 1781, 1822 are *Terrier*, 1583 are NCWills; 1622 and 1640 are *Amc*; 1628, 1634[2], 1724[2], 1736, 1752, 1755, 1756[1], 1757[1], 1768, 1769, 1773, 1788, 1797, 1800, 1807, 1826, and 1845 are *Td'E*; 1636, and 1718 are MC; 1679, 1690, 1708, 1715, 1725, 1728, 1732, 1740, 1742, 1756[2], 1757[2], 1759, 1772, 1776, 1779, 1781, 1784, 1787, 1792, 1795, 1796, 1802, 1805, 1806, 1807, 1809, 1812, 1817, 1825, 1826, 1829, 1838, 1840, 1848, 1854, 1872 are *Deeds*; 1738, 1745, 1792, 1829, 1840 are *TSJ*; 1747 are *MiscDon 140*; 1774 are *EnclA*; 1823 are *BRA 513*; 1846 are *NW*; 1849 are *TA*; 1852, 1857, 1872, 1874, and 1880 are *Padley*; 1863 are *TSJ* (Plan).

(a) Bagmore Cl 1774, Bagmore 1846 (perhaps *v.* **bagga, mōr**[1]); Barn Cl 1880; the Barthorpe Closes (*v.* the Milking Hill *infra*); Beck Crooks Furlong 1769 (cf. *the becke* 1628, *v.* **bekkr**); Brick Kiln Cl 1781, Brickkiln Cls 1796 (cf. the same name in Market Rasen f.ns. (a) *supra*); Brigg Rd 1774, 1797, 1800, 1829 ('the road to Brigg'); But-Stile 1772; Claxby Rd 1774 ('the road to

Claxby' *supra*); Common House Cl 1880; Cooper Cl 1774 (*Cowper close* 1583, *Cooper Close* 1606, *Cowperclose corner* 1612, *Cowper Close* 1724, 1726, - *close* 1634, 1638, named from the *Cooper* or *Cowper* family, cf. John *Cowper* a1537 *MiscDep 43, v.* also Moor Cl *infra*); Cooping Cl 1849; Cottage Pasture 1880; the Cow Cl 1840, Cow Cl 1863; Cowham 1772; the Cow Pasture 1774; Dave-Acres 1772; Dean's Paddock 1872 (named from a family well-evidenced in the parish, cf. Richard *Dene* 1563 *Inv*, Robert *Deane* 1604 *BT*); Dovecoat Cl 1776, Dove Coat - 1781, 1792, Dove-Cote - 1863 (*Dove Coat Close Dungeon Close or Plumtree Close* 1740, *Dovecoat Close* 1725, 1728, 1742, *v.* **douve-dote** and cf. *Plumb Tree Close* in (b) *infra*); formerly called or known by the name of Draker Sikes but now Davenport Cls 1788, 1807, Draper Sykes afterwards Draker Sykes but now or lately Davenport Cl 1826, that Close ... called ... Davenport 1807 (*draper Sike Close* 1742, *Draper Sike* - 1736, successively named after the *Draper* and *Davenport* families, with **sík**); Great East Cl 1806, The Great - 1807, 1812, 1825, the Great - 1826, 1829, 1854, Far -, Middle -, Near East Cl 1849; East Dale Btm 1806, The East - 1807 (*v.* **botm**); the East Fd 1756[1], 1756[2], 1757[2], 1805, East Fd 1757[1], 1768, the - 1805 (*the east feildes* 1601, - *easte Feild* 1606, - *Eastfeild* 1612, - *east feild* 1628, *East Feild* 1634[1], - *Feilde* 1638, *the East feild* 1715, 1724[1], one of the great fields of the village); East mdw 1849; The Eleven Acres 1857; Far Btm 1839 (*v.* **botm**); Great Far Cl 1839, The (Great) Far Cl 1840, 1852, Far - 1849, 1880; Fifteen Acre Cl 1863; Fish pond Cl 1863; the Folley 1774 (described in the document as "Ancient Inclosure", *v.* **folie**); Fourteen Acre 1863; Furrdale 1769; Gammill Green Cl 1795 (1715), Gamble Green 1822 (cf. *gamble thinge, Gamble greene thinge* 1628 (*v.* **þing** 'property, possessions'), *Gamble Close* 1628, *Gammill-Green Closes* 1708, *Gammil Green Close* 1715, cf. *gamble furres* 1601, 1612, *Gamble Furres* 1606, - *Furs* 1698, 1707, 1724, *Gamle furrs* 1634, *Gamlefurrs* 1638 (*v.* **furh**), from the surn. *Gamble* or *Gam(m)ill*, cf. Geoffrey *Gamell'* 1332 *SR*; if the forms relate to the same piece of land, it appears that earlier arable, indicated by *Furrs* forms, has been laid down to grass, *v.* **grēne**[1] and **grēne**[2]); Garing Cl 1773 (from dial. *gairing* (EDD s.v. *gair*) 'a triangular piece of land which cannot be ploughed with the rest of a field', recorded only from L. It is probably a derivative of ON **geiri** and cf. *A garinge called half an Acre* in Linwood f.ns. (b) *supra*); Gatehouse, Garden & Railway 1849; Gibbons Thing 1792 (*Gibbons Thinge* 1634[2], - *thing* 1640, *Gibbon's Thing* 1738, *Gibbons* - 1745, from the surn. *Gibbon*, common in the parish, cf. John *Gybbon* 1570 *Inv*, Thomas *Gybbon* 1595 *BT*, and **þing** 'property, possessions'); Glebe Cl 1822; Graper Sike 1823, 1845 (cf. *Graper Hill Moor* 1603; this name is found also in Market Rasen, *v.* **sík**); Grass Cl 1849; the Grave Pit (sic) 1774; Great Cl 1806, 1829, the Great cl 1825, - Cl 1826, 1838, 1854, 1872 (area 11a.1r.8p.); the Green Hill 1774, 1802 (*the greene* *hill* 1628, *v.* **grēne**[1], **hyll**); Grimsby Road 1774; Hall Cl 1880 (1697, *ad aulam* 1259, 1332[2] both (p), *ad Aulam* 1327 (p), *atte Halle* 1370 (p), *Haule close* (sic) 1638, *the hall Close* 1664); Gt -, Lt High Field 1774, North -, South High Field Cl 1838, High field Cl 1840 (*the hygh feild* 1601, - *weste highe field* 1606, - *heigh feild* 1612, 1634[1], - *high feild* 1628, *the high Feild* 1698, - *High Fields* 1707, - *high Fields* 1707, 1724[1], cf. *ye high feild dike* 1628); the High Field Cls (*v.* the Milking Hill *infra*); Hill Cl 1839, 1863; Holly-tree Cottage 1880; The Holt 1852 (*v.* **holt**); Home Cl 1839, 1849, 1863, The Home - 1857; Horscotts flg 1772 (the reading is uncertain); The Horse Cl 1840, Horse - 1863; Huttings Cl 1863 (*Mr Benjamin Hutting* is named in the document); the King's Mear 1805 (*v.* **(ge)mǣre**); Kirkheadland Lane 1839 (*the church headland* 1601, *the Kirke headland* 1606, *the kirke headland* 1612, *the Kirke headland* 1634[1], 1638, - *headlande* 1662, *Church headland* 1698, *the Kirkheadland* 1708, *ye Kirk headland* 1724, *v.* **kirkja**, **hēafod-land**, the variation between *Kirk* and *Church* being noteworthy); Kirkin Bank 1772; Lair Pits 1772 (*v.* **leirr** 'clay'; this is probably the same as *Lairepittes* in West Rasen f.ns. (b) *infra*); Linwood Road 1774; The Little Yard 1822; Mare-Acres Flg 1772; Meadow Cl 1849; the Meer 1772 (*v.* **(ge)mǣre**); Middle -, South Melbourn flg 1772 (this is to be identified with *Medelberg* 1290, *Methelbergh'* 1329, *mealbarugh* 1577, *Middle melbrough* 1601, - *melborowe* 1606, *midlemelborough* 1612, *Midlemelborow* 1634[1], *Middle Melber* 1638, 1698, *midle melber* 1662, *middle-Melbrough* 1707, *Middle Melbrough* 1724, *south melbrough* 1601, - *molborowe* (sic) 1606, *southmolbory* (sic) 1612, *Southmelborn* (sic) 1634[1], *South Melber* 1638, - *melber* 1662, - *Melbrough* 1698, 1707, 1724, 'the middle hill', *v.* **medal**, **berg**, a Scand. compound); Middle Cl 1880; Middle Dale 1863; Middle Moor 1846 (*v.* **mōr**[1]); the Milking Hill 1784, formerly ... the Milking Hill but now the High Field Cls 1809, close formerly ... 1817, - Cl 1826, formerly ... the Milking Hill but now - 1812, close formerly ... the Milking Hill afterwards the High Field but now better known by the name of the Barthorpe Closes 1840 ("formerly in the occupation or tenure of John Barthorpe" 1840); Moor Cl or Cowper Cl 1752, the Moor Cl or - 1755, Moor Cl 1781 (cf. Moor Fm *supra* and *v.* also Cowper Cl *supra*); the Near Cl 1809, 1817, the nearClose (sic) 1812, Near Cl (2x) 1849; East -, West Neebly's Cl 1857 (*nebly* a1575, *neeblay* 1601, *Neebley* 1606, 1638, *Nebley* 1612, 1634[1], *Neebly* 1662, *Neeblay* 1698, 1707, 1724, obscure); The Nine Acres 1839, the - 1840, Top Nine Acre 1863; North Cl 1849; the North Fd 1759, Great -, Little North Fd 1774 (*the north fildes* 1601, *the North feild* 1679); North East Fd 1772, 1802, the North - 1829 (*the North and Northest fieldes* 1610); the North-East Pasture 1787, the North East - 1774, 1797, 1800; Pt of North Home Cl 1849; the North West Field 1769, 1772, 1774 (*the North west Feild* 1638, cf. West Fd in (a) *infra*); Occupation Road 1849; the old Acre-Dike Fence 1784 (cf. *le Akerdik* 113, a common name in North L, *v.* *ye acredikes* PN

L **2** 13, cf. *Hakerdik* in West Rasen f.ns. (b). An early form for this name has been noted from Berks, *on þa æcer dik* 956 (c.1200) BCS 924 (S 605) in the bounds of Abingdon, PN Brk 735, where it is simply translated 'acre ditch'. A further example, Acre Ditch 1767, is recorded from Shelden, PN Db 165. The frequency in L suggested a possible ODan origin, but the Berks example clearly shows that it must be derived from OE ***æcer-dīc** though in L the form certainly has been Scandinavianized); Orchard 1849; Orchard Cl 1863; Orvises Bottom Nine Acres 1863 (named from the *Orvis* family, cf. John *Orvis* 1840 *TSJ*); Paddock 1863; Pasture Cl 1849; Patrick Nooking 1772 (cf. *Patricke headland* 1662, named from the *Patrick* family, cf. Symon *Pattrick* 1606, with **nōking** and **hēafod-land**); Pingle 1822 (cf. *pingle Close* 1606, 1634[1], *v.* **pingel**); Plantation 1849; (East) Ploughed Cl 1806, the Ploughed Cl 1829, - ploughed cl 1825, The East ploughed Cl 1812, the East Ploughed Cl 1825; (East -, West of) Railway, Railway Cl 1849 (2x); Ramper Cl 1863 (from dial. *ramper* found in L for 'a raised road or way, the highway' (EDD s.v.), *v. The Rampart* PN L **1** 92); Run-Gates 1772; Sand Lane 1863; The Seven Acres 1839, 1840, Seven Acres 1863; the Short Lane 1772; The Six Acres 1857; the East End of Skinnams 1759, at the - 1802, Skinnills forlong, Skinhils Moor 1772, Skin hills Road 1774; South Cl 1849; the South Dale Drain or Beck 1774, South Dale 1863 (*the south dale* 1601, *the South Dale* 1707); the East -, the West South Fd 1759; the South East Fd 1769, 1774, 1809, 1812, 1817, 1829, 1840, South East - 1772 (*the South east fielde* 1662, and cf. the East Fd *supra*); the South East Pasture 1774; the South West Fd 1772, 1774, 1779, 1781, 1792, 1809, 1812, 1817, the South West (sic) 1776, the South-West Fd 1784, the south Westfield 1822, South West Fd 1829 (*the South west fielde* 1662, cf. West Fd *infra*. The names North -, South East Fd and North -, South West Fd are paralleled by similar names in West Rasen, and such divisions of the great fields of the village have not so far been noted elsewhere in L); Sow Pig Lane 1774, 1781, 1792, Sow pig Lane Fd 1776; a piece of Ground known by the name of strife moore 1822 (probably self-explanatory); One land lying on Switch and Spurr End 1769 (cf. *one land of swich & spurre* 1662; Mr John Field points out that both these terms are names of grasses. *Switch* is one of the numerous expressions for couch grass (*Agropyrum repens*); spur-grass, according to NED s.v. *spur*, is *Glyceria distans*, i.e. reflexed meadow grass, a rather local species, generally found on sandy pastures and waste places in counties near the sea. Spur-grass does not appear to be a nuisance of the same kind as couch-grass, and so the pairing of the names still lacks explanation); Tealby Road 1774 (leading to Tealby *infra*); Long Ten Btm, - Ten Cl 1839; Thirteen Acres 1863; Thornton's Cl 1863; Todds Hills 1863 (named from the *Todd* family, cf. William *Todd* 1725 *BT*); Great Top Cl 1806, the Great - 1826, 1838, Little Top Cl 1806, the little - 1807, 1812, 1825, 1829, The Little -

1812, 1825, Little - 1826; Top Moor 1846; Topham's Parish 1806, Tophams - 1809 (cf. *Tupham Manor* 1718, i.e. Tupholme, and *v.* Middle Rasen *supra*); North -, South Town End Cl 1880; Twelve Acres 1863; Far -, Middle -, Near West Cl 1849; West Dale 1863; West Fd 1756[1], the West - 1756[1], 1757[1], 1768, 1774 (*in occidentali campo* 113, - *campo occidental'* c.1407, *the West feildes* 1601, *the Weste Feild* 1606, *the west feilde* 1612, *the west feild* 1634[1], *the West feild* 1628, 1715, *the west Field* 1698, *the West Field* 1707, 1724, *the high west Field* 1662, one of the open fields of the parish, cf. the North Fd and the East -, the west South Fd *supra*); West mdw 1849; White Layers Flg 1769, White Lairs flg 1772; the Willow Tree Wath 1772 (*v.* **vað** 'a ford'); Windings 1772 (for a discussion of this name *v.* PN L 2 173); the Wood Hill 1822; Yorkshireman cl 1822.

(b) *beane oxlandes* 1601, *Bean Oxlands* 1698, 1707, *bean* - 1724, *beane oxland* 1628, *beanoxlands* 1638, *east beane oxland* 1606, *Eastbean oxlands* 1612, *east beane Oxlands* 1662 (*v.* **bēan, ox-land**; MED s.v. *oxe-land* n. 3a defines the word as 'a place where a cow is found, also, an area of land that can be worked with an ox', and compares *ox-gang(e)* which it explains as 'A measure of land equivalent to that which can be plowed (sic) by an ox in a season, varying in size from about 8 to 30 acres'); *Belewelle* a1290 (*v.* **wella**; the first el. is uncertain, but may well be the ME surn. *Bele*); *Belwether hole* 1601, 1612, 1634, *bellwether* - 1606, *belwether* - 1638[1], 1662, *Belwether hole* 1698, - *Hole* 1707, *Bellwetherhole, Belwether headland* 1724 (no doubt from the ME surn. *Belwether* and **hol**[1] 'a hollow'); *Bencelllanges* (sic) 1259 (obscure); *Blailand'* 1259, *blealandes* 1601, *Blealands* 1612, *blay landes* 1606, *Blay land* 1634[1], *bla lands* 1638, *Blea lands* 1698, *Blealands* 1707, 1724, *longe blalandes* a1575, - *blayland* 1628, - *Blalands* 1662, *short blaylandes* 1628 (probably like Blayfield in Normanby le Wold f.ns. (a) from ME *bla, blo* (ON **blá(r)**) 'bluish grey, lead- or ash-coloured, dark', *v.* MED s.v., perhaps extended in meaning to 'cold, cheerless', in this case with **land**); *bouly* 1601, *Boul...r* 1698 (obscure); *Bratts* 1606 ("one furlong soe called"), 1674, *the Bratts* 1638, *Brats* 1671, *fore Leas of Brates* 1703 (*v.* **brot** 'a small piece of land', and cf. *Brottes* in West Rasen f.ns. (b)); *John Brombye garth end* 1628 (*v.* **garðr**); *Byellyorhouslanges* (sci) c.1260 (*v.* **lang**[2]); *Campoxelandes* a1575 (cf. *beane oxlandes supra*); *Chantree Moore* 1628 (*v.* **chaunterie**; the Chantry of St Mary, West Rasen, founded a1376, held "4 marcates of land in *Middlerasyn*" and presumably this is the Chantry referred to; this is no doubt the same as *Chauntry Moore* under the Moor Close in Market Rasen f.ns. (a)); *the Church Mill* 1747 (it was next to the Water Mill); *Clint* a1290, *the Clintes* 1612, *y*[e] - 1628, *the topp of the Clintes* 1634, *the' Clint hooles* 1601, *Clint Holes* 1698, 1707, *Clinthole* 1724[1] (*v.* **klint** 'a rocky cliff' etc.); *the comons* 1601, *the Commons* 1606, 1671, 1698, 1724[1]; *the corne feild* 1622; *cracough stigh* 1601, *Cracoughs stigh* *moor* 1662, *Craugh Stigh* 1698, 1707, 1724[1] (*v.* **stīg**) (probably the same as *Crakhou* in West Rasen f.ns. (b) *infra* and cf. Craycow in Kirkby cum Osgodby); *the dayle*, *y*[e] *dayle nooke* 1628, *dale Nook* 1698, *the Dale Goat* 1698 (*v.* **deill** 'a share of land' with **nōk, gotu**); *dangattes* 1601, *dangate* 1606, 1662, *Dangates* 1612, *dangates* 1634[1], *dangats* 1628, *Dangates* 1707, 1724 (this is almost certainly the same as Dam Gates (*daynegate, dangattes* 1601) in Linwood parish f.ns. (a), where a probable etymology is suggested); *dead landes* 1601, *- land* 1606, *- lands* 1634[1], *deadlands* 1612, *deadlans* 1638 (probably the same as *dedelandes* in West Rasen f.ns. (b)); *Dead Leys* 1707, *- Leyes* 1724 (*v.* **dēad, lea** (OE **lēah**), in the pl.); *Thomas Denis Close corner* 1612 (named from *Thomas Dennes* 1601 *Terrier*); *Dungeon Close* (*v.* Dovecoat Cl in (a) *supra*); *Est' Estlanges* 1259 (*v.* **ēast, lang**[2] 'a long strip of land'); *Esthymare* (sic) 1235; *Farnwell* 113 (*v.* **wella**); *farswell* 1601, 1634, *Farswell* 1612, 1638, *Nether Farswell* 1662, *Farsewell* 1698, *lower Farswell* 1707, 1724 (*v.* **wella**; the forms are too late for any certainty, but Dr John Insley suggests that the first el. *may* be a ME byname or surn. derived from OE **fearr** 'an ox, a bull'); *in campis de Rasne* 1235, *the feildes of Midle Rasyng* 1537-8, *the feilde of Midell Raison* 1606, *Middle Rasen Field* 1703, 1707, 1709 (*v.* **feld**; presumably this is the same as Gt -, Lt High Field *supra*); *Germaynland* m13 (m14) *Drax* (from *Germ' fil' Walteri* who made a donation to Drax Priory in the relevant document); *gildhous* m13 (Ed1) (*v.* **gildi-hūs**); *Goldhwitenest* (sic) 113 (Mr John Field wonders whether this is not an error for *Godwite-* and points out that the bird-name *godwit* (*Limosa spp.*) is recorded in NED from 1544, where it is noted "Formerly in great repute, when fattened, for the table." He comments that the location of nesting sites would be worth recording); *gooseinges* 1601 (doubtful reading), *goose myers* 1612, *gosemiers* 1634[1], *Gosemires* 1638, *goose myers* 1662, *Goose Mires* 1698, *Goosmires* 1707, *Geesmires* (sic) 1724 (*v.* **gōs, mýrr** 'a mire, swampy ground'); *hayrecotts* 1606, *Hare Coates* 1612, *hare coates* 1662 (self-explanatory, *v.* **hara** 'a hare', **cot** 'a shed'); *Hemletwath Close* (sic) 1575; *the high lees* 1628 (*v.* **lea**); *Hilstye* 113 (*v.* **hyll, stīg**); *houlwath* 1601, *houl Wath* 1606, *houlewath* 1612, *holewath* 1634[1], *the Howlwath* 1638, *Howlwarth* (sic) 1698, *Howlwath* 1707, 1724 (*v.* **vað** 'a ford'); *Howeforlong* (uncertain reading) a1575, *how furlong* 1601, 1638, *How Furlong* 1707, 1724[1] (*v.* **haugr, furlang**); *Howhil furlonge* 1606, *Howhill forlonge* 1612, *howhill -* 1634[1] (from the sequence of names in the Terriers, this is almost certainly the same as the preceding name, from **haugr** 'a mound, a hill' with **hyll**); *huggons close* 1662 (from the surn. *Hug(g)on* common in this parish, cf. Christopher *Huggon* 1588 *BT*, Thomas *Huggon* 1641 *LPR*); *Hundeley* 1259 (probably from OE *hunda* gen. pl. of **hund** 'a hound, a dog', and **lēah** 'a wood, a glade in a wood'); *the Inge gate* 1628 (*v.* **eng, gata**); *y*[e] *Kiln*, *M*[r] *Dixon's Kill* (sic) 1724 (*v.* **cyln**); *Laming Farme* 1725, *Lamming Farm* 1728, 1742 (named from the *Lamming* family, cf. Robert *Lammynge* 1550 *Inv*; John

Lam(m)ing is named in the 1725 and 1742 documents); *little brough* 1628; *Littelhoubi* (sic) c.1260, *Litelhouberh* 113 (*v.* **lýtel, haugr** 'a mound, a hill', with **beorg,** and cf. *Scarhouberh infra*); *long leyes* 1601, 1606, *long leas* 1612, *the long leas* 1662, *the Long leyes* 1698, - *Leys* 1707 (*v.* **lang¹, lea** in the pl.); *Lound toft* 1369 (*v.* **lundr, toft** 'a curtilage, a messuage'); *the low feildes* 1601; *lyneoxlandes* 1601, *line oxland* 1628, *Lyme Oxlands* 1698, 1707, 1724¹ (*v.* **lin** 'flax', **ox-land,** cf. *beane oxlands supra*); *Magestyhe* 1259, *Maggesty* a1290; *Magote More* 1492; *Mares* 113 (*v.* **(ge)mære** in the pl.); *Market Rasin meere* 1662 (*v.* **(ge)mære** 'a boundary'); *Matyns thinge* 1495, *mattynes lande* a1575 (from **þing** 'property, possessions' and **land** and the surn. *Matyn,* cf. *illud mesuagium ... nuper Joh'is Matyns* 1492 *MC*); *Maweacres* 1601, *Mawe Akers* 1612, *mave acres* 1662, *Maveacres* 1698, *Mave acres* 1707 (presumably from the ME surn. *Mawe* and **æcer,** though so far the surn. has not been noted before Mathew *Maw* 1774); *Middelestlanges* 113 (*v.* **east, lang²** and cf. *Est' Estlanges supra*); *middle howles* 1628; *the Motton Meadow* 1698, 1724, *Motton Meadow* 1707; *manor called Nevell Fee* 1495, *Nevills Manor* 1636 (Robert *de Nevill'* is recorded as holding one third of a fee in Middle Rasen 1242-43 Fees and *v.* also *manor called Panell Fee infra*); *the new dale* 1634¹, 1638 (*v.* **deill** and cf. *Neudayle* in West Rasen f.ns. (b) *infra*); *the newe ditch* 1606, *the new dike* 1612; *the noonings* 1601, *the nooneinge* 1606, *one plott of tethering ground called the nooneinges* 1612, *the Noneing* 1634¹, *the noonings* 1662, *yᵉ Noonings* 1698, 1707, *the -* 1724¹ (*v.* **eng,** the first el. being uncertain); *North Bek de Media Rasen* 1369 (p), *Northebek* a1376 ChancCert (*v.* **norð, bekkr**); *North hall* 1399; *Oven laine house* 1690, *Oven lane House* 1732; *manor called Panell Fee* 1495, *Paynells Manor* 1636, *Pagnells and Nevills Fee* 1718 (Ingram *Paynel* is recorded as holding half a fee in Middle Rasen 1242-43 Fees. For the family and their holdings in West Rasen, Drax YWR and East Quantoxhead So, *v.* Clay and Greenway, *Early Yorkshire Families,* 68-9, *v.* also *Nevell Fee supra*); *George Parker's Close* 1747; *Plumb Tree Close* 1742 (perhaps named from the *Plumtree* family, cf. William *Plumtre* a1537 *MiscDep 43,* and *v.* Dovecoat Cl in (a) *supra*); *Rasen gates,* - *hedge* 1628; *Ratunrau* 1364 (is described as a messuage in the charter; this is a name frequently applied to a dilapidated building or buildings, *v.* **ratoun, ráw**); "messuage called" *Rawlins* 1558-79 (from the surn. *Rawlin,* cf. Richard *Raulyn* 1369); *Scarhouberh'* 113 (cf. *Littelhoubi supra;* the etymology of *Scar-* is uncertain); *Scatter witt close* (sic) 1698, *Scatterwitt* 1707, 1724; *One other Mill or Kiln ... for the Bolting of Oats comonly called the Shooting Mill* 1747; *Sinder* 113, *Sinderes* a1575; *the South Dale* 1707 (cf. *Southdayle* in West Rasen f.ns. (b) *infra*); *Stacehepole* 1259 (*v.* **pól** 'a pool'); (*de*) *Stannes de Middelrasen* 1375 (p); *yᵉ stigh* 1628 (*v.* **stíg**); *Stokberughdale* c.1407 (no doubt from *Stokberh'* in West Rasen f.ns. (b) *infra*); *Surdaile* 1628 (*v.* **súr, deill**); *thirty acres* 1679; *Thorestanges* 1259 (the first el. is the pers.n. ON *Þórir,* ODan

Thorir, Thori, hence 'Thori's poles of land', *v.* **stong**); *Thornhoustlanges* 113 ('the east long strips', *v.* **austr, lang**², with **þorn** prefixed, and cf. *Estlanges supra*); *thorne oxlandes* 1601, 1606, *Thorne Oxlands* 1612, 1662, 1698, *thornoxlands* 1638, *Thorn* - 1707, 1724¹, *Under Thorne oxlands headland* 1612, *Thorne oxland headland* 1634¹ (*v.* **þorn, ox-land,** and cf. *beane oxlandes supra*); *Thorpe meere* 1662 (*v.* **(ge)mǣre**); *toothill* 1601, *Toot-hill* 1698, 1707, *Tooth hill* (sic) 1724 ('the look-out hill', *v.* **tōt-hyll**); *the towne furlong* 1601, *y*ᵉ *Town Furlong* 1698, *the* - 1707; *Vickerbrigg* 1612, *Vicar brigg* 1638, *the Vicar Bridge* 1698 (self-explanatory; the Scand. forms in *brigg* are worthy of note); *viam de Westrasen* 113 ('the road to West Rasen'); *West Rasen meare* 1628 (*v.* **(ge)mǣre** 'a boundary'); *Wharledikes* 1707 (perhaps from **hwerfel, hvirfil** 'a circle' and **dīk**); *Wranglandis* c.1407, *wrong landes* 1601, *Wronglands* 1698, *Rongelands headlande* 1662, *Wronglands headland* 1707 (*v.* **vrangr** 'crooked', **land** and cf. *Wranglandes* in West Rasen f.ns. (b) *infra*).

West Rasen

WEST RASEN
    Magna Rasna 1150-54 *AddCh, Mikelrasen* 1367 Ipm, "Great"
       Rasen 1361 Fine
    Westrasen(') 1175-88 (1464) Pat, 1202 Ass, 1205 OblR, 1206
       Ass, 1212 Fees, 1218 Ass, 1226 FF, 1247 RA iv, 1250
       FF, 1254 ValNor, 1263, 1268, 1272 FF, 1275 RH, 1287
       Ipm, 1321 Pat *et freq* to 1427 ib, Cur, *West Rasen*
       1200 Cur, 1265 FF, 1282 Pat, 1290, 1291 Cl, 1299
       Ipm, 1303 FA, 1306 Pap, 1312 Pat, 1316 FA *et freq,*
       W Rasen 1576 Saxton, *West Rasene* 1290, 1305
       Cl, 1305 Fine, *Westerasene* 1220 Cur
    Westrase 1185 RotDom, 1187, 1188, 1190, 1191 P, 1196
       ChancR, 1202 Ass, 1227 ClR, *-rasse* 1197, 1198, 1199
       P, *-rasa* 1186 ib
    West Rasne 1200 Cur, 1242-43 Fees, 1274 *FF, -rasne* 1205 P,
       1208 FF, 1239 Ch, 1510 LP i
    Westrasyn(') Hy3 (1409) Gilb, 1381, 1385 Pat, 1405 RRep,
       1410, 1413 Cl, 1421 Pat, 1424 IBL, 1446 Pat, 1466
       Fine, 1548 ChancCert, - *Rasyn* 1291 Tax, 1389 Pat,
       1402, 1428, 1431 FA, 1441 Cl, 1495 Ipm, 1526 Sub,
       - *Rasin* 1386 Pat, *Weste Rasyn* 1510 Ipm
    Westrasun 1236, 1237 Cl, *-rason*(*e*) 1373 Peace, 1502 Ipm

West Raseyn 1350 Pat, 1469 Pap
Westraysen 1501 Ipm, *West* - 1583 NCWills i, *-rayson* 1502
    Fine, *-raison* 1568-70 *MinAcct,* - *Raison* 1634 VisitN,
    - *Raisen* 1583 NCWills i
Westreason 1556 Pat, a1567 LNQ v, 1576 LER, - *reasen* 1556
    InstBen

For the etymology of Rasen, *v.* Market Rasen *supra.* It is distinguished from Market and Middle Rasen, apparently first as *Great, Mickle, v.* **micel, mikill,** then as *West.*

TOFTLEY'S FM, *Toftleys F^{m}* 1828 Bry, cf. *Toftlaye* 1323, *-law* 1326, 1329, *-lawe* 1339, *-lagh* 1331 all *MC, toftley* 1577, *Toftley* 1596, - *Leys* c.1677 all *Terrier,* the second el. is probably **lēah** 'a woodland glade', 'meadow, pasture', though this is rare in L. Topography suggests that the sense here is 'meadow, pasture'. The early forms for the name are paralleled by those for Wheatley, PN YW 1 36 and for Healaugh, ib 4 240-41, and see also ib 7 84, paragraph 20. The farm is near Toft next Newton and so the name probably means 'the meadow, pasture near or belonging to Toft'. The later spellings have apparently been influenced by *leas, leys* common in North L.

BROKENBACK, *brokenbak* a1244 *MC, Brokenbak* 1337 Cl, *Broken Back* 1804 *EnclA, Brokenback Hill* 1824 O, 1830 Gre; *brokenback* is not recorded in NED and MED only gives *broke(n)-bak(ked),* adj., in the sense 'hunchbacked'. This p.n. is clearly a compound noun and must represent OE **brocen-bæc* from OE **brocen** 'broken' and **bæc** 'a back', used topographically of something resembling a back. The mound, now covered with trees and undergrowth, is described by Mr Ian George, Sites and Monuments Record Officer, City and County Museum, as flat-topped, about 26' across at the top, 7' high, with a ditch 15' broad and 3' deep and with a causeway on the west side. Its overall diameter is about 100'. It stands in the midst of medieval ridge and furrow to which it appears to be related. It may be noted that on 1824 O it is marked as a windmill, but there is no visible evidence of this today. The origin of the mound is unknown, though it is clearly artificial and could well be a round barrow.

COCKTHORNE FM, cf. *Kokthornes* e13 *MC, CokkeThornes* (sic) c.1285 *ib, Cotthorn* (with *-tt-* for *-ct-*) c.1300 RA iii, *kakthornes* (sic, with *Cockthorns* in margin in a later hand) 1577 *Terrier, Cockthornes* 1608 *MC, Cockthornes* 1804 *EnclA*, 1828 Bry, 1830 Gre; *cockthorn* does not seem to be recorded as a species of thorn, so presumably this is simply 'the thorn-bushes where cocks (perhaps woodcock) are found', *v.* **cocc**^{2}, **þorn**. THE DALE, *ad uallem* 1320 *MC, le Dale* 1337 Cl, *the dale* 1577 (17) *Terrier*, self-explanatory, *v.* **dæl, dalr**. DALE BRIDGE is *Wood Pasture Bridge* 1824 O, 1830 Gre, cf. Wood Pasture in f.ns. (a) *infra*. FIELD HO (lost, TF 044 905), *Field Ho.* 1828 Bry, 1830 Gre, cf. *campis de Westrasen* 1329, 1330, 1372, 1389 *MC, - de Westrasin* 1343 *ib, campo de Weste Rase* 1349 *ib, campos ... de West Raison* 1596 *ib, West Raisin field* 1671, 1674 *Terrier* (Kirkby cum Osgodby), self-explanatory, *v.* **feld**. GRANGE FM is apparently a late example of the use of **grange** found often in L in the sense 'a homestead, a farm-house, esp. one standing by itself remote from others', recorded in EDD from L, cf. Croxton Grange, PN L 2 100. HOLME FM is *Smiths F^m* 1828 Bry, presumably named from a family long-established in the parish, cf. Walter *Smyth* 1337 Cl. It is possible that the modern name is represented by *Hulme* 1203 FF, *holm* e13, *Holm* a1244, *Holm(e)* p1290 all *MC, le Holm* 1337 Cl, for the farm is situated beside the R. Rase. However, there is a HOLME HILL, 1824 O, 1828 Bry, 1830 Gre, on a spur of land between the R. Ancholme and the R. Rase, so that it is difficult to decide to which the early forms belong, though the contexts suggest that it is the former. Both are clearly derived from ON **holmr** 'an island of land, a piece of higher ground amidst marsh, etc.'. INGS FM, cf. *ye Inges* 1624 *MC, the Ings* 1671 *ib*, 1804 *EnclA, v.* **eng** 'meadow, pasture'. MILL, 1828 Bry, cf. *Milnehill', milnehill* a1244 *MC, Milnehul* 1337 Cl, *milne hyll* a1575 *MC, v.* **myln, hyll**. POPLAR FM. RECTORY FM is *Glebe F^m* 1824 O, 1830 Gre, cf. *the Rectory House* 1864, - *Parsonage houses* (sic) 1674, *The Parsonage House* 1700 all *Terrier*. SEDGE COPSE, SEDGECOPSE FM, *Sedge Copse* 1828 Bry, *Setcopp* a1287 *MC, Over-, Netherseccecuppe* (*-cc-* for *-tt-*) 1337 Cl, *Setcop, Settcope* 1577 *Terrier, Set Cop* 1824 O, 1830 Gre. The early forms quoted under the Sedcock in Kirkby cum Osgodby f.ns. (a) belong here. The meaning is 'the flat-topped hill', *v.* **set-copp**, for the hill above the Copse has a distinctively flat profile. SOUTH PARK FM, *Sth Park F^m* 1828 Bry, cf. *South Park* 1804 MC, self-explanatory.

WESTLANDS COTTAGE. WESTLEAS. WEST RASEN BRIDGE, also known as PACKHORSE BRIDGE. According to Bridg 36 "West Rasen Bridge, also known as Bishop's Bridge, as it is supposed to have been built by Bishop Dalderby early in the fourteenth century, has three segmental arches spanning a distance of nearly 20 yards. The width is only 4 feet". In P&H, 420, it is described as "C14. With three arches". No reference was found to the bridge in a search through Bishop Dalderby's *Register* (*BR 3*, LAO) and it is strange that no early forms at all have been found for the name. However, several references have been noted to a bridge: "a selion upon *Holm* at the bridge" 1337 Cl, *le Horebrigge* 1337 ib (probably 'the grey bridge' *v.* **hār, brycg,** with the second el. in a Scandinavianized form) and *inter Gardinium W.... & pontem puellarum* 1366 *MC* ('the maidens' bridge', the charter being partially illegible). Unfortunately, it is impossible to tell whether any of them refer to the existing bridge. WOOD HO. WOODSIDE FM is *Kingerby Wood F^m* 1828 Bry, named from Kingerby Wood in Kingerby parish *supra.*

Field-Names

Forms dated e13[1] are e13 (Ed1) *Bart*; e13[2], a1272, 1286, 1348, 1351, 1354, 1608 and 1804[1] are MC; e13[3], a1244, a1275, c.1285, a1287, a1290, 1320, 1326, 1329, 1330, 1331, 1339, 1343, 1349, 1356, 1366, 1372, 1389, 1400, 1403, 1430, 1434, 1438, 1457, 1492, 1495, 1571, a1575, 1588, 1596, 1603, 1621, 1624, 1639, c.1677, 1774 and 1850 are *MC*; 1210-15 are RA iv; 1327 and 1332 are *SR*; 1337[1] are Ipm; 1337[2] are Cl; 1351 are Pat; a1376 are ChancCert; 1568-70, 1589-91 and 1605-6 are *MinAcct*; 1577, 1674, 1679, 1700, 1788 and 1864 are *Terrier*; 1583 are NCWills; 1607 are *Rental*; those dated 1804[2] are *EnclA*.

(a) Great Barram 1804[1] (*Barholm, barthom* (sic) a1244, *Bareholm* 1434, 'the piece of higher ground amid marsh where barley is grown', *v.* **bær[3], holmr**); Bishop Bridge Rd 1804[2] (leading to Bishop Bridge in Glentham LWR); Chatterton's Walk 1774 (from the surn. *Chatterton* and **walk** 'land used for the pasture of animals, especially sheep'); the Church Rd 1804[2] (perhaps the same as *Kirkelane* 1434, *v.* **kirkja, lane**); all that s[d] Piece or Parcel of Old Inclosed Land ... called Ding Dong ... to pay the clear Rents and profitts to some poor Parishoner of the said Parish of West Rasen for Ringing a Bell in the said Parish Church at a certain Hour during some part of each and every year for ever 1804², a piece of Land called Ding Dong 1850 (evidently an endowment dating from before 1804); the Grove 1804²; the Kilngarth 1804², the Kiln Garth Cl 1788 (*one Killne house and the Killne garthe* 1621, *Kilne Garth* 1671, *the Kill garth* 1674, *the Kilgarth* 1679, v. **cyln, garðr** 'an enclosure, a small plot of ground', as elsewhere in the parish); the Ley Cl 1804² (v. **lǽge** 'fallow, untilled', **clos(e)**); the Marsh (1700), the Marsh Foot Way 1804² (*In Marisco de Westrasen* a1244, *de marisco* 1327, 1332 both (p), *le Mersh* 1337², *Mersch* 1343 (p), *in marisco* 1372, *le Merske* 1400, *Close called the Marsh* 1674, *closse commonly called the Marsh* 1679, self-explanatory, v. **mersc**; the 1400 form is in a Scandinavianized form); the Mownray 1788, The - 1804² (*peice or parcell of meadow called ... Momeray* 1671, *Close called Mownray* 1674, *closse called Monwray* 1679, *Mon Ray* 1700; the first el. is uncertain, the second is probably **vrá** 'a nook, a cattle shelter'); the Mulberry Cl 1788, The - 1804², farm called Mulberry 1804¹ (*mulberry Close* 1671, *the Mulberry Close* 1674, *closse called Mulberry* 1679, *Mulberry Close* 1700, self-explanatory); New Ings 1774 (v. **eng** 'meadow, pasture' in the pl., as elsewhere in the parish); New Fm 1804¹; the North East Fd 1804² (*le North Est fieldes* 1596, cf. the South East Fd and *in Estcampo infra*); the North West Fd 1804² (cf. the South West Fd *infra*); Sand Lands 1804² (*Sandlandes* 1320, *sandlandes* 1577, *Langsandlandes* c.1285 and cf. *le Sandes* 1337², *Langsandes* 1320, v. **sand, land**); the South East Fd or Alltmᵗ of Land or Ground 1804² (*le Sowth Est fieldes* 1596, *the South East feild* 1671, cf. the North East Fd *supra*); the South West Fd 1804² (*le sowth west fieldes* 1596, *South West feild* 1671, cf. the North West Fd *supra* and *in Westcampo infra*. The names North -, South East Fd and North -, South West Fd are paralleled in Middle Rasen *supra* and similar divisions of the great fields of the village have so far not been noted elsewhere in L); Stephenson's Walk 1774 (from the surn. *Stephenson*, cf. Edward *Stephenson* 1851 *Census*, and **walk** and cf. Chatterton's Walk *supra*); Toft Rd 1804² (the road to Toft next Newton, an adjoining village); the Town St. 1804²; the West St. 1804²; Wood Pasture 1774, 1804², the Upper Wood Pasture 1864 (cf. *Wdefurlanges* e13¹, *Wodefurlanges* 1337², *Wod End* p1290, *Rasing Woode* 1537 LP xii); David Young's Home Cl 1804².

(b) *Akerdyk* (v. *Hakerdyk infra*); *aldhaithe* e13³, *aldharth* (sic) e13¹, *Aldeheythe* (*hole*) 1320, *Aldhaythe* 1389, *awdith* (sic) 1577 (v. **ald** 'old, long-used' and probably **heiðr** 'a heath, uncultivated land'); *le Barlocces* (sic in transcription) 1337² (perhaps an error for *Warlottes*, the same as *Le Warlotes infra*); *le Bek* 1337², - *Bek'* 1339, *the bek(e)* 1577 (v. **bekkr** 'a stream', as elsewhere in the parish); *Beclangfurlanghes* a1244, *Beklanges* 1337². *Becklands* 1603 (v. **bekkr, lang²** 'a long strip of land', as elsewhere in the parish, with **furlang** and in the 1603 form **land** for **lang²**); *Belthinge* 1589-91, 1605-7, 1607 (named from the family of

John *Bell'* 1574 BT with **þing** 'property, possessions'); *del Berch* (sic) 1286, *le Berkes* (sic) 1331, - *Bergh* 1337², *the barugh, Little Barugh, litle Baruge* 1577, *le Barghe* 1596 (*v.* **be(o)rg** 'a hill, a mound', as elsewhere in the parish); *Bergathe* a1290 (*gathe* is found in L as a spelling of *gate; Bergathe* is perhaps to be compared with *le Beregate* in South Kelsey and Owersby f.ns. (b) *supra; v.* the former for a discussion of the name); *Blaberth* 1286, -*bergh* 1320, 1329, 1337², 1351, 1389, *blebar, blaber* 1577 (*v.* **be(o)rg**; the first el. is probably ON **blá(r)** 'dark, blue, livid' in the extended sense 'cold, cheerless', cf. Blabers in Newton by Toft f.ns. (a) *supra*); *Blaberghoustlanges* a1244, 1343, *Blaberkestlanges* p1290, *Blabergh estlanges* 1320, - *Estlanges* 1337² (from the prec. with **austr** 'east' varying with **ēast, lang²**); *Bleber oxlandes* 1577 (from *Blaberth supra* with **oxa, land**, cf. *beane oxlandes* in Middle Rasen f.ns. (b) *supra*); *blakemild* a1244 (*v.* **blæc, mylde** 'soil, earth' and probably identical is *Blakvildos* (sic) 1337², though this form seems to be corrupt); *pratum bē marie* 1330 ('the Blessed Mary's meadow' and cf. *Chauntre Hous infra*); *Braythow* a1287, *Est braithowe* 1337² (*v.* **breiðr** 'broad', **haugr** 'a hill, a mound', a Scand. compound and cf. *Westerbradhou infra*); *Braelheng* (sic) 1337² (*v.* **eng**, *Brael-* may be an erratic spelling for *Braith-*, cf. Breathings in Walesby f.ns. (a) *infra*); *Brottes* e13¹, e13³ (*v.* **brot** 'a small piece of land'; cf. *Bratts* in Middle Rasen f.ns. (b) *supra*); *Bynlandes* 1337² (Dr John Insley suggests that this is perhaps 'the cultivated, occupied strips of land', from OE *b ne* 'cultivated, inhabited, occupied' and **land** in the pl.); *Byrygh* 1337² (probably from **byrg**, dat.sg. of **burh**, 'a fortification, etc.'); *calfe marfar* 1577 (*v.* **calf, marfur** 'a boundary furrow'); *mesuagium vocat' Chauntre Hous* 1492, *the Chauntry howse* 1588, *messuag' voc. le Chauntry House* 1568-70, - *Chauntrey House* 1589-91, *mes' voc' le* - 1605-7, *mess' voc' le Chauntrey howse* 1607, *the Chauntrey howse in Rayson parcell of the said Chauntrey of the blessed Marie in Weste Rayson* 1588, cf. *the Chantry landes* 1577 and *Chantree Moore* in Middle Rasen f.ns. (b) *supra*; for the Chantry of St Mary, West Rasen *v.* AASR 36, 267-69); *Clakebergh'* 1210-15, -*bergh* a1244, *le Clakbergh* 1337² (*v.* **be(o)rg**; the first el. is probably the Scand. pers.n. *Klakki*, a side-form of *Klakkr*; rather than **klakkr** 'a lump, a hill'); *Clederhouwe thynge* 1495 (named from the family of John *Clederhowe* 1492 with **þing** 'property, possessions'); *Cokething* 1568-70, 1589-91, *Cookthinge* 1605-7, *Cokethinge* 1607 (probably from the family of Agnes *Cok'* 1337² with **þing**); *Coleputtes* 1337² (*v.* **col¹** 'coal', **pytt**); *cotgarth* 1389 (*v.* **cot, garðr**); *Cowper hedland* a1575 (from the surn. *Cowper* well-evidenced in the district, cf. Cooper Cl in Middle Rasen f.ns. (a) *supra*, and **hēafod-land**); *Crakhou* p1290, *Crakehow* 1337², *Crakoe* 1596 (probably 'the mound, hill where crows are found', *v.* **kraka, haugr**, a Scand. compound; later forms presumably occur in Middle Rasen f.ns. (b) s.n. *cracough stigh supra*); *Craklincroft* e13³, *Crakelincroft* a1244 (*v.* **croft**; the first el. is perhaps a pers.n. or byname

*_Crakling_, an AScand derivative of ON _Krakaleggr_ (a nickname applied to Olaf Tryggvason, _v._ LindB 217-18 s.v. _Krakabein_), cf. AScand _Spracaling_ corresponding to ON _Sprakaleggr_, for which _v._ Reaney, s.n. _Sprackling_, as suggested by Dr John Insley); _crosberhille_ a1244 (_berhille_ is apparently 'the barley hill', _v._ **bere, hyll** and _cros-_ perhaps means 'across, lying across' rather than **cros** 'a cross'); _dauencroft_ a1244 (_v._ **croft**; the first el. is the pers.n. _Dawen_ (<ME _Dawe_), a West Midland form, _v._ Reaney, s.n. _Dawn_); _dedlandes_ a1290 (_v._ **dēad** 'barren, uncultivated', **land**; later forms of the same name are probably found in Middle Rasen f.ns. (b) _supra_); _Dikfurlonges, -furlanges_ a1244, 1320, _Dykeforlanges_ p1290, _Dykfurland_ 1337², _Dikforland_ 1348, _Dikefurlangs_ 1372, _Dykefurlonges_ 1434, _dikfurlonge_ 1577 (_v._ **dīk** 'a ditch', **furlang** and cf. _Langdikfurlanges_ and _Skortedikfurlonges infra_); _Diking_ 1203, 1203 (Ed1) _Barl_, 1603, _Dikheng'_ p1290, _Dikheug_ (_-u-_ = _-n-_) 1337² (presumably from **dīk** and **eng**); _Doderholm_ 1320 (_v._ **holmr**; the first el is probably ME _doder_, the parasite plant _Cuscuta_, _v._ NED s.v. _dodder_); _Douthlandes_ 1320 (_v._ **land**; the first el. is probably, as suggested by Dr John Insley, ON _dauðr_ 'dead', referring to barren soil, cf. _deadlandes supra_ with the cognate OE **dēad**. The two names may well, therefore, refer to the same strips of land with a remarkable interchange between OE **dēad** and ON _dauðr_); _Dowlandes_ 1337², _-landys_ 1389, _dowlandes_ 1577, 1596 (perhaps from **dūfe** 'a dove', with the dial. form _dow_, and **land** in the pl.); _the east Inges_ 1639 (_v._ **ēast, eng**); _Elleuestang_ 1320, _Eller stong_ 1577 (from **stong** 'a pole' also a measure of land, with an uncertain first el., which might perhaps be the ODan pers.n. _Elef_); _le Estbenland_ 1330, _Est benalandes_ (sic) 1337², _East Beanlandes_ 1596 (_v._ **ēast** and **bēan** 'a bean', **land**); _in Estcampo de Westrasin_ e13³, _- Westrasen_ e13¹, _In Estcampo_ a1244, _in Orientali Campo_ 1330, _in campo oriental'_ 1339, _super meram campi orientalis_ 1434 (_v._ **(ge)mǣre** 'a boundary'), _the est field, the east feild_ 1577, _- of West Raysen_ 1638 (one of the great fields of the village, _v._ **ēast, feld**); _estholme_ 1343 (_v._ **ēast, holmr**); _Estlanges_ e13¹, e13³, a1244, (_v._ **ēast, lang**² 'a long strip of land', and cf. _Nordlanges, Westlanges infra_); _le Estmar(e)_ 1330, 1337² (_v._ **ēast, (ge)mǣre**); _le Farthynges prat'_ 1403 (literally 'the Farthing's meadow', but the sense of _Farthynges_ here is unclear); _Fassed thing_ 1568-70, _- thinge_ 1605-7, 1607 (from **þing** 'property, possessions' and presumably a surn., perhaps _Fasset_, for which _v._ Reaney s.n. _Fawcett_); _Faunelcroft_ 1337² (this is probably a misreading for _Fauuelcroft_ from the ME surn. _Fauuel_, from ME _favel, fauvel_ 'flattery, insincerity, guile, intrigue', _v._ MED s.v., and **croft**); _in Campis de Westrasen_ e13¹, 1330, _in campis_ - 1329, 1331, 1372, _in campo de Westrasne_ 1339, _in campis de Westrasin_ 1343 (the open field(s) of the village, _v._ **feld** and note _the feildes mer_ 1577, _v._ **(ge)mǣre**); _Frampistlegate_ a1244 (_v._ **gata**, the first el. being obscure); _Fraunselgate_ (sic) 1330, _Fraunceysgate_ 1337², _Fraunceis_ ..... 1348 (the rest of the name is illegible) (from the family of Alice _Fraunces_ 1430 and **gata**); _Galchdaile_

a1244 (v. deill 'a share of land', the first el. perhaps being galga 'a gallows'); atte Gote 1337^2 (p.n.) (v. gotu 'a water-channel, a stream'); Goukegarth 1366 (probably 'Gauk's enclosure', from the ON pers.n. Gaukr and garðr, rather than ME gauk, gouk 'a cuckoo', from ON gaukr); viridem meram 1434 ('the green boundary', v. grēne, (ge)mǣre, perhaps, as Mr John Field suggests, alluding to a grass-grown boundary lane); le Greites 1337^2 (v. grjót 'gravel, stones', with reference to stony or gravelly places); Grene Toft, - toft 1210-15, a1244, Grentoft 1389, Nethergrenecroft (sic) 1337^2, Nether-, Overgrenetoft 1337^2, Overgrene Toft 1339 (v. grēne, toft 'a curtilage, a messuage' with neoðera, uferra 'upper'); Hakerdik a1244, Akerdyk 1337^2 (cf. the old Acre-Dike Fence in Middle Rasen f.ns. (a) supra); le Halle dayle p1290 (v. hall, deill 'a share of land', as elsewhere in the parish, and cf. "an old hall on the north side of the church" 1337^1, atte Halle 1389 (p)); high furre Close 1671; Holgatehil e13^2, Holgathill' 1389 ('the road running in the hollow', v. hol^1, gata with hyll); Hornething 1568-70, -thinge 1589-91, 1605-7, 1607 (presumably from the surn. Horne and þing); Hundyk 1337^2 (v. hund 'a hound', dík); le Hyulam 1286 (obscure); Inedbriggehill (sic) a1244 (from brycg in a Scandinavianized form with hyll; the first el. might just be ME inede 'needy persons', v. MED s.v. inēden); Jolane thynge 1495 (named from the Jollan family, cf. Jollan' fil' bereng' e13^3, Thomas Jollan 1337^2, John Jollayn 1350 MC, and þing); Kauenesholm e13^3, Cavenesham (sic) 1372^2 (perhaps 'Kafni's higher land in marsh' from an unrecorded Scand. pers.n. *Kafni and holmr, a Scand. compound. The same pers.n. is the first el. of Caenby LWR, but here has a gen.sg. -es. Dr John Insley, however, suggests that the first el. is rather a hypocoristic ME form of a pers.n. based on ONFr cauf 'bald' and compares this with the surn. Cavell, for which v. Reaney s.n.); Kirtelakre a1244, Kirkelacre (sic), Kirtilacres 1337^2 (the first el. may be ME kirtel 'a kirtle', used as a byname or an occupational name, with æcer, akr); "a house called" Knighteschaumbre 1337^1, 1337^2 (cf. la Chaumbre de Westrasen 1351 (p), v. chaumbre; the sense of Knight here is uncertain); Lairepittes a1244, Layrpittes a1272, a1275, -putes 1337^2 ('the clay pits', v. leirr, pytt; it is probably the same as Lair Pits in Middle Rasen f.ns. (a) supra); Langedayle, Langdaile a1244, le Landayl (sic) p1290, Longedayles 1337^2, Langdale 1457 (v. lang^1, deill); Langdikfurlanges a1244, Langedyk furlanges 1337^2, cf. Dikfurlonges supra and Skortedikfurlonges infra); super viam linc' 1330 ('the road to Lincoln'); Litelbergh' 1320, Lutelbergh 1337^2 (v. lytel, be(o)rg); Litteldikgate a1244 (v. lytel, litill, dík with gata); Littlehow 1337^2, Lytilhow 1389, Littelhougate 1337^2 (v. lytel, litill, haugr 'a mound, a hill' with gata); an Maniwile del berec aput meridiem (sic) e13^3, maniwillthes (sic) a1290, Maniwylghes, le maniwilwghes 1337^2 ('the many willows', v. manig, wilig, and cf. Maniwilghes in Newton by Toft f.ns. (b) supra, no doubt the same name); the manor house of West Rasen 1608; in pratis de Westras' p1290, - de Westrasen 1330 ('West Rasen meadow'); *Meadlegate headinges* 1577 (*v.* **middel, gata,** and for *headinges, v.* PN L 2 14); *Micheldayle* a1244, *mikkeldayled* (sic) 1330, *Mikeldaylside* 1337² (*v.* **micel, mikill, deill,** with **hēafod** and **side**); *Mikelegate* 1203 FF, 1203 (Ed1) *Barl, Mikelgate* e13³, 1337², *Mikkelgate* a1244, *mikilgate* a1290, *Mikelgat* 1337² (*v.* **mikill, gata,** a Scand. compound); *In mora* a1244, *Le More* p1290, 1337², *in mora* 1400, *Inthemorforlang* (sic) 1348 (*v.* **mōr**¹ with **furlang**); *Morwanges* 1434 (*v.* **mōr**¹, **vangr** 'a garden, an in-field' as elsewhere in the parish); *Motkantoft* a1376 (named from the family of Thomas *Motekan* 1337² with **toft**); *Murwine marfur* 1577 (the second el. is **marfur** 'a boundary furrow'; the first is presumably a surn., possibly derived from the OE pers.n. *Mōrwine*, though the form is too late to allow any degree of certainty); *Nedermurnewang* a1244, *Overmirnewang* 1337² (*v.* **vangr**; *murne/mirne* is obscure); *Neudayle* 1337², *Nudayll'* 1389 (*v.* **nīwe, deill**); *Neutonforlonges* 1434, *Newton Yngfurlong* 1577 (named from Newton next Toft, the adjoining parish, with **eng** and **furlang**); *Nordhalanges, Nordherhalanges* e13³, *Nordhalanges* a1244, *le Northeralanges* 1330, *Northanlanges* 1337² (*v.* **norð, norðan, norðerra** and **lang**², but the significance of -(*h*)*a*- is unclear; cf. *Sudherhalanges infra*); *Nordlanges* a1244 (*v.* **norð, lang**² and cf. *Estlanges supra, Westlanges infra*); *North dayle* 1434 (*v.* **norð, deill** and cf. *Southdayle infra*); *in boriali campo* p1290 ('the North Field'); *Northforland* 1348 (*v.* **norð, furlang; furland, forland** are frequent spellings of *furlong* in North L); *Nyne Rygges* 1577 ('nine strips of land', *v.* **hryggr,** in ME 'a cultivated strip of land, a measure of ground'); *Oldehayes* 1337² ('the old enclosures', *v.* **ald, (ge)hæg,** the latter being rare in north L); *viam que dic' venell' Radulphi Pesse* p1290 ('the way which is called Ralph Pesse's lane'); *Petreshull* 1339, *Petersell* 1596 (presumably this is 'Peter's hill', *v.* **hyll**); *Phipelfure* 1210-25 (*v.* **furh**; the first el. is an unrecorded hypocoristic form of the ME pers.n. *Philip,* cf. Reaney s.n.); *Pottershull* 1337² (*v.* **hyll**; the first el. being the occupational name or surn. *Potter*); *Prestferthing* (from **prēost** 'a priest' and **feorðung** probably in the sense 'a fourth part'); *Rachelwang* a1272, a1275 (from the fem. pers.n. *Rachel* and **vangr**); *Randbylande* 1389 (from the *Randby* family (from Ranby LSR), cf. Roger *de Randeby* 1337², and **land**); *Raulwang* 1337² ('Raul's garden, in-field', from the OFr pers.n. *Raoul,* itself from West Frankish *Rādulf,* and **vangr**); *Rigdalleswathe* 1577 (*v.* **hryggr, deill** with **swæð** 'a track', later 'a strip of grassland'); *Rischeng* a1244, *Reskhenges* 1290, *Riskholm* 1337¹, 1337² (from a Scandinavianized form of **risc** 'a rush' with **eng** and **holmr**); *pratis vocatis Santmargreteynges* 1403 ('St Margaret's meadows', *v.* **eng** in the pl.); *Shortres* p1290, *Scettres* (sic) 1337² (obscure); *Skortedikfurlonges* a1244, *Shortdikefurlong* 1596 (*v.* **sceort** and cf. *Dikfurlonges* and *Langdikfurlanges supra*; the form dated a1244, like others of that date in this parish, is partially Scandinavianized); *le Sike* p1290 (*v.* **sīk**); *Skouelakres* a1244, *scouelacres* a1287, *Nedhreskouelakres* a1244 (*v.* **scofl, æcer, akr** with **neoðera** 'lower', denoting a narrow piece of land as broad as a shovel; again forms in *Sk-*, *Sc-* are Scandinavianized); *Smerholm* 1337² *v.* **smeoru** 'fat', **holmr**, denoting rich higher ground amidst marsh); *Snypdale* a1244 (*v.* **dalr**; Dr John Insley suggests that the first el. may be the ON byname *Snípr*, which seems to correspond to Norw *snipa* 'miser, unsociable person', *v.* LindB 346); *Southdayle* 1434 (*v.* **sūð, deill** and cf. *Northdayle supra*); *Southebergh'* 1343, (*le*) *Southbergh'* 1354, 1389 (*v.* **sūð, be(o)rg**); "the South field" a1272, *in australi campo* p1290, - *Australi* - 1329, "the South field of Westrasen" 1351, 1354, *in campo australi* 1372, *the Southfield* 1577 (*v.* **sūð, feld**; one of the great fields of the village); *le Southmere* 1434 (*v.* **sūð, (ge)mǣre** 'a boundary, land on a boundary'); *Stickething* (sic) 1568-70, *Sykes thinge* 1589-91, 1605-7, *Sikes Thinge* 1607 (the four names appear to refer to the same property, so the most likely explanation is that *Sticke-* is an error and the name is derived from the *Sykes* family and **þing**, though the surn. has not been noted in the sources searched); *Stokberh'* p1290, *Stokebergh* 1337², *Stocberk'* 1339, *Stockebergh'* 1343, *Stockebergh'* 1343, *Stocberksyke* 1339 (*v.* **stocc** 'a tree-trunk, especially one left standing, etc.', **beorg** with **sīk**); *Stonelacres* 1337² (probably 'the stony hill', *v.* **stān, hyll** with **æcer**); *Sudherhalanges* e13³, *le Southalang*, *Sutheralanges* 1337² (*v.* **sūð, sūðerra, lang²** and cf. *Nordhalanges supra*); *Toftfeild* 1577 (named from the adjoining village of Toft in Newton next Toft *supra* and **feld**); *Toftwranglandes* a1272, a1275 (similarly named from Toft and cf. *Wranglandes infra*); *Turfgreynes* 1583 (*v.* **turf, grein**, probably in the sense of a small valley or valleys forking off another where turves were obtained); *forar' de Twell'* a1244 (perhaps a headland (*v.* **hēafod-land**) named after a family called *Twell*); *twelve ende* 1577; *Twenti* e13³, a1244, *le Twenty* 1337² (referred to as a plot) (this is clearly the numeral, OE *twentig*, perhaps referring to the size of the plot of land); *Wadhou* 1210-15, *-how* 1337², 1349 (perhaps 'the mound, hill where woad grows', *v.* **wād, haugr**; the form *Wadehōt* (sic) a1244 may belong here); *le Warlotes* 1343 (*v.* **warlot** and for a full discussion of this name *v.* PN L 2 67); *de molendina aqua* a1275 ('the watermill'); *Watfures* e13¹, *-foures* 1320 (*v.* **vátr** 'wet', **furh** 'a furrow'); *Wellesty* 1337² (*v.* **wella, stīg** 'a path'); *Westdikforlanges* 1330 (*v.* **west** and cf. *Dikfurlonges supra*); *the west end of the towne* 1577; *in West campo* e13¹, *In Westcampo* e13³, a1244, *in ocidenta' campo de Westrasen* 1400, *the west feild* 1577 (*v.* **west, feld**; one of the great fields of the village); *Westerbradhou* a1244 (*v.* **brād, haugr** with **westerra** 'more westerly' and cf. *Braythow supra*); *Westhallegarth* 1337² (*v.* **west, hall, garðr**); *Westholm* 1337² (*v.* **west, holmr**); *Westlanges* a1244, 1330, 1337², *-langs* 1339, *le Westlanges* p1290, *Westlang* 1337², *Langwestlanges* a1244 (*v.* **west, lang²** 'a long piece of land' with **lang¹** 'long' and cf. *Estlanges, Nordlanges supra*); *Whytputteland* 1337² ('the white pit', *v.* **hwīt, pytt**, presumably denoting a chalk pit, with **land**); *le Wilghes* 1337² (*v.* **wilig** and cf. *Maniwile supra*); *Whitpintelland* a1244 (from the ME nickname

Whitepintel 'white penis' and **land**); *Wormething* 1568-70, *Wrenetinge* (sic) 1589-91, *Wrenthinge* 1605-7, 1607 (all four forms apparently refer to the same property, so as with *Stickething supra*, the first *Wormething* is presumably an error; the others are derived from the family of John *Wrenn* 1438 and þing 'property, possessions'); *wra* e13[1], e13[3], *Wra* a1244, *le* -, - *Wro* 1337[2] (*v.* vrá 'a nook, a secluded spot, a cattle shelter', cf. dial. *wro*); *Wranglandes* 1337[2], *Wronglandes* 1577, *Wrenglandes nooke* 1603 (*v.* vrangr 'crooked', **land** 'a selion' with nōk and cf. *Wranglandis* in Middle Rasen f.ns. (b) *supra*); *Wypdayl* 1337[2] (*v.* deill; *Wyp-* is obscure); *Wysmangarth* 1337[2] (from the ME surn. *Wisman* and garðr).

Stainton le Vale

STAINTON LE VALE

    Stainton(') 1086 DB, 1208 Cur, 1210-15 RA iv, 1212 Fees,
        1228 Welles iii, 1240 (13) *Alv*, 1266 *MiD*, 1275 RH,
        1301 Ch, - *super Waldam* 1296 RSu, - *in le Hole* 1576
        LER, 1585 SC, - *in ye Hole* 1692 *MiD*, - *in y*[e]
        whole 1693 *ib*, - *in the Hole* 1695 *ib*, 1713 *Holywell*,
        1782 *Tur*, - *Le Hole* 1695 *MiD*, 1702 *Terrier*, - *Lee Hole*
        1707, 1709, 1712, 1715 *ib*, - *le hole* 1768 *MiscDep*
        108, - *le Hole* 1824 O, 1830 Gre, - *Vale* 1808 *Td'E*,
        - *le Vale* 1828 Bry, *Staintone* 1086 DB (3x), l12, e13
        MiD, -*tona* 1136 (1312) Whit, 1143-48 (15) ib, 1174-79
        (c.1240) ib, *Wald Staintona* e13 (13) *Alv*
    Stayntona Hy1 (c.1400) Whit, l12 (1409) Gilb, -*ton*(') eHy2,
        1240, 1245, 1250 all (13) *Alv*, 1272-73 Misc, 1279
        RRGr, 1281 RSu, 1314, 1315 Ipm, 1316 FA *et passim*
        to 1535 VE iv, - *iuxta Keuermund* 1288 *Ass*, - *iuxta*
        Hirford 1297 *MiD*, - *juxta Irford* 1346 Orig, - "by" *Irford*
        1346, 1350 Pat, 1394 Fine, 1399 Pat, - *iuxta Thoresway*
        1322 *MiD*, - *super Waldam* 1299 RSu, - "upon Wold"
        1318 Pat, *Waldestaynton*' 1376 *FF*, - *in le Hole* 1440,
        1480 Pat, 1497 HMCRep, 1542-43 Dugd vi, 1546 LP
        xxi, 1547, 1551 Pat, 1562 InstBen, - *in the Hole* 1567
        Pat, - *in le valy* 1530 Wills ii
    Northstaynton 1274 Ipm, 1414, 1415 Fine, 1424 IBL, 1431
        FA, *North Staynton* 1428 FA, 1591 *Inv*, - *Steynton* 1352
        Ipm
    Steintuna c.1115 LS, 1150-55 Dane, -*ton*(') 1182, 1190, 1191,

1192 P, 1202 Ass, 1206 Abbr, 1207, 1208 Cur, 1219 Ass,
a1221, 1225 Welles iii, 1559 InstBen, *-tona iuxta Yreford*
1288 *Alv*
Steynton(') 1190 (1301) Dugd vi, 1242-43 Fees, 1254 ValNor,
1277 Ipm, 1277 RRGr, 1281 QW, 1291 Tax, 1291 *Alv*,
1338 Ipm, 1455 Fine, *-tona Iuxta Binbrok'* 1284 *Alv*
Stantune a1135 (c.1240) Whit, *-ton* 1428 FA, *- juxta Bynbroke*
1538-39 Dugd vi, *- iuxta Bynbroke* 1546 *AOMB 217*, *- in
the Hole* 1539-40 Dugd vi, 1607 Camden, 1610 Speed,
1656, 1672 *Holywell, - in le hole* m16 *Cragg, - in y^e
Hole* 1695 Morden, *Waldestanton* 1208 (13) *Alv*
Staunton(') 112 *MiD*, 1245 (13) *Alv*, 1526 Sub, *- in Valle*
c.1300 RA iii, *- en le Hole* 1547 Pat, *- in le Hole* 1645
Holywell

This is no doubt a partial Scandinavianization of OE **Stantūn*
'the farmstead, village on stony ground', *v.* **stān, tūn,** with ON **steinn**
replacing the cognate OE **stān,** comparable to Stainton LWR,
Stainton by Langworth and Market Stainton LSR, and topographically appropriate. It is described as being near Binbrook, Kirmond,
Orford and Thoresway and upon the Wold (*v.* **wald**). The
commonest affix is some form of *in the Hole* 'in the hollow', *v.* **hol**[1],
again an exact description of the situation. Though two forms in
Valle and *valy* have been noted c.1300 and 1530, the modern *le Vale*
does not become common till the 19th century.

ORFORD (now represented by ORFORD HO)
Erforde 1086 DB, *-ford* 1414 WillsPCC
Iraforda c.1115 LS, *Ireforde* c.1115 ib, 1562 Pat, *-ford*(')
1150-60 Dane, c.1189 LAAS v, 1200 Cur (p), e13
HarlCh, e13 RA viii, 1220-23 *HarlCh*, 1202 Ass, 1204
Cur (p), 1205 FF (p), 1219 Ass, 1254 ValNor, 1316
Pat, *-fford* 1552 PrState, *Irreford* 112 MiD, 1291 Ipm
Hireford a1150, c.1155, 1155-60, 1160-66 Dane, *-forda* R1
(1318) Ch, 1212 Fees, *-fort* 1160-66 (1409) Gilb
Hyreforth c.1150-55 Dane
Yreford(') 1180, 1181, 1188 all (p) P, a1189 LAAS v, 1190,
1191, 1192, 1193, 1194, 1195 all (p) P, 1202 Ass (p),
1204 Cur (p), 1281 QW, *-uord* c.1200 RA iv (p)
Irfordie c.1155 Dane, *-forda* 1228-32 (1409) Gilb, *-ford*(')

1281 Ipm, 1316 FA, 1327, 1332 *SR*, 1341 Orig *et passim*
to 1679 *Terrier*, - *iuxta Bynbrok'* 1288 *Ass*, - *alias Urforth*
1539 LP xiv, -*forde* 1556 AASR xxxvii, - *alias Urforthe*
1544 LP xix, -*forth* 1526 Sub, - *alias Urforthe* 1567 Pat,
- *alias Urforth* Eliz ChancP
Urford 1536 LP xi, 1620-32 WillsA, 1634 *Terrier*
Orford 1662, 1709, 1715 *Terrier*, 1824 O *et passim*, - *House*
1740 *Cragg*, 1828 Bry

According to Ekwall, DEPN s.n., this is either 'the ford of
the Irish', from OE **Iraford*, or 'the ford of Yra', from an
unrecorded OE pers.n. **Yra* and **ford**. For the latter, he compares
Irchester, DEPN s.n., (PN Nth 192) and suggests there that the first
el. is an OE pers.n. *Ira* or **Yra*, noting that the former is recorded
as the name of a moneyer. (The editors of PNNth take the first el.
to be the pers.n. *Ira*). It is likely, however, that this is in fact an
ON pers.n. since the forms *Iire*, *Ire* and *Irra* occur on coins struck
at York during the reigns of Cnut and Æthelstan II on which, *v.*
Veronica Smart, *Sylloge of the Coins of the British Isles* 28,
Cumulative Index of Volumes 1-20, British Academy 1981, p. 48b.
Ekwall's alternative suggestion that the name means 'the ford of the
Irish' from OE *Iras* (gen.pl. *Ira*) and **ford**, seems doubtful too, for
Iras seems to be a late development in OE - BT points out that
"the people of Ireland are often spoken of as Scottas in Old English
sources". Further, such a name compounded with **ford** has no
parallel noted so far. Now, ON **Íri** (gen.pl. **Íra**) occurs not
infrequently compounded with **bý** and as a hybrid with **tūn**, but a
hybrid with **ford** seems improbable, unless **Íri** has replaced some
earlier first el., though it should be noted that Smith, EPN 1 304,
accepts **Íri** as the first el. of Irford (sic). Dr John Insley comments
that Ekwall's OE pers.n. **Yra* "is a step in the right direction, but
the etymology of this pers.n. is not discussed". He goes on "We
would rather seem to be concerned with an OE pers.n. **Ȳra*, which
is to be connected with the rune-name *ȳr*. R.I. Page, *An Intro-
duction to English Runes*, London 1973, p. 85, favours 'bow' as the
meaning of the rune-name and thinks that *ȳr* originated as an
i-mutated variant of the rune-name *ūr* 'aurochs' (*ib* 84; for this
rune-name, *v.* Page *ib* 73-74). The etymologically identical pers.n.
el. *Ūr* occurs in OHG (Alamannic *Ūro*, *Ūr(w)ald*, *Ūrolf*. Dr Insley
concludes "Orford is therefore best interpreted as 'Ȳra's ford'" and

this is no doubt the most plausible interpretation of the name.
The comparatively late (17th century) development to Orford
does not seem to have any parallel. It should be noted that forms
with prosthetic *H-* are common enough in p.ns. and f.ns. in north L,
especially in names with **eng** as first el.
Orford is once described as near Binbrook.

ASH HOLT. BLACK HOLT. CHERRY HOLT. COWDYKE
PLANTATION, cf. *Cow dike Bottom(s)* 1852 *Plan,* self-explanatory,
v. **cū, dík, botm.** DALES BOTTOM PLANTATION. GOODY
ORCHIN PLANTATION, cf. *Goody Hotchen* 1828 Bry, *-orchin* 1852
Plan, obscure. HALFMOON PLANTATION, named from its
shape. LUD'S WELL, presumably to be identified with *riuulum de
ludewell* 112 *MiD,* though no other forms have been noted. It is
probably 'the loud spring', *v.* **hlūd, wella,** though an OE pers.n.
Luda is formally possible as first el. MANOR HO. MILL (lost,
TF 165 947), ... *q'd' molendino* 1297 *MiD, quod'· molendino aquatico
in villa de Staynton'* 1300 *ib, the water milne* 1612, *the Mill* 1625,
1634, 1638 all *Terrier, Mill* 1828 Bry, cf. ... *stagni molendini* 112 *MiD,
the mile banke* 1601, *the milne banck(e)* 1606, 1612, *milne Bank*
1700, *the milln yard* 1638 all *Terrier, yᵉ mill dick banke,* - *Milne
Miers* 1693 *MiD, Mill Mires* 1695 *ib, the mill mire* 1709, *mill dam*
1707, *the Mill Acres* 1709, *the millstead* 1712 all *Terrier,*
self-explanatory and *v.* **banke, geard, dík, mýrr, damme, (dammr),**
æcer and **stede.** NELSON'S SHIP PLANTATION, cf. *Nelsons Ship*
1854 *Plan,* NELSON'S WOOD, *Nelsons Wood* 1824 O, 1830 Gre,
1854 Plan, *presumably* commemorating Admiral Lord *Nelson,* though
it should be noted that *Nelson* is a local surn. cf. Sarah *Nelson* 1814
BT. NIMBLETON PLANTATION, cf. *Nimbleton* 1854 *Plan.*
NURSERY RIDE PLANTATION, cf. *Nursery Bottom* 1854 *ib.*
ORFORD BRIDGE. ORFORD PLANTATION, 1828 Bry, - *Plant.ⁿ*
1830 Gre. PRIORY FM, *the* - 1854 *Plan,* and it is *Abbey F.ᵐ*
1828 Bry, commemorating the former Priory here. RASEN
PLANTATION, 1824 O, cf. *Raisen Hutt* 1606 *Terrier,* presumably
from a local family called *Ra(i)sen.* SMITHFIELD PLANTATION,
cf. *the Smithfeild, Smith feild walke* 1652 *Rad, Smithfield Lane* 1828
Bry, *Smith field Lane* 1854 *Plan;* perhaps named from the *Smith*
family, well-recorded in the parish, cf. John *Smith* 1596 *Inv.*
SOUTH FM, 1842 White, 1854 *Plan,* self-explanatory. SPRING
RIDE PLANTATION, cf. *Spring Bottom* 1854 *ib.* STAINTON

HALL is *Stainton Vale Ho.* 1824 O, 1830 Gre and is *Hall* 1828 Bry, cf. also *Stanton West hall* 1618 *Deeds*, *West Hall*, 1692, 1694 *MiD*. STAINTON PLANTATION, 1828 Bry. THORGANBY CORNER PLANTATION, from the adjacent parish of Thorganby. TRUSOLES PLANTATION, presumably from a surn. TUNNEL PLANTATION, *Tunnel* is perhaps a surn. here.

Field-Names

Undated forms in (a) are 1854 *Plan*. Spellings dated 112, e13, 1297, 1300, 1689, 1690, 1692, 1693, 1694, and 1695 are *MiD*; 1155-60 are Dane; R1 (1318) are Ch; 1272-73 are Misc; 1288 are *Ass*; 1327 and 1332 are *SR*; 1343 are NI; 1392 are Works; 1414 are Fine; 1536 are LP xi, 1539 are ib xiv. The remainder are *Terrier*.

(a) Barley Cl; Bottomside Cl; Brick yard Cl; Bulley hill; Caistor Lane (self-explanatory); Chapel hill Cl; Church Fd; (New -, Old) Cinque foil (probably for *Sainfoin*, v. **sainfoin**); Cogdale hill (*Cockdayle Hill* 1634); Cow Cl (2x); Croxby Bottom (from the neighbouring parish of Croxby and **botm**, and cf. Croxby Top *supra*); East Fd (*the East feilde* 1612, y^e *East feild of Orforth* 1662, - *field of Irford* 1679); Fish Pond; Flint Hill (v. **flint**. **hyll**, cf. *the buske of flints* in (b) *infra*); New -, Old fold Cl; Forty acres; Free hold hill; Goar Cl (v. **gāra**); Gooks Cl; Hairy Cl (sic); Hill Plat (*de la Hul* 1272-3, *Uponyehull'* 1288, *super Montem* 1327, *del Hull'* 1332, *del Hill* 1343, *atte Hill* "of" *Northstaynton* 1414 all (p), y^e *Hill* 1692, *close ... called the Hill* 1694, v. **hyll**, **plat**[2] 'a plot of ground', as elsewhere in this parish); Homes cl (y^e *Holmes* 1693, *the Holmes* 1695, cf. *holmesike* e13, v. **holmr** 'higher land in marshes', as elsewhere in this parish, and **sik**); Honey Hills, - holes (probably 'land with sticky soil', v. **hunig**); House Cl; First -, Second Intake (v. **inntak**); Jusby hill; Kermond Plat (from the neighbouring parish of Kirmond le Mire and **plat**[2]); Legsby Plat (v. **plat**[2]); Lime Kiln Plat (v. **plat**); Long Mdw; Marle Plat (v. **marle**, **plat**[2]); The Moors (*the comon more* 1606, y^e *more* 1612, *ye Moare* 1689, *the mores* 1692, y^e *Moore(s)*, *Stainton moore* 1693, *the More*, - *Moore* 1695, *the Moor* 1707, y^e *southe more* 1601, *the south more* 1606, - *Southe More* 1625, - *South moor* 1638, - *Moor* 1700, *the south moor* 1712, *the Neither More close* 1634, *the nether more* - 1638, *the upp More* - 1634, - *Close* 1638, *Moor Closes* 1689, v. **mōr**[1]); New Cl; Nine Acres; Orford hill; Pasture Cl; Pleck (several) (from ME *plek(ke)*, v. **plek** 'a small plot of land'); Red Hill(s); Rye Plat (v. **ryge**, **plat**[2]); Old Seed walk (v. **walk**); Seventy acres; Sheargrass hill, Skear Grass (sic) (perhaps 'the boundary grassland', alluding to a turf boundary, *v.* **sccaru, gærs,** though it is not on a known boundary); Bottom -, Top Snape hill (*Snapehill* 1689, 1690, *Snaphill* 1693, *v.* **snæp** 'a boggy piece of land', **hyll**); Swinthorp hill; Thorney Plat (*v.* **þornig, plat²**), Thorns Cl; Trencher Pce (possibly named for its shape, though this is in fact very irregular); (Far -, Middle) Wang Wong (*v.* **vangr** 'a garden, an infield'; *Wang* may be for *Wrang, cf. Wrangland* 112, *v.* **vrangr** 'crooked'); Wanty Cl; Warren house Plat (*v.* **plat²**); Water Mdw; Far -, Middle West Hills; Willow Garth (*v.* **garðr**); Wood (cl, - Fd) (*yᵉ woodd close* 1601, *the woode, the wood close* 1606, *the Wood close, the wood laine* 1612, *the Woodd* 1625, *Wood close* 1692, *the wood* 1709, *the wood end* 1638, *one pennill at -* 1707, *one pinnell by the wood* 1709, *- by the wood side side* 1712, *v.* **wudu**; *pennill* and *pinnell* are no doubt from **pingel** 'a small enclosure').

(b) *Adellof gard* (sic) e13 (the text also refers to *vno tofto in Staintona quod Adellulful* [sic, for *Adellulfus*] *Hwitel tenuit*, it is probably 'Adelulf's enclosure', *v.* **garðr**, the first el. being the Continental Germanic (Frankish) pers.n. *Adelulf, -olf,* which is known from Flemish sources. The byname *Hwitel* belongs to OE *hwitel* 'a blanket, a cloak'); *arniw well* (sic) e13 (the division of the two words is at the end of a line; it is perhaps 'Arni's spring', *v.* **wella,** the first el. being then the ODan pers.n. *Arni*); *the common becke* 1601, *- comon becke* 1606, *yᵉ common beck* 1612, *the becke* 1625 (*v.* **bekkr**); *Atte Bynne* 1327, *atte Binne* 1332 both (p) (this is apparently from OE **binn** 'a manger, stall', ME *binne* 'a stall or stable', *v.* MED s.v.; presumably the sense is 'a stable'); *Blaikesdale* e13, *Blacksdalebottom* 1694 (probably from the Scand. pers.n. *Bleikr* and **dalr** 'a valley' and **botm,** the 17th cent. forms showing Anglicization of the first el.); *ad Pontem* 1327, *atte Brigg* 1343 both (p), *Bridge mires* 1692 (*v.* **brycg, mýrr**); *the buske of flints* 1601, *the bushe of flyntes* 1606, *the Buske of Flintes* 1612, *the Bushe of Flints* ... (tear in MS) 1625, *the Busk of flints* 1634, *Buskey Flints* 1692, *the Busky Flints* 1700, 1707, *one acre and halfe in the Flints called Busky* 1712 (alluding to a piece or pieces of scrubland in an area apparently known as *The Flints,* cf. Flint Hill in (a) *supra, v.* **flint, buskr**); *Cherrywell dales* 1692 (from **chiri, wella** with **deill** 'a share, a portion of land'); *yᵉ church hyll* 1601, *the Church Hill* 1606, 1634, 1709, *- Hyll* 1625, *- hill* 1638, 1700, *the Churche-hill* 1612, *Church-hill* 1718 (*v.* **cirice, hyll**); *Colecroft* R1 (1318) (probably an error for *Tolecroft, v. Tolecrofth infra*); *yᵉ Cowhills* 1693, *Cowhills* 1695; *Crewgate* 1692 (probably from dial. *crew(e),* which EDD defines as 'a small yard or enclosure, a pen, a fold for cattle or sheep' and **gata** 'a road'); *Croxbie gate* 1638, *Crosby - * 1692, *Crosby Gates* 1712 ('the road to Croxby', *v.* **gata**); *Anthony Dewick close* 1625; *dockedeil* 112 (*v.* **docce, deill**); *Duffe Coate Close* 1692 (*v.* **dūfe, cot**); *Fissehorne* e13 (the second el. is apparently **horn** 'a horn' in a transferred topographical sense, such as 'a horn-shaped piece of land', 'a horn-shaped hill'; the first el. is obscure); *Furrehill* 1536, 1539 (*v.* **hyll**; the first el. may be **fyrs** 'furze', but in N Lincs *furre* is often from ME *furre* (OE **feor** 'far')); *gatewelledale* e13 (*v.* **gata, wella** with **deill** or **dalr**); *Gosmore* 1693 (*v.* **gōs** 'a goose', **mōr**[1]); *M*[r] *gylbys Woodd* 1601; *hericholm* 112 (*v.* **holmr**, the first el. being the OE pers.n. *Hereric*); *Horse hills* 1692, *y*[e] *Horsehills* 1693; *in The Hyrne* 1327, *in the hirn'* 1332 both (p) (*v.* **hyrne** 'a corner, an angle of land, etc.'); *Kele* 1332 (p), *del Kel* (sic) 1343 (p) (perhaps this is from **kjolr** 'a keel, a ridge', if it is a local surn.); *Lailand* 112 ('uncultivated land, an untilled piece of land, open ground', from ME *leiland*, cf. MED s.v. *lei*(*e*)); *Langedale* e13, *Longdales* 1692 (probably 'the long valley', *v.* **lang**[1], **dalr**); *Langhove* e13 (*v.* **lang**[1], **haugr**); *Langraue hille* e13 (perhaps from **lang**[1] and **grāf** 'a grove, a copse', with **hyll**); *laxedale* e13 (probably 'Leaxa's valley', *v.* **dalr** and for the first el. cf. Laxton PN YE 254, PN Nt 54 and PN Nth 168); (*de*) *Litegate* 1272-73 (p) (*v.* **lītill**, **gata**); *Litelhov*, *litalhou* e13 (*v.* **lītill**, **haugr** 'a (burial) mound'); *Lone-Wells* 1692; *y*[e] *Lowffeild* 1692, *the Lowfeild* 1694; *y*[e] *meadows* 1693; *Meddleys Farme* 1694 (presumably *Meddley* is a surn.); *midle furlonge* 1638, *y*[e] *middle furlong* 1662; *the mires* 1712 (the same as *Mill Mires* under *Mill supra*); *the North Feild* 1625, - *feild* 1634, *the North feild*, - *north field* 1638, *Northfeild* 1689, *Northffeild*, *the north ffeild* 1692, *y*[e] *North feild*, *Stainton Northfeild* 1693, *north feild* 1694, *the North Field* 1695, - *field* 1700, *the north field* 1707 (one of the open fields of the parish, *v.* **norð**, **feld**, cf. *the South feild infra*); *novtstigel*, *novtstigeldale* acre e13 (from **naut** 'cattle', **stigel** 'a stile' perhaps in the sense 'a steep ascent', with **dalr** or **deill** and **æcer**); *ovstdaleclif* e13 ('the eastern valley', *v.* **austr**, **dalr** (a Scand. compound) with **clif**, cf. *a valley called the West Dales infra*); *y*[e] *parsonage ground* 1601, *the homestall of scite* [i.e. site] *of the parsonadge* (sic) 1606, *the parsonage land*, - *pingle* 1612, *one Pinnell* 1638, *one stong called the Pinnells in close* 1700 (the last four references relate to the same feature, *v.* **pingel** 'a small enclosure'); *the parsonage howse* 1606, - *house* 1638, *a Parsonage House* 1634, *There is no remains of a Parsonage House, only a little yard wall'd about* 1718; *the parsons breeke* (sic) 1638 (*v.* **brēc** or **bræc**[1]); *pindaleclif* e13 (*v.* **clif**); *Rauendale*, *Rauendalestige* e13 (from **hræfn**, **hrafn** 'a raven' or the derived pers.n. and **dalr**, and so comparable with East and West Ravendale LNR, over five miles north-east, with **stigr** 'a path'); *Reylie* 1692 (possibly 'rye meadow', *v.* **ryge**, **lēah**); *Roundhill* 1694, *Round Hill* 1689; *Scrubb Furrs* 1692, - *furrs* 1694 (*v.* **scrubb**, **fyrs**, but note that *furrs* is freq. the pl. of **furh** 'a furrow' in north L); *Mr Sheffeild Closse* 1606; *the south Feild* 1625, - *feild* 1638, - *Field* 1700, *the South feild* 1634, - *Feild* 1689, 1692, - *Field* 1709, *Southfeild* 1690, 1693, 1695 (*v.* **sūð**, **feld**; one of the common fields of the parish, cf. *the North feild supra*); *South Hills* 1707; *the South Meare* 1638 (*v.* **sūð**, (**ge**)**mǣre** 'a boundary, land on a boundary'); *Attestanes* 1297, *Atte Stanes* 1327, *atte* - 1332, 1343 all (p) (*v.* **stān**); *ex p'te occidentali vie Lincolnie* e13 ('the way to Lincoln'); *the street furlong* 1709 (*v.* **strǣt**; the reference is no doubt to High Street, as in Tealby *infra*); *the Swallow Holes* 1709, *Swallow-Hole* 1718 (a *swallow-hole* is a cavity through which a stream disappears underground); *Thorpedale* e13 (*v.* **dalr** 'a valley'; it is odd that no other reference has been noted to *Thorpe* 'a secondary or outlying settlement'); *Toft dike* (*banke*) 1693 (*v.* **toft**, **dīk**); *Tolecrofth* 1155-60 (probably 'Tóli's croft', from the ON pers.n. *Tóli* and **croft**; this form is found in an original charter and *Colecroft supra*, from a 14th cent. copy, is no doubt an error for *Tolecroft*); *Trussedale, trussedale West graine* e13 (the forms in *trusse-* are comparable to the early spellings of Trusley, PN Db 613, where it is suggested that they are from **trūs** 'brushwood', but it is pointed out that the almost universal -*ss*- is puzzling. Perhaps the first el. is rather OFr *trousse* 'a bundle', here probably used as a byname; the second el. is no doubt **dalr** 'a valley'. *West graine* is certainly 'the west branch or fork', *v.* **west**, **grein** and cf. dial. *grain* 'a small valley forking off from another'); *Turnesdale Close* 1692; *the 20 acres* 1692; *Wellgraine* e13 (*v.* **wella**, **grein**, though the reading is uncertain); *Weeste dayles* 1601, *West dailes* 1606, *the West Dailes* 1612, - *Dayles* 1625, *West daile* 1612, *a valley called the West Dales* 1634, *the west dayle* 1638, *the west Dale side* 1700, *the west Dale* 1709, *the west dales* 1712, *West Dales* 1718 (probably 'the western valley(s)', *v.* **west**, **dalr**, though forms in *dayle*(*s*) suggest **deill**, cf. *ovstdaleclif supra*); *Wells Close* 1692; *Were'dale* 112 (obscure); *the west cloase* 1638, *y*^e^ *West Close* 1707 (in Orford); *the West feilde* 1612, *Irford west field* 1638, *the west field of Orforth* 1662, *y*^e^ *west field* (Orford) 1679, *the west field* 1712 (*v.* **west**, **feld**; the references dated 1612 and 1712 almost certainly refer to Stainton); *Westffurs* 1689, *the West Furze* 1692, *ye west fure* 1694 (for the second el. cf. *Scrubb Furrs supra*); *y*^e^ *west hall yards and closes* 1693; *les West Milles* 1392 (*v.* **west**, **myln**); *Wrangland* 112 (*v.* **vrangr** 'crooked', **land**).

Tealby

TEALBY

Tavelesbi (3x) 1086 DB, *Tauelesbi* (4x) 1086 ib, 1136-40
(1464) Pat, c.1150 (1409), c.1160 (1409) Gilb, 1166-79
YCh vi, Hy2 (1409) Gilb, 1195 P (p), 1196 ChancR (p),
lHy2 RA ii, 1203 Cur, 1204 P, 1210-12 RBE, 1218 Ass,
- *hundred* 1086 DB, *Tauelesbia* 1100-8 (1356) YCh vi,
1212 Fees, *Tawelesbi* 1202 Ass, *Tauelsby* Hy2 (1409)
Gilb, *Taulesbi* 1206 Cur

Tablesbeia 1094 France

Tablesberiis (sic) 1090-1100 (1402) YCh vi, 1100-8 (17) ib vi,
1147-53 ib vi

Tauellesbury lHy2 (1409) Gilb

Taflesb' 1185 Templar

Tauelebi 1086 DB, 1209 P, -*by* 1202 Ass, *Tauelby* 1208 Cur,
1210 FF, 1220 Cur, *Tavelby* 1589-90 Lanc, 1707 *Terrier*

Tafleby c.1190 RA iv

Teflesbi (2x) c.1115 LS, -*by* a1168 Semp (checked from MS),
1187 (1409) Gilb

Teuelsby c.1150 (1409) Gilb, *Teuelesby* (3x) Hy2 (1409) ib,
a1183 RA iv, 1194, 1195 P, 1219 FF, c.1221 Welles,
-*b'* 1235 IB, *Tevelesby* 1254 ValNor

Tevellesby 1206 OblR (p)

Tiuelesby 1201 Cur, *Tyuelesby* 1226 FF

Tefleby 1187 (1409) Gilb

Teuelebi 1219 Ass, -*by* 1220 Cur, *Teuleby* 1250 FF, *Teuelby*
eHy3, 1228-32 (1409) Gilb, 1242-32 Fees, 1257 FF,
1276 RH, 1284 Fine, 1298 Ass, 1313 Ch, 1316 FA *et
freq* to 1481 AD, -*be* 1507 Ipm, -*bye* 1610 Speed,
Teuylby 1310 *Extent*, 1381 Peace, *Tevelby* 1291 Tax,
1299, 1310 Pat, 1311 Orig, 1322 Pat *et passim* to 1723
SDL, - *als Teylby* 1618 *Td'E,* -*bie* 1554 InstBen, 1576
LER, 1609, 1631 VisitN, -*bye* 1557 Pat, 1675 Ogilby,
Tevilby 1296 AD, 1327 Pat, 1534, 1544, 1646, 1766,
1811 *Td'E,* - *alias Tealby* 1629 *ib,* 1702 *Foster,* -
otherwise Tealby 1757, 1820 *Td'E,* -*bye* 1549 Pat,
1538-39 Dugd vi, 1552 Pat, -*bie* 1552 *Td'E,* -*be* 1548
ib, Tevylby 1529 Wills ii, 1546 LP xxi, 1552 Pat, -*bye*
1539 LP xiv, 1547 Pat, *Tewelby* 1428 FA, *Tewylby*
1436 Fine, *Tewilbie* 1552 *Td'E*

Teilebi 1210 P, *Teilby* 1553 *Td'E,* 1556 *Mad, Teylby* 1504
LouthCA, 1529 Wills ii, 1537-38 Dugd vi, *Teylebie*
1551 Pat, *Taylebye* 1551 ib

Teleby 1252 FF, 1385 Peace, 1393 Works, 1396 Peace, 1462,
1465 Pat, 1535 VE iv, Eliz ChancP ii, *Telby* 1384
Peace, 1404 Pap, 1507 Ipm, 1529 Wills ii

Teelby 1375, 1384 Peace, - "alias" *Thelby* 1395 Pat, 1428
FA, 1526 Sub, *Tealby* 1526, 1624, 1649, 1660 *Td'E et
passim,* - *alias Teavelby* 1667, 1716 *ib,* - *als Tevilby*

1681 *Td'E,* 1723 *NW,* - *otherwise Teavelby* 1742, 1763,
1765 *ib,* 1773 *NW,* - *otherwise Tevelby* 1770 *NW,* 1783
Td'E, -otherwise Tevilby 1787 *NW,* - *alias Teavilby* 1806
Td'E, Tealeby 1705 *ib*
Tylby 1533 *Td'E*
Tyvelbye 1561 Pat
Teavelby 1622 *Td'E,* 1625 Heneage, 1634 *Td'E,* 1638 VisitN,
1652, 1659 *Td'E,* - *als Tevylby* 1626 *ib,* - *als Tealeby*
1666 *ib, -bie* 1623 *ib*
Thaueleby 1219 Cur
Thevelbe 1252 Ch. *-by* 1281 QW, 1285 RSu, 1291 Tax, 1295
Ipm, 1296 RSu, 1305 Pat, 1312, 1324 Cl, *Theuelby* 1252
(1389) Pat, 1261 RRGr, 1336 Pat, *Thewelby* e13 (1409)
Gilb *Theulby* 1303 FA
Theleby 1240 FF

This is a difficult name. Ekwall, DEPN s.n., compares
Tealby with Tellisford So, Thelsford PN Wa 250, and Tablehurst PN
Sx 329, though the spellings for Thelsford suggest that it has a
different etymology. A full list of forms for Tellisford is not yet
available, while those for Tablehurst do not parallel the variants in
the collection above. It is probably safest to take Tealby on its
own.

Ekwall suggests that the first el. seems to be an OE **tæfli,*
*tefli, derived from OE *tæfl* 'a chess-board', probably in the sense 'a
plateau', and compares Dutch Tafelberg and German *Zabelstein.*
Smith, EPN 2 174-75 s.v. **tæfl(e),** notes Ekwall's comments and adds
that "it may be recalled that ME *tavele, tevele* 'to contend with dice,
etc.' was also used in a more general sense 'to argue, strive' and
some such application in p.ns. to 'land in dispute' (cf. þrēap), as
proposed in Sx 329, is not out of the question". Interestingly, he
links Tablehurst and Tellisford and "possibly" Thelsford to the el.
tæfl(e), tefle, but makes no mention of Tealby here. As an alternative meaning for the el., he suggests 'the flat stones forming the
track of a ford', but goes on to say that it is in any case difficult to
distinguish it from the "OE pers.n. *Theabul*". This, as Dr John
Insley points out, is a misconception, because the pers.n. form
Theabul, which is contained in the witness list of an original charter
of 697 (S 19), stands for OE *Þēoful,* a pers.n. which is also attested
in *The Earliest Life of Gregory the Great* (ed. B. Colgrave, Kansas

1968, 102) in the form *Teoful.* OE *Þeoful -ol* is also the best explanation for the first el. of Thelsford, PN Wa 250, 319.

Fellows-Jensen, SSNEM 74, following Ekwall, notes that the "related ODan *tafl* n. is apparently used of a square-shaped piece of land in Danish p.ns. ... and it may be the Danish word that was originally found in T." She adds that forms in *Te-* would then show the influence of an OE **tefli.*

The collection of forms here certainly suggests that *Taveles-, Taueles-,* with medial *-s-,* represent the earliest sequence from DB to the early 13th century, though these are represented only sporadically later. On the other hand, *Tefles-, Teueles-* appear first in c.1115 and these and subsequent developments are found till the mid 13th century. Forms in *Th-* can be ignored from an etymological point of view, occurring from the early 13th century to 1336, perhaps due to AN influence, *v.* Feilitzen 93-94.

It should be noted that forms with medial *-s-* predominate from DB to the mid 13th century, after which they cease to appear. An occasional spelling without medial *-s-* occurs in DB and in 1187 and c.1190, then frequently in the early 13th century, and after 1254 they alone are represented in this collection.

The forms in *Tables-,* however, are not easily explained, but the second el. in *-beriis,* which occurs three times is probably not significant, for in each of the documents in which they occur, similar spellings are found for the second els. of both Roxby and Scawby LWR. It may be significant that one of the three documents containing this form is of French origin, as is also that with the spelling *Tablesbeia.* The unique *Tauellesbury* is hardly sufficient evidence to suggest that OE **burh** 'a fortification' has been replaced by ON **by** 'a farmstead, village'.

Dr John Insley comments "Ekwall's OE **tæfli, *tefli* is an unlikely formation. OE *tæfl* is a feminine noun in the ō-declension, being a loan of Latin *tabula.* It merely denotes a gaming-board, not specifically a 'chess-board', as stated by Ekwall. Its genitive singular ending is *-e,* hence Ekwall has to invent the variant **tæfli, *tefli* in order to explain what he presumably assumed to be genitival *-es* in the early forms of Tealby. Final *-l* of OE *tæfl* is syllabic, and *i*-mutated side-form *tefl* occurs in early glosses (e.g. Erfurt: *tefil*; Corpus, Leiden: *tebl*). ODan *tafl* 'a gaming board, board game' occurs in Danish p.ns. with the sense 'square or rectangular piece of land', though Lindroth, *Ortnamn på -rum* 79f.

presumes that *Tavl* in the Danish p.ns. Tavlgaarde and Tavlov has the sense 'piece of raised ground, eminence', and is to be compared with Halland dialect *tavel* 'small bank' (*v. Danmarks Stednavne* iii.lx; viii.127-28; ix.174)".

"None of these Danish p.ns. is a genitival compound. If Tealby contained OE *tæfl, tefl,* we would expect it to be morphologically similar to Danish names of the type Tavlov, from ODan *Taflhøgh, v. Danmarks Stednavne* viii.127-28, and be uninflected. The medial *-s-* in the early forms of Tealby is an obstacle to parallels of this kind. If we interpret this medial *-s-* as having genitival significance, we come up against semantic questions, and a pers.n. *Tæfl,* *Tefl* lacks parallels in other Germanic languages and is semantically and etymologically implausible. It would seem therefore, necessary to re-think the etymology of Tealby and to get rid of the preconceptions engendered by Ekwall's discussion."

Dr Insley continues "In Domesday Book, Tealby is a settlement of some consequence and has all the appearance of an ancient village in terms of its geld assessment and its importance as a jurisdictional and administrative centre. [It should be noted that it is referred to as *Tauelesbi hundred* in DB f.276a]. I would suggest that Tealby contains the name of the East Germanic *Taifali,* OE *Taflas/*Tæflas.* Detachments of *Taifali* are recorded in Britain by the early 5th century *Notitia Dignitatum* (ed. O. Seeck, reprinted 1962, 130), and it is *possible* that they retained their separate identity for some time in the post-Roman period. This was certainly the case in Gaul, where Gregory of Tours mentions an insurrection of the *Taifali* in the vicinity of Champtoceaux (Maine-et-Loire) c.561, and where their name is contained in the p.ns. Tiffauges (Vendée) and Tivauges (Semur) (*v.* E. Schwarz, *Germanische Stammeskunde,* Heidelberg 1956, 104). Tealby would be a parallel formation to French names of this type, being originally merely the simplex of the tribal name, OE *Taflas/*Tæflas.* The second el. ODan **by** would have been added to the simplex form after the Scandinavian occupation of the area. Simplex tribal names as p.ns. are rare, but an exact parallel is provided by Wales from OE *W(e)alas* 'Welshmen', PN YW 1 155-56. The alternation of mutated and unmutated forms *Taflas/*Tæflas* reflects the fact that original uncertainty as to whether we are concerned with *-il* or *-ul* caused the emergence of parallel forms with and without *i*-mutation, cf. OE *tæfl/tefl.*" He goes on to point out that "in OHG a parallel can be seen in the forms taken by the pers.n. el. *Wandil-*, from the tribal name of the Vandals, cf. such forms as *Uuentil(a), Uuandalgarius, Uuentilger, Uuantalmar, Uuentilmar* in the confraternity book of the Swabian abbey of Reichenau (p. 167)."

Clearly, Dr Insley's interpretation of the meaning of Tealby is easily the best so far offered.

BAYONS MANOR, 1813 *Td'E*, 1824 O, *Bayon's Manor House* 1842 *Td'E*. P&H 185 comment "The house is the expression of the lineal aspirations of its builder *Charles Tennyson* (d'Eyncourt), uncle of the Poet Laureate. Tennyson, the politician, could never acquire a peerage, so he invented a quasi-Gothic lineage for himself, discovered a dim d'Eyncourt connexion, and surrounded himself with a panoply of baronial objects. Before 1835 Bayons was a plain bay-fronted Regency 'cottage'". They add "Bayons is now in total decay and never looked better". It was, in fact, demolished in 1964. The name, however, goes back at least to the early 16th century and there are many references to the *manor* - "the manor of" *Bayons* 1514 LP i, - *Baions* 1520 ib iii, *manerium ac tenement' de Bayons et Tevelby als' dict' Bacons et Tevelby* 1546, *Mannor of Bayons* 1570 (17), *Bayons & Tevelbie mannor* 1583, *Manerium de Tevilbie vocat Bayons et Tevilbie Mannor* 1602, *Maner' de Tevilby vocat Bayons et Tevilbye* 1610 *et freq* to *Manor of Beacons otherwise Bayons in Tealby* 1792 all *Td'E* and note also *terre pertinent' Bayshalls* 1533, *lands belonging to Bayshall* 1548, *Beacons als Bayons hall* 1669, *the Mannor of Bayes Hall* 1712, *the manor or Lordship ... of Beacons otherwise Bayons, Beacons otherwise Bayons Hall* 1783 all *Td'E*. The *de Baiocis* family held land in Tealby from at least 1202 Ass and cf. also Richard *le Bayeus* 1284 Fine, Juliana *Bayous* 1327 SR, John son of Robert *Bayus* 1346 FA, William *Bayus* 1428 FA, while the Bishop of Bayeux held Tealby in DB.

ASH HOLT, *Wood or Ash Holt* 1842 *Td'E*. BETTS PLANTA-TION, named from a local family, cf. John *Betts* 1669, Thomas *Betts* 1779 both *ib*. BARNSFIELD PLANTATION. BAYONS PARK, *Park, Deer Park* 1842 *ib*. BEDLAM PLANTATION, presumably a derogatory nickname, *v*. Field 17 s.n. Bedlam. BROGGERY PLANTATION, cf. *Furlong called Brogery* 1793 *Td'E*. BULLY HILL, 1637 *ib*, 1824 O, *Bulley hill* 1854 *Plan*, cf. *Bulloe furlong* 1631, *Bull hill* 1793 both *Td'E*. CASTLE FM, 1842 White, *The*

Castle F. 1824 O, so-named from its appearance. CHAPEL HILL, 1795 *EnclA*, cf. *the Chapel Homestead and Chapel Hill Close(s)* 1804 *NW*, and *Messuage Comonly called or knowen by the name of St Thomas Chappell* 1630, *that Mesuage called or knowen by the name of St Thomas Chappell* 1637 both *Td'E*, *St. Thomas Chappell* 1661 *NW*, *All that peece and parcell of Ground ... upon which formerly stood a Messuage Tenement or Farme House ... comonly called or known by the name of St Thomas Chappell* 1706, *Chappel House* 1787 both *Td'E*. CHURCH LANE, 1797 *EnclA*, cf. *ye Church close* 1707 *Terrier*. CHURN WATER HEADS. COWLANE, 1807 *Td'E*, cf. *Cowlane Close* 1726 *ib*. DAIRY FM. DOVECOTE HO, cf. *dovecote close* 1627, (*The*) *Dovecote Close* 1637, 1671, *the Dovecote close* 1653, 1663, - *Dovecoate Close* 1705, 1742 all *Td'E*, *Dove Court Close* 1773 *NW*, *Dove coat Close* 1795 *EnclA*, and *Temple Garth ... together with the dovecote* 1663 *Td'E*, self-explanatory. For *Temple Garth*, *v*. Temple Garth Farm in f.ns. (a) *infra*. GIBBONDALE PLANTATION, cf. *Gibbon Dale* 1793, - *Blades* 1630 both *Td'E*, named from a family well-evidenced in the parish, cf. Anthony *Gybbon* 1592 *Inv*, Francis *Gibbon* 1660 *Td'E*, and probably **dalr** 'a valley'. *Blades* appears to be from OE **blæd** 'a blade (of grass)', presumably in some transferred topographical sense. THE GROVE, *Groves* 1842 White. HAMILTON HILL, *Hammalton* -, *Hamelton* -, *Hambleton Hill* 1536 LP xi, *Hamyldon How* 1537 iv xii (*v*. **haugr**), *Hambleton Hill* 1603 *MC*, *Hamlinton hill* 1675 Ogilby, *Hambleton Hill* 1795 *EnclA*, 1804 *NW*, from **hamol** 'crooked, scarred, mutilated' and **dūn** 'a hill'. The sense is probably 'the scarred hill', a hill which appears to have been mutilated and which is topographically appropriate. This is a common name found with various modern spellings in at least ten counties. HAWKHILL PLANTATION. HIGH STREET, *apud Stratam intra villam de Teelby et Staynton* 1375 Peace, - *inter villas de Theuelby et Staynton'* 1376 ib, cf. (*the*) *Street Furlong* 1671, 1637, 1743, 1768, 1793, *street furlong* 1631. *Furlong called Street and Dale* all *Td'E*, the name of the ancient trackway from Horncastle to Caistor, found in many parishes in this Wapentake. It is called Middlegate Lane from Caistor to South Ferriby. KILLICK'S PLANTATION, named from the family of Thomas *Killick* 1803 *Td'E*. KING'S HEAD INN, *King's Head* 1842 White. LARCH PLANTATION. LOW MOOR FM. MANOR HO, *the* - 1795 *EnclA*. NORTH MOOR FM, cf. *the North Moor* 1787 *NW*, 1795 *EnclA*, *Tealby North Moor* 1804

Td'E, cf. Tealby Moor *infra*. NORTH WOLD FM. PAPERMILL
LANE, cf. *One Paper Mill* 1746, 1748 *Td'E*, *that New erected
Mesuage or Tenement and Paper Mill thereto adjoining called the
Paper Mill or Pad Mill* 1752 *ib*, *Paper Mill* 1749, 1763 *ib et freq*, *a
Mill lately used as a Paper Mill* 1845 *ib*; for the alternative name,
which is recorded earlier, cf. *Pad Mill hill* 1617, *pad mill hill* 1631,
Furlong called Padmill-hill 1768, *Pad Mill Hill* 1842 all *Td'E* and
Pad(*e*) *becke* 1617, *Pade beck* 1631, *Padebeck* 1743, *Pade Beck* 1793
all *ib*, *Padebeck Close* 1795 *EnclA*; *Pad*(*e*) *becke* appears to mean
'the stream where toads are found' *v*. **padde, padda, bekkr** and if
this is to be associated with *Pad Mill*, perhaps the latter is a
shortened form of *Pad Beck Mill*, though the forms are too late for
any certainty. PASTURE HO, cf. *pasture hill* 1617, 1631, *Pasture
Hill*(*s*) 1793, 1843 all *Td'E*. PEBBLE COTTAGE. PICKARD'S
PLANTATION, no doubt named from the family of John Kirkby
Picard (sic) 1810 *Td'E*. SOUTH MOOR FM, cf. (*the*) *South Moor*
1787 *NW*, 1795 *EnclA*, *Tealby South Moor* 1804 *NW*. SWEET
HILLS, no doubt a complimentary nickname. TEALBY GRANGE,
grang' de Teulby 1381 Peace, "grange in Tealby" 1537-39 LDRH,
Tevilbye, firma grang' 1538-39 Dugd vi, *Tevelby Graunge* Eliz
ChancP, *Tealby grange* 1650 *Td'E*, *Grange Farm* 1792 *ib*; it was a
grange of Sichills Priory. TEALBY MOOR, 1803 *NW*, *the common
moore* 1631, - *Common Moor* 1743 both *Td'E*, - *More* 1697 *NW*, -
Moor 1724 *Td'E*, 1795 *EnclA*. TEALBY THORPE 1704 *NW*, 1824
O, *a place called Thorpe* 1697, 1804, 1815 *NW*, *Thorpe* 1792 *Td'E*,
1794 *NW*, 1795 *EnclA*, cf. *Thorp*(*e*) *Close* 1724 *NW*, 1757, 1793
Td'E, 1795 *EnclA*, 1804 *NW*, *Great, Little Thorping* 1707 *Terrier*, -
Thorpe Ings 1773, 1795 *NW* (*v*. **eng** 'meadow, pasture'), *Thorpe
Lane* 1792, 1811, 1816, 1824 *Td'E*; it is earlier *Westhorpe* 1665 *Td'E*,
cf. *Westropp feild* 1617 *ib*, 1686 *NW*, *West Thorpe feilde* 1627, - *Feild*
1637 both *Td'E*, 1697 *NW*, - *field* 1667 *Td'E*, 1704 *NW*, - *Field*
1768, 1792, 1794 *NW*, 1795 *EnclA*, 1820 *Td'E*, *Westhorp feild* 1631
ib, 'the outlying farmstead to the west, belonging to Tealby', *v*.
west, þorp. TEALBY VALE. TEALBY WATERMILL,
watermilne 1663, 1671, *Watermilne* 1663, -*mill* 1652 all *Td'E*, 1686
NW, 1734 *Td'E*, *Water mile* (sic) 1705 *ib*, *Water Mill now used as a
Paper Mill* 1768 *NW*, *Water Corn Mill* 1795 *EnclA* and is *Mill* 1824
O. Earlier references to a mill in Tealby include *molendinum de
Tauelesbi* 1136-40 (1464) Pat, *unum molendini quem habuimus super
aquam de teuelesby* Hy2 (1409) Gilb (i.e. on the R. Rase), *unum*

molendinum in villa de Tefleby 1187 (1409) ib, *Tevilbye, firma molend', Tevilbye, firma molend' aquat'* 1538-39 Dugd vi, "a water-mill ... in Tevylbe" 1541 LP xvi, each of which denoted a mill owned by Sixhills Priory. The first reference under 1538-39 suggests that this is a different mill from the watermill. Indeed, certainly from DB onwards, there were several mills in the village - - *molendino de Willelmi de Baiocis* 1202 Ass, "two water mills in Teylbye which belonged to Willoughton", i.e. "the commandry of Willoughton, Lincs, and St John's of Jerusalem" 1543 LP xviii, and *Fower mylnes in Teavelby* 1627 *Td'E*. Note also the various f.ns. in *Mill* in f.ns. (a) *infra.* THORPE MILL, 1828 Bry, 1849 *Td'E.* VICARAGE, *Vicarage House* 1795 *EnclA.* WARREN WOOD is north of Lynwode Warren. WASS'S BECK PLANTATION is named from the family of Thomas *Wass* 1842 *Td'E*, 1842 White. WHITEHOUSE FM.

Field-Names

Forms dated c.1150 (1409), and 1187 (1409) are Gilb; 1327 are *SR*; 1374, 1375, 1383, and 1384 are Peace; 1526, 1534, 1544, 1548, 1552, 1617, 1623, 1627, 1630, 1631, 1634, 1637, 1645, 1648, 1650, 1652, 1653, 1656, 1659, 1660, 1663, 1665, 1666[1], 1667, 1669[1], 1671, 1694, 1703, 1705, 1706, 1712, 1723[1], 1724[1], 1726, 1727, 1734, 1742, 1743, 1744, 1746, 1748, 1749, 1750, 1756, 1757, 1762, 1763, 1768[1], 1776, 1778, 1783, 1785, 1786, 1787[1], 1792[1], 1793, 1796, 1797, 1800, 1802, 1806, 1807, 1811, 1816, 1818, 1820, 1821, 1822, 1826, 1827, 1836, 1842, 1849, are *Td'E*; 1529 are Wills ii; 1537-8 are *AOMB 409*; 1556 are *Mad*; 1649, 1661, 1666[2], 1669[2], 1679, 1686, 1697, 1701, 1704, 1723, 1724[2], 1760, 1768[2], 1769, 1770, 1773, 1787[2], 1792[2], 1794, 1804, 1815 are *NW*; 1675 are VisitN; 1707 are *Terrier*; 1790 are *LTR*; 1795 are *EnclA*; Forms common to 1768[1] and 1768[2] are marked 1768*.

(a) The back-lane-close 1783, Back lane Cl 1795, Close ... formerly called the back lane Close and lately called the Church Close 1826 (self-explanatory); Baily Garth 1792[2] (*Baley Garthe* 1537-8, *Baylye garth* 1617, *Baily* - 1631, from the surn. *Bail(e)y*, cf. *Jennet Baylye* 1592 BT and **garðr** 'an enclosure' as elsewhere in the parish); Bardill 1768[1], 1793, Flg called Bardill, - Bardhill 1768[2], Bard hill 1793, Bardale 1795 (1617), Lt Bardill 1768[2], lt bardill 1792[1], Lt Bardale otherwise Bardill 1821 (*Bardaile* 1631, *Bardell, Bardale* 1743); Barnapit 1792[1], Barnapit Cl 1768[2], 1795 (*Barna pitts* 1617, *Barrey pittes* 1631, *Bernay Pitts* 1743, *v.* **pytt**); Barn Garth 1795 (*v.* **bere-ærn**, **garðr**); Batemans Cl 1795, Bateman Cls 1804 (from the surn. *Bateman*); Beadle Garth 1773 (*v.* **garðr**; the *beadle* was the parish law officer and the land may have been that assigned to him by virtue of his office); Beam Croft End (sic) 1793; Bean Plotts 1768[1], Beanplotts 1768[2], Bean Plotts, - Crofts 1793 (*ye Beane platt* 1617, *Beane* - 1637, *beane plat* 1631, *Bean Plott* 1743, *upper beane croft* 1666, *v.* **bēan, plat**[2], **plot, croft**); the Beck 1768[2], 1792[2], 1793, 1804, 1815 (1697, *v.* **bekkr**); Beetle Garth 1795 (perhaps an error for Beadle Garth *supra*); Bell Cl 1785, 1795, Bellgapp, Bell Sike 1793, - Green 1795 (*Bel sike* 1617, *Belsike* 1631, *Bell Syke* 1743, *v.* **sík, gap**; *Bell* is probably a surn. found in the parish, e.g. Francis *Bell* 1641 LPR); Benniworth Yds 1793 (from the surn. *Benniworth*, cf. William *Bennyworth* 1685 *BT*); Blades 1778, the East Blades Cl 1787[2] (*the Blades* 1630, 1661, *Close of pasture called the Blades* 1706, *Blades 2 Carr close* 1707; this is apparently from the pl. of **blæd** 'a blade, a leaf (of grass etc.)' used in some topographical sense or as a synonym for 'grassland'); Bold Garr or Gale 1768[1], Boldgar 1793 (*Boulgare* 1617, *Bowle gaite* 1637, *v.* **gāra, gata**); Bragate, Bray Gate 1793 (the forms are late, but this *may* be 'the broad road', from ON **breiðr** and **gata**, with a similar development to that in Brayford, PN L 1 17-18, but it could well be associated with a place called Bray *infra*); Braithing 1768[1], 1793, Flg called - 1768[2] (*furlonge called Brathin* 1617, *Braithing, furlong called brathing* 1631, *a Furlong called Brathing* 1743, perhaps 'the broad meadow', a Scand. compound from **breiðr** and **eng**, cf. a Close called also Brathing 1762 in the f.ns. (a) of Claxby *supra*); Flg called Branderith 1793, Brandnish (sic) 1842 (perhaps from the surn. *Brandreth*, but ME *brand-erth*, *brend-hurth* 'a field that has been burned over in preparation for tillage', *v.* MED s.v., is perhaps more likely); Bransby Garth 1795 (probably from the surn. *Bransby* and **garðr**); Braw Sike (sic) 1793, Bray Sike 1842, Bray Syke Cl 1792[2], - Sykes Cl 1804, Brae Syke Cl 1792[2], 1794 (*Bray sike* 1617, *Braisyke* 1631, *Braysicke* 1637, *-side* 1743, *close ... called Brasike* 1659, *brasike close or Church Close* 1660, *Bray Syke Close* 1697, - *Sike closes* 1717, presumably cf. the foll. and *v.* **sík** 'a ditch etc.'); a place called Bray 1792[2], 1794 (1697; perhaps from **brēg** 'a brow of a hill', cf. Bray PN YWR 5 126); Brookes Garth 1792[1], Brook(e)s - 1795, Brooks - 1818 (named from the *Brookes* family, cf. William *Brooks*, mentioned in 1795, and earlier Thomas *Brookes* 1739 *Td'E*); Broughton Pingle 1795 (named from an ancestor of Banks *Broughton* 1842 White); Buck Toft 1795 (*Buck toftes* 1617, *Bucktofts* 1631, 1743, *Bucktoft* 1659, *Bucktofte* 1660, *v.* **toft** 'a curtilage'); a Close ... called Buckliffs 1783 (sic, probably an error for Ruckliffe *infra*); Bull-Cock-Yard 1768[2], Bullcock's Garth 1770, Bulcock Garth 1795 (probably named from the family of Thomas *Bulcock* 1641 LPR); Burnt House Garth 1793, Burnthouse Yd 1795 (the interchange of **garth** and **yard** is noteworthy); Butlers Cl, - Garth 1795 (named from the family of James *Butler* 1592 *Inv*); Butter cakehill 1768[1], Butter-cake-hill 1768[2], Butter Cake (Hill) 1793 (perhaps a complimentary name, but probably alluding to soft, greasy soil, having the consistency of buttered bread, cf. Buttercake PN Ch 3 299, ib, 5 398); Caistor Thorns 1793 (*Caistor thorne* 1637; perhaps from a family name, since Caistor is some distance away, or perhaps on the road to Caistor, cf. *Caster gate* 1617, *Caister* - 1631, - *Gate* 1743, v. **gata**); Flg called Cake-Dale 1768[2], Upr Cake Dale 1793 (*Cake dale* 1617, 1743, - *daile* 1631, - *doale* 1637, cf. *Butter cakehill supra*); Candle head (sic) 1793, Cawdle End 1793 (*Caudle head Close* 1669[1], *Cawdel hill close* 1623, perhaps from **cald** with **hyll**, the compound being opaque, so that the addition of *hill* in the 1623 form is regular); the Carr 1768[2], 1792[2], 1793, 1795, Carr 1778, Tealby Carr 1804 (*del Kerr' de Teuelby* 1383, 1387 (p), y^e *Car* 1617, *the Carre* 1637, - *Carr* 1631, 1649, 1661, 1666[1], 1669[2], 1686, 1697, 1704, 1706, 1743, and cf. *est Carr* 1529, v. **kjarr** 'brushwood', 'a marsh, etc.'); Carr Cl(s) 1795, - Cls 1804, the low Carr Cl 1783, Low - 1787[2] (*Carr close* 1623, *the Carr Close* 1669[1], v. **kjarr** 'a marsh'); Causie Flg 1768[2], Causey Flg 1768[2], 1793, the Causeway 1795, 1811, Tealby Causeway 1796, 1826, Causeway House 1787[1], - Flg 1795 (*Cawsy -, Cawsye furlonge* 1617, *Cawsey Furlong* 1631, 1743, *Cawsye furlong* 1631, *Cawsey Close* 1627, 1637, *the cawsie close* 1653, *the Cawsie close* 1663, - *Close* 1671, *the Causey Close* 1705, *The* - 1742, v. **caucie**); Cavilla Hill 1768[1], Cavilla-hill 1768[2], Cavilla hill 1793 (perhaps to be identified with *Cawell hill* 1631, 'cabbage hill', from ME *caul, cawel* (OE **cawel, caul**) 'cabbage, cale, colewort, rape etc.', v. MED s.v. *caul* and **hyll**); Chapel Blades 1795, - Blades Cls 1804 (cf. *Chappel dale, - Garth* 1630, - *dayle, - garth* 1661, *Chappel Dayle, - garths* 1706, - *close* 1707, v. **deill, garðr, clos(e)** and Blades *supra*; for Chapel v. Chapel Hill *supra*); The Chapmans Cl (from the surn. *Chapman*; William *Chapman* is named in the document); Cherry Holt 1795; Church Cl 1795 (v. also The back-lane-close and the early forms of Braw Sike *supra*); Church-hill 1768[2], Church Hill 1793 (1637, *kirkhill* 1617, *the Church hill* 1631, 1743, *the churchill* 1631; it is worthy of note that Scand. *kirk* is found once as an alternative to *church*); Clarks Cl 1795 (named from the *Clark(e)* family, cf. Humphrey *Clarke* a1537 *MiscDep 43*, 1553 *Td'E*); Clays 1768*, 1793 (v. **clǣg** in the pl., presumably with reference to clayey places); Cliff House 1776; Clifton's Cl 1783, Clifton Btm 1793 (named from the *Clifton* family, cf. John *Clifton* 1734, with **clos(e)** and **botm**); Close 1795; Club Bank 1768[1], - Banks 1768[2], 1793 (cf. *Club Furlonge* 1617, - *furlong* 1631, - *Furlong* 1743); Cold Leas 1783 (v. **cald** 'cold, cheerless, exposed', **lea** (from **lēah**) in the pl. 'meadows'); the Common 1760, 1792[2], the late Common 1804 (cf. *voc le Coem Closses* 1556); the Common Hill 1787[2]; Coney Green Cl 1849 (presumably a variant form from **coninger** 'a rabbit warren'); Copp hill side 1793 (*Copp hill* 1617, *Cophill* 1617, 1631, 1743, v. **copp** 'a summit, a peak' or 'a mound, a ridge of earth', **hyll**); (North) Corn Fd 1778, the Corn Fd 1792[1] (*the Corne feld* 1552); Cowfold Hill 1804 *Td'E*; Cow Lane

1807 (*Cowlane Close* 1726); the Cowpasture 1768*, 1796, 1816, the Cow Pasture 1795, a place ... formerly called the Cow pasture 1826, the Cow pasture 1842 (cf. *cowclose* 1659, *cow close* 1660, *the Cow Close* 1704); Crabtree Nooking 1793 (perhaps from the surn. *Crabtree* and for *Nooking v.* The Nookings *infra*); Cracks 1768[1]; Cross Green 1800 (presumably self-explanatory); Cumber Lane 1793 (cf. *Cumberlands* 1686, *a comon called Cumberhill* 1552, *Cumber Hill* 1712); Cutlers Cl 1785 (presumably from the surn. *Cutler* and **clos(e)**); The Dale, High Dales, Gt -, Little - 1795, Dale Mires 1795, 1842 (*Close of pasture called daylemyre* 1665, cf. *close ... called the Dale close* 1660, *the Dale Close* 1724, *the high dale closes* 1665, *v.* **dalr, mýrr**); Close of meadow or pasture Ground formerly called New Close and now or late called Damsdike Cl 1750, Damsike 1760, Dam Syke Cls 1795, - Sykes Cl 1804, close ... formerly called or known by the name of the New Close but since the Dam sike Close or Closes 1820 (*Close of meadow or pasture formerly called the New Close and now called or knowne by the name of Dam sike Close* 1703, *v.* **dammr** (ME **damme**) 'a dam, a pond', **sík**, once confused with **dík** in 1750); Dial Garth 1795, 1804, 1815 (*Dyall Close* 1704, - *Garth* 1723[2], *v.* **dial, garðr**, probably an enclosure with a sundial cut in the turf); Flg called Dozmons 1793 (obscure); Drapers Cl 1795, 1842 (from the surn. *Draper*, cf. William *Draper* 1659); The Gt -, the lt Dunsgarth 1783, Flg called Dun Scolds 1793, Ft -, Lt Dunsgoe 1795 (the form *Dunsga* is found in a copy of *EnclA* 1795), Dunsgate 1842 (*greate Dunsgarre, little dansegare* (sic) 1623, *Le Dungare als Dunsgarth* 1630, *Duns Garres als duns Garthes* 1634, *the great -, the little Dunsgarrs Close* 1669[1], *v.* **gara, geiri** 'a gore', apparently confused with or replaced by **garðr**; the first el. is uncertain); East Fd 1756 (*ye east feild* 1617, *the* - 1631, *the East feild of Tavelby* 1637, *ye east and south feilds* 1648, *the East Field* 1743, one of the great fields of the village); Fallows 1793 (*the Fallow* 1552, *ye Fough* 1617, *the* - 1631, 1743, perhaps *v.* **falh**)); Feelsons Yd 1795 (presumably from the surn. *Feelson*); Flg called Fennel 1793; The Field 1794 (*the felde* 1529, *campis de Tevilby* 1534, *the Feildes of Tevilby* 1637, *Tealby field* 1744, *v.* **feld**); First Cl 1795; Flints 1768[2], Flints 1793 (*y[e] Flints* 1617, *the* - 1631, *the Flynts* 1637, *a place called the Flints* 1743, presumably this is self-explanatory); the Fools 1768*, Flg called Fouls (sic) 1768[2], Meadow Grd by the Fools, Meadow Grd called long Fools 1793 (it may be more than coincidence that forms in *fools* begin in 1768, while others in *foots* are recorded from 1617 till 1743 -- *y[e] foots, - long foots* 1617, *the Foots, - long foots* 1631, *the Foots* 1743, *Footts 20 Carr* (sic) 1707, *Limefoots* 1743; it is possible for some now unknown reason that the feature known as *y[e] foots* became *the Fools* and continued to be so called till, at least, the end of the 18th century. If *foots* is indeed the earlier name, it presumably reflects the pl. of **fōt** 'the foot (of a hill, etc.); the 40th Bottom 1768[1]; The Forty Acres 1793, a Little Cl called Forty Acres 1804 (possibly an ironic name for a very small field); Furhill Cl 1795, the - 1804 (*Fur hill close* 1660, *v.* **hyll**); the Furze Cl 1773 (*v.* **fyrs**); Furze Hill 1795; Golding-lithe 1768², Golding Lithe Flg, - Lythe (Hill) 1793 (*v.* **hlið²** 'a slope' and cf. *Gouldinge infra*); Grass Clifts 1768¹, - Cliffs 1768², Grass Cliffe, the Mills Flg called Glass Cliffs (sic) 1793 (*girsy cliffe* 1617, *grassey* - 1631, *v.* **gærs**, -**ig³**, **clif**); Great Cl 1795; the Green 1795, 1797 (*Grene de Teuelby* 1374, *atte Grene* - 1375, 1401 *FF* all (p), *v.* **grêne²** 'a green, a grassy spot'); Half Moon Cl 1795, the half moon cl 1804 (doubtless so called from its shape); Hall Garth 1795 (*the hall garth Close* 1669¹, *v.* **hall, garðr**); Hang Dogg cl 1757, - Dog Cl 1790 *LTR*, 1795, Hang-Dog-Cl 1769 (a derogatory name for poor land); high Barn Pitts 1793 (perhaps the same as Barnapit *supra*); the East and West high Fds 1792², 1794 (*the East and West high Fields* 1724¹); The Hill Cl 1783, Hill - 1795 (*Hill de Teuelby* 1428 *FA* (p), *Hill close* 1660, *v.* **hyll**); Hinging -, Inging Banks 1793 (*v.* perhaps ON **hengjandi** 'hanging' and **bankc**); Hint Flg 1793; the Home-Cl 1773, Home Cl 1795, 1821 (situated near a homestead); The Homestead 1793, 1802, the - 1804, 1815; Home Yds 1768², - Yd 1802 (cf. the Home-Cl *supra*); Hordale 1795 (perhaps *v.* **horu** 'dirt', **dalr**); the horse Cl 1783, Horse Cl Head 1793, Horse Cl 1795, 1842 (*the horse close* 1669¹); How Toff (sic) 1793, How Toft Cl 1795 (perhaps from the surn. *How,* cf. Edward *Howe* 1600 *BT,* and **toft** 'a curtilage, a messuage'); Hunger Hill 1793 (a common derogatory name for infertile land); Hurd-Dale 1768², Hurd Dale 1793; The Ings Flg 1793, Ings 1842, Gt -, Lt Ings 1793, the Ings Corner 1795 (cf. *the little Ings Close* 1669¹, *Ings Close* 1707); Intake Cl 1795 (*the intax close* 1653, *Intackes* - 1656, *the Intackes* - 1663, *the Intaks Close* 1671, *the Intax Close* 1705, *the Intake* - 1742, cf. *the intackes* 1649, *the Intak* 1704, *v.* **inntak**); Jennyhole Fd 1795, Jenny Hole - 1804, - hole Fd 1815 (*genny hole* 1666¹); Kersey Garth 1795 (*v.* **garðr**); Ketley Cl 1792², 1794, Ketlock - 1795, 1804, Kettock - (sic) 1842 (*Ketley close* 1724²; these refer to the same piece of land and *Ketley* etc. is here a surn., cf. Edward *Kettley* 1684 *BT*); Kirmongate 1768², Kirmond Gate 1793, Far Kirmond - 1842 (*kirmon gate* 1617, *kermond* -, *kirmond* - 1631, *Kirmond gaite* 1637, - *Gate, Kirmon* - 1743, 'the road to Kirmond le Mire (LSR)', *v.* **gata**); Lady Garth 1795 (*v.* **hlæfdige, garðr**); Lamas Blades 1795 (*v.* **lammas** and Blades *supra*); Lawn 1842; Lea Cl 1773, Ley(s) - 1795, the Ley - 1802, 1804, 1822, Ley - 1806 (*the Lea close* 1653, 1663, 1742, - *Close* 1671, 1704, 1705, *Lea Close* 1723², *the great Ley close* 1686, *v.* **lea** (from OE **lêah**) in the sense 'a meadow, an open pasture'); Leather Mill 1792¹, the Leather Mill 1795; Lincoln Cross Cl 1795, 1802; Little Cl 1795; the Long Acres 1793 ('the long selions', *v.* **lang¹, æcer**); The long Cl 1783, Long Cl 1795, 1815, The long Cls 1795, the Long Cl 1804; Long dale Hill 1793, Longdale - 1842; longgars 1793 (*v.* **lang¹, gara**); the Lotts 1793 (*certeyne meadowe called Lotts being Eight in Number* 1669², *meadow called Lottes* 1686, *v.* **hlot**); the Low

Fd 1793, the East -, the South -, the West Low Fd 1792[2] (*the East Lowe feild* 1686, *East Low Feild* 1697, *the South -, the West Low Field* 1724[2]); Mare and Foal 1842 (the same name has been noted as Two lands called Mare and Foal in Nettleton, *v.* PN L 2 243, where it is noted that these were perhaps two adjoining strips of unequal size, and as *one broode land ... called by the name of the mare and foale* in North Kelsey, *v.* PN L 2 192-3 and there it is suggested that it was a foaling strip of land; there is no indication in the present example of the extent of the field); the Marsh or Mire Cl 1783, The Marsh Cl 1793, 1795, Marsh - 1842, the Marshes 1768[2], 1793 (*marshe close* 1623, *close called Marsh* 1666[1], *the Marsh Close* 1669[1], y^e *Far Marshes* 1617, *the Far marshes* 1631, *the Farr Marshes* 1743); Great Michael Hill Cl 1750, 1762, 1806, Micklethe (sic) 1768[2], Mickle the Hill (sic) 1793, Mickle Hill 1795, Mickle hill Cl 1776, Mickle Hill Cls 1804, 1815 (*Mickle lyth hill* 1617, 1631, - *lythe Hill* 1637, *Great-hill Close* 1669[1], *Great Michael Hill Close* 1703, *v.* **mycel, hlið**[2] 'a slope, a hill-side'; **hyll** has been added tautologically in most forms; *Michael* is found elsewhere for *Mickle* from **mycel**, and there is further tautology in the 1669[1] form, in which the significance of *Michael* has been missed and *Great* added, if this refers to the same feature); the Middle fd 1768[1], - Field 1768[2], the Middle or South Fd 1793, Middle Fd 1795; Middle Moor 1787[2], 1795, the - 1804, 1820 (cf. North Moor Fm *supra*); Middle Stone 1793; the Mill Alley Gutter 1793, The Mill Cl 1793, Milne alias Lea Cl 1792[2], 1794 (*Mills close* 1627, *Mill Close* 1637, *the milne close* 1653, 1663, - *Milne Close* 1671, *Milne Close als Lea Close* 1697, *the Millne Close* 1705, *the Miln Close* 1742, cf. Lea Cl *supra*); Mill Gate 1768[1] (1743), Low -, Short Mill Gate 1768[2], 1793, long -, short - 1793, the Long Mill Gate Cl 1821 (*Mill gate* 1617, 1631, *short Mill* - 1617, *Myllne gaite* 16327, *v.* **gata**); Mill Ings 1773, the Mill Ings 1787[2] (*v.* **eng**); The Mills Flg (*v.* Grass Clifts *supra*) (for these f.ns. in Mill cf. Tealby Watermill *supra*); the Moor Cl 1792[2], North Moorgate 1795 (*v.* **gata**) (named from Tealby Moor *supra*); Nettle Dale 1768[1], - dale 1793, Nettledale 1768[2] (*Netle dale* 1637, *v.* **netel, dalr**); the New Cl 1760, New Cl 1768[2], 1792[2] (1697, *the New close* 1704, cf. Damsike Cl *supra*); the nine Leys 1795, - Nine Leys 1826 (*Nyne Leys* 1627, - *Leyes* 1637, *the nyne leas* 1653, 1663, *the nine leas* 1705, *the Nine Leas* 1742, 'nine allotments of meadow or pasture' *v.* **lea** (from OE **lēah**) in the pl., cf. Ten Leys *infra*); The Nookings, Short - 1793, Nookings, Nooking Cl 1795, the Nookings Cl 1804, Nookins Cl 1826 (*the Nookins* 1649, *the Nookeings* 1679, - *nookeinges* 1701, *Close ... called Nooking* 1723[2], from dial. *nooking* 'a nook', *v.* **nōk** and PN Nt 288 s.v.); the North Fd 1768*, 1793, the two North Fds 1783, the High Northfield 1785, High -, Low North Fd 1795, the low North Fd 1821 (*in Campis borial'* 1556, *the North Field* 1743, *Upper North Field* 1723[1], *the two north fields* 1734, one of the open fields of the village); Northings 1778, the North Ings 1792[2], 1794 (*Northynges*

1529, *northings* 1707, *the North Ings* 1724[2], *upper northen* 1617, *upper north Inges*, *upper northing* 1631, *upper Northings* 1743, *v.* **norð, eng** 'a meadow', cf. South Ings Cl *infra*); the North Plot (sic) 1849 (cf. the South Plots *infra*); North Willingham Road 1804 (self-explanatory); Nunfirkins 1795 (*Nunne Furkin* 1627, - *Furkyn* 1637, *close of pasture* ... *knowne by the name of Nunfirkin* 1665, *v.* **nunne** 'a nun'; the etymology of the second el. is obscure); West Overgates 1757, the - 1787[2], West over Gates 1778, West Over Gates 1795, a place called West Overgates 1804 (*West overgate* 1630, *the west overgates* 1661, *West Overgate Close* 1661, 1706, *a place called Westovergates* 1706, *west overgates* 1707, from **uferra** 'higher, upper' and **gata**); the Ox pasture 1756; Paddock 1842; Padman Hill 1793 (perhaps a mistake for Padmill Hill under Papermill Lane *supra*); Parsons Cl 1795; Peat Hills 1793; the pen Green 1783 (*the comon' called pengren'* 1548, *pen greene* 1666[1], *Penngreene close* 1623, *the upper Pengreene, the low Pen-greene Close* 1669[1], probably from **penn[2]** 'a pen', **grēne[2]** 'a green, a grassy spot'); Perkins Yd 1795 (*Perkins Garth* 1726, named from the family of *Gemm' Perkins* 1784 *Td'E* and **garðr;** the variation between *garth* and *yard* is paralleled in the Stack Yd *infra*); the Pinfold Cl 1842 (*v.* **pynd-fald** and cf. the pen Green *supra*); the Pingle 1792[1], 1842 (1704), Pingle 1795 (*v.* **pingel** 'a small enclosure'); Pleasure Grds 1842; Tealby Ploughed Ings 1804 (*v.* **eng**); Priest Wong 1768[2], Priest Wong 1793 (*v.* **prēost, vangr** 'a garden, an infield', cf. *Preestholme* 1617, *v.* **holmr**); Pudding Poke Nooking 1793 ('land with sticky soil', cf. Pudding Poke PN Db 297 and for Nooking, cf. The Nookings *supra); Rasen Cl 1795, 1842 (adjacent to the boundary with Market Rasen); Rasen Poor Cl 1804 (land endowed for the benefit of paupers in Market Rasen); Redhill 1768*, Red Hill 1793; Risby Garr 1768[1], - Gurr (sic) 1793 (*v.* **gara** 'a triangular plot'), Risby Mear 1768, 1793 (*Risby meare* 1617, - *maire* 1631, 'the boundary (of Tealby) with Risby in Walesby parish', *v.* **(ge)mǣre,** cf. *Risby Hedge* 1617, - *hedge* 1631); a place or close formerly called the Coroner Robinson's close or Mr Robinson's close 1792[2], 1794 (*Coroner Robinsons Close or Mr Robinsons Close* 1697, named from a member of the *Robinson* family, well represented in the parish, cf. John *Robynson* 1554-7 *Inv*); The Rownsey dale Cl 1783, Rounce Dale End 1793, Rounsey Dale Cl 1795 (*Rouncedall'* close 1623, *Raunsdale Close* (sic) 1669[1], *Raunsadale* (sic) 1704, *v.* **deill;** the first el. may be a form of the surn. *Runcie*, for which *v.* Reaney s.n.); Ruckcliffe 1762, Ruckliffe 1768*, 1793 (1669), Rockecliff, Rockecliffe Cl 1795 (*v.* **clif**, the first el. *may* be **hrōc** 'a rook'); Rye Croft 1768[1], Rie Crofts 1793 (*Ryecrofts* 1617, *a place called Rycrofts* 1631, *Ry croft* 1743, *v.* **ryge, croft**); Saltergate 1793, Salter Gate 1842 ('the road used by salt-merchants, *v.* **saltere, gata**); It Sandons (sic) 1768[1], Sandams Cl 1792[2], 1794, 1815 (1697), Sandhams 1795, Sandham(s) (Cl) 1804 (*Sand holmes* 1649, *Sand Holmes, the Sand Holm's* (sic) 1704, *Sandholmes* 1723[2], *v.* **sand, holmr** 'higher dry ground in the marshes'); Sandhill 1768*, Sand Hill 1793, Sandhills 1842 (v. sand, hyll); Scotch Marshes 1768[1], Scotch Marsh 1768[2], 1793 (Scot marshe, - Marshes 1617, Scot marshes 1631, Scott Marsh(es) 1743), Scott Flg 1768*, 1793 (Scott Furlonge 1617, 1743, - furlonge 1637, Scot furlong 1631, both appear to be from the surn. Scot(t) with mersc and furlang, with Scot(t) replaced by Scotch in the former); The Screed 1795, 1804, the - 1815, the Screed Cl 1804 (the little Screed 1630, 1661[1], the long - 1630, 1706, the longe - 1661, from ModE screed 'a narrow strip or tongue of land'); Sewels Cl 1795, Sewell(s) Cl 1804, 1815 (from the surn. Sewell); Shepherd Hut 1793 (perhaps from the Shepherd family, cf. John Shepherd 1734 BT); Simons Yd 1795 (no doubt named from the family of Thomas Simons 1796); Skittah 1768[1], Long -, Flg called Short - 1768[2], long Skittah, - Skillah (sic) 1793 (a place ... called long Skitta 1617, long Skittey 1631, short Skitta 1617, - skittey 1631, long Skittay 1743, perhaps 'the dung-hill', from skitr, haugr); Sleight Bur Hill 1793 (perhaps 'the smooth, level hill' from sléttr and berg, with hyll added because the meaning was unknown, but the form is too late for certainty); Small Thornes 1768[1], Small-thorne Banks 1768[2], small Thorns 1793 (Small thorne 1617, 1631, a Place called Small thorn 1743, v. smæl perhaps in the sense 'thin', þorn); Sound Thorn (sic) 1793; the Two South Fds 1757, the South Fd 1768[2], 1795 (in Campis austr' 1556, in australi campo de Tealby 1623, in Australi Campo de Tealbie als Teavelbie 1630, the two south fields 1634, ye east and south feilds 1648, The south Feildes and Moore (sic) 1669, one of the great fields of the village); the South Flat 1792[1]; South Ings Cl 1750, Southings 1778, the South Ings 1787[2], 1792[2], 1793, 1804, South Ings 1795 (le South Inges 1627, the - 1637, South Ing 1650, Sowth Ings 1666[1], South Inge 1686, the South Ings 1694, 1697, the South Ings Close 1703, the South Inge 1704, South Ings (Close) 1707, v. sūð, eng, cf. Northings supra); the South Plots 1849 (v. sūð, plot, cf. the North Plot supra); Spencers Garth 1750 (1703, named from the family of Thomas Spencer 1633 Inv and garðr); Spiritual Hill 1793 (unexplained); the Square Pce 1793 (alluding to its shape); the Stack Yd 1793, the Stack Garth Cl 1795 (v. geard, stakgarth; the variation between yard and garth is noteworthy, cf. Parkins Garth supra and The Tythe Barn Yd infra); Stainton Gate 1768*, - Gate (Btm) 1768[2], - Gate Flg 1793, Stainton hole Btm 1768[1] (v. gata, botm, cf. Stainton hill 1617, 1631, 1743, referring to the neighbouring parish of Stainton le Vale, earlier Stainton le Hole); Stamp Cls 1787, 1820, Stamps Cl 1795, - Garth 1842 (named from the Stamp family, cf. Mathew Stampe 1641 LPR); Stevensons Garth 1795 (named from the Stevenson family and garðr; Joseph Stevenson is named in the same document); Stone Pitts 1793 (Stone pitts 1617, the stone pitts 1631); Stony Gars 1793, the Stoney Garth 1795; Flg called Sun-burnt-hill 1768[2], Sun burnt hill 1793, - Burnt Hill 1842 (perhaps because of a southern aspect); Sweet Jack Cl 1757, 1797, Sweet Jack 1795, the Close called Sweet Jack 1804, the close ... called or known by the name of Sweet Jack 1836 (*Close ... called Sweet Jack* 1723, presumably a complimentary f.n. for a productive piece of land); Syke Cl 1804 (*v.* **sik**); Syren Dale Cl 1792[2], Syron Dale Cl 1794, Cypher Dale 1795, Syren or Cypher Dale (sic) 1815, - Dale Cl 1804 (*Ciperdale* 1686, *Cypher Dale or Syper Dale* (sic) 1723[2], *Syper Dale Close* 1697, obscure); Tame Hole 1793; Temple Garth Fm 1776, 1778, 1792[1], Temple Garth 1795, Temple Fm 1790 *LTR* (*Temple Garth* 1627, 1637, 1663, 1671, 1705, - *garth* 1645, 1653, 1663, 1742, *v.* **temple, garðr**, cf. *Temple Ings* 1653, 1663, - *Inges* 1671, *the Temple Ings* 1705, 1742, *Temple Ings close* 1666; the reference is to the Willoughton Preceptory of the Knights Templars, cf. *parcell of the landes & possessions of the late Commandry of Willoughton* (sic) 1544; *Temple garth* in 1663 is referred to as a *capitall Mesuage*); Ten Leys 1795, the Ten - 1826 (*Tenne Leyes* 1627, *Ten Leyes* 1637, *the ten leas* 1653, *The Tenn Leas* 1705, 1742, *the ten leas close* 1663, - *tenn leas close* 1671, *the ten leazes close* 1663, *v.* **lea** in the pl. (from **lēah**) and the nine Leys *supra*); Tenter Garth 1795 (cf. *the Tenter close* 1660, *v.* **teyntour**); Thornes 1768[1], the Thorns 1792[1] (*v.* **þorn**); Thorne Hills 1793, Thornhill 1842 (*v.* **þorn, hyll**); Thorn Tree btm 1793 (*Thorn tree bottom* 1744, *v.* **botm**); Flg at the Town End 1793, Townend Gap 1795; the Town Street of Tealby 1826; Gt Turnhill 1795, Turnhill 1842, Great Turnhill Cl 1804, 1815 (cf. *close of meadowe or pasture ... called great Turnells* 1659, *great Turnells* 1660, *the Turney hills Close* 1669[1], *little Turnil* 1707); the Turnpike Road leading from Market Rasen to Louth 1804; The Tythe Barn Yd 1793, Tithe Barn Garth 1821 (for the variation between *yard* and *garth*, cf. the Stack Yd *supra*); Underfull Hill 1793, Underfall 1795; Waghing Stoop Cl 1768[1], Wainstoop 1795, Wainstoop Cl 1804, 1815; Wailesby Lane 1807, Walesby - 1811, 1816 (leading to Walesby); the High Walk Mill 1795; Wardle hole 1793 (perhaps named from the family of Robert *Wardall* 1699 *BT* and **hol**[1]); Warlotts 1768[1], Flg called - 1768[2], the warlotts 1793 (*North warlottes, south warlotes* 1617, *the South Warlotts* 1631, *South Warlotts* 1743, *Wharelattes* (sic) 1637, cf. *farre warlottes* 1631, *v.* **warlot**, cf. *Warlottes* in Toft Newton f.ns. (b) *supra* and ye Warlots in South Kelsey f.ns. (a) *supra*); Washdike Cl 1756 (1727, *Washdike close* 1659, 1660, *v.* **wæsce, dīk**, for washing sheep); Water Dike 1793 (*Water dike* 1617, - *Dike* 1637, *the water dyke* 1631, *v.*, **wæter, dīk**); Water Gap 1778 (*v.* **wæter, gap**); Waterworth Pit 1793; West Cl 1768[2], - Cl(s) 1795, the West Cl 1820, Great West -, 1827; West Fd 1756, 1778, the - 1787 (y[e] *west feild* 1617, *the* - 1631, - *field* 1743, *the West Feylde of Tavelby* 1637, *Westerfield* (sic) 1723[2], one of the great fields of the village); West Lodge Paddock 1842; Wheat Cl 1763, 1795 (1746, 1748, 1749), the Wheat Cl 1768[1] (*the wheat close* 1707, self-explanatory); Widow Nooking 1793 (perhaps an error for Willow Nooking *infra*); Wilkinsons Yd 1795 (presumably named from the *Wilkinson* family, cf. Daniel *Wilkinson* 1842 White); Willingham

Hedge 1787[2], 1793 (alluding to North Willingham parish); Willow Nooking 1768[2] (cf. Widow Nooking *supra* and The Nookings *supra*); the Windings 1793 (cf. *two wyndinges* in Binbrook f.ns. (b) and *Windinges* in South Kelsey f.ns. (b) *supra*); Winter Hill 1793 (*Wynter hill* 1631, from the surn. *Winter* or *Wynter*, cf. John *winter* 1617 and *the heires of Wynter* 1631); Wold Cl 1795 (*le Would close* 1630, *the Would close* 1634, *the would close* 1665, *the would* or *Littlehill Close* 1669[1], *the Would* 1686, *the Wold* 1704, *v.* **wald**); Wood Fds 1792[2], 1794 (1697); Wood Flg 1768[2], 1793 (1743, *wood Forlonge* 1617, - *forlong, woodforlong* 1631, *v.* **wudu, furlang**); the World Cl End 1793 (perhaps for the Wold Cl -, *v.* Wold Cl *supra*); Wronglins 1768[1], Long -, Short Wronglin's (sic) 1768[2], Short Wrong Lande 1793, long -, short Wrong lings (sic) 1793 (*shortwrong landes* 1617, *Wrong lands* 1637, *v.* **vrangr, land**, cf. *Short Wrong dale* 1743, *v.* **deill**); Yawd Syke Cl 1795 (*yaud Sike* 1707, *v.* **sik**; the etymology of *Yawd* is uncertain).

(b) *Ardens Close* 1688, 1700 *Deeds* (from the family name *Arden*, cf. Christopher *Ardayne* 1590 *BT*); *Barnard Staynges* 1529 (from the *Barnard* family, presumably an ancester of Thomas *Barnard* 1629 *Td'E*); *Bartle Close* 1724 (presumably from the surn. *Bartle*); *Bayes hill Closes* 1623 (perhaps cf. Bayons Manor *supra* for which forms in *Bay(es)-* have been noted); *Baysicke* 1637 (*v.* **sik**, the first el. being perhaps comparable to the prec.); *Meadow called the Bearthor* 1743; *the blinde layne* 1631, *the Blind* - 1743 (i.e. a cul-de-sac); *Brame landes* 1617, *braime landes* 1631, *Brame Lands Alias Rutcliffe* (sic) 1743 (*v.* **land**; the first el. is IME *brame* 'briar or bramble'; for the alternative name *v.* Ruckliffe in (a) *supra*); *Brame well* 1617, *braime* - 1631 (from the same first el. as in the prec. and **wella**); *Brigg Lees* 1707 (*v.* **brycg**, in a Scandinavianized form, with **lea** (from **lēah**) in the pl.); *Broadsikes* 1649, *Broad Sikes* 1704 (*v.* **brād, sik**); *Brownes House* 1703 (named from the *Browne* family, cf. Thomas *Browne* 1577 *Inv*); *Cafforth layne* 1529; *Calfe Close* 1688, 1700 *Deeds*; *Carworth close* 1652 (presumably *Carworth* is a surn. here); *Catgrene* 1327 *SR*, *Cattegren* "of Tealby" 1341 *RIL*, *Catgrene de Telby* (sic) 1384 all (p) ('the grassy spot, green, where cats are found', *v.* **cat(t), grēne**[2], but this may not be a local name); *the Chequer* 1669[2] (*v.* **cheker**, from the chequered appearance of the ground); *Comberhowe* 1327 (p), *Comerawhill* 1529 (perhaps to be associated with Cumber Lane *supra*); *the commen way* 1529; *le Corner Close* 1667; *Cowdale bottom,* - *goulding,* - *lyth* 1617, *Cowdaile bottom,* - *lyth* 1631, *Cowdaleith* 1743, *v.* **cū, deill** with **botm** and **hlið**[2], *v.* also *Gouldinge infra*); *le dale mill* 1623 (*v.* **dalr, myln**); *long* -, *short durdale* 1617, *long* -, *short durdaile* 1631, *long* -, *Short Durdale* 1743 (perhaps from IME *dure* 'hard' and **deill**); *Flatlandes* 1617, 1631, *Flatt Lands* 1743; *Fostlee Dale* 1743; *Foston Blades* 1630, 1661, 1706 (named from the *Foston* family, cf. John *Foston* 1571 *Inv*, and *v.* Blades in (a) *supra*); *Frisby* *hedge* 1743; *Gilden garth* 1650, 1694 (named from the *Gilden* family, cf. Godfrey *Gilden* 1641 LPR, and **garðr**); *Glead howe* 1617 (though the form is late, this is likely to be 'the kite hill', *v.* **gleoda, haugr**, cf. Gleadow PN L 2 10); *Gouldinge, -ings* 1617 (probably 'the meadow, pasture where gold flowers grow', *v.* **golde, eng**); y^e *gutter* 1617 (from ME *gōter, gotter, gut(t)er* 'a gutter', *v.* MED s.v.); *Hayward Close* 1650, 1694 (from the occupational name or derived surn. and **clos(e)**); *the High Laith Barn and Close* 1675 (*v.* **hlaða** 'a barn', and note *Barn* itself has been added); *Holbray sike* (perhaps cf. Braw Sike in (a) *supra*); *Innamed Closes* 1686 (obscure, unless it is to be associated with **innām** 'a piece of land taken in or enclosed'); *Kendilgarthes* 1562, *Kendall garth* 1623, *Kendal house or garthes* 1665 (no doubt from the surn. *Kendal*, the earliest member of the family so far noted being Edmond *Kendall* 1604 *BT*, and **garðr**); *Kilnehowse close* 1623 (*v.* ME **kilne-hous** 'the building in which a kiln is housed' and **clos(e)**); y^e *kircher* 1617, *the kercher* 1631 ('the kerchief', possibly an allusion to a small piece of land, cf. the frequent *Handkerchief* f.ns. with this sense); *Lyarrdaole* (sic) 1637 (Dr John Insley suggests that the first el. seems to be ME *liard*, normally used to denote a horse spotted with white or silver grey, but here probably a ME byname or surn; the second is obscure, though it may be corrupt for *dayle* from **deill**); *Lyne toftes* 1617, *Lynetoftes* 1631 (*v.* **lin** 'flax;, **toft** 'a curtilage'); *Molton meadowes* 1660 (presumably from a surn.); y^e *Peele* 1617, *the peell* 1631, *the Peel* 1743 (perhaps named from its shape; a *peel* is the shovel used by bakers to insert bread into and withdraw it from the oven); *Read hill* 1631; *super aquam de Teuelsby* c.1150 (1409), - *de Teflesby* 1187, *the river of Tealby* 1675 (the 12th century references are to a mill of Sixhill Priory on the R. Tealby, now the R. Rase); *Scammadyne Close* 1627, *Scamadyne* - 1637, *Skammadyne close* 1653, *Scammadine* - 1663, *Scamadine* - 1663, - *Close* 1671, 1705, 1742 (from the surn. *Scammadyne*, cf. Jane *Scammadyne* 1627); *Scurles close* 1660 (possibly from the surn. *Scurle*, cf. Reaney s.n. *Squirrel*); *streetern dale* 1617, *strethorns dayle* 1631 (probably named from High Street *supra*); *Talbots hedge* 1669[1] (named from the *Talbot* family, cf. Richard *Talbot* 1529, Bryan *Talbot* 1602 *Inv*); *Tevilby meare* 1666[2] (*v.* **(ge)mǣre** 'a boundary'; it is on the boundary with North Willingham); *Claus voc Turpyhnge* 1556 (possibly from the surn. *Turpin*); *Vicar Dale* 1630, *Viccar dale* 1661 (*v.* **deill** and cf. Vicarage *supra*); *Walnut garth* 1666[1] (*v.* **garðr**); *Water gareing* 1617, *Water gayring* 1631, *Water Gareing* 1743 (*gareing* has been noted as an appellative in *a gareing merefurr* 1577 *Terrier* in South Kelsey and as *a garing* 1611 *TLE* in Owersby, and is presumably a late -*ing* derivative of ON **geiri** 'a triangular plot of land'; *water* has probably the sense 'wet' here); *place called water wheele* 1617, *water wheele, water wheell pitts* 1631, *a certain place called Water whell* (sic) 1743 (self-explanatory, and cf. *Wheel Pits* 1743); *Wayneflet close* 1627, - *Close* 1637, *waineflet closes* 1653, 1663, *Wainflet Close* 1671, *the Wainflett Close* 1705, *the Wain Flett Close* 1742 (unless *Wayneflet* is a surn., it is identical with Wainfleet LSR 'the stream which a wagon can cross', v. **wægn, flēot**); *West Crofte* 1552 Pat; *the Worn Flatt Close* 1742; *Wrawby Close* 1669[1] (from the surn. *Wrawby*, cf. Oliver *Wrawbie* 1611 *Inv*); *yow toftes* 1614 (v. **toft**).

Thoresway

THORESWAY

 Toreswe (2x) (sic) 1086 DB

 Toresweia c.1115 LS, 1212 Fees, -*ueia* 12 *RevesInv,* -*waye* 1260 Cl

 Toreweia c.1115 LS, 1223 Pat

 Thoresweie 1187 (1409) Gilb, 1199 FF, -*weiam* 1198 (1328) Ch, -*weia* 1200-23 *HarlCh,* -*weye* 1242-43 Fees, 1252 Ch, 1288 Ipm, 1298 Pat, 1303, 1306 Cl *et passim* to 1375 ib, -*wey* 1239 RRG, 1303, 1316 FA, 1336 *HarlCh,* 1356 Orig *et passim* to 1549 Pat, *Thoriswey* 1242-43 Fees, *Thoryswey* 1242-43 ib, 1428 FA

 Thoreswaia Hy2 Dane, -*waya* 1335 Cl, -*way* 1275 RH, 1288 Ipm, 1290 Abbr, 1303 FA, 1327, 1332 *SR,* 1334, 1340 Ipm, 1341 NI, 1346 FA, 1349 Pat *et freq, Thoresway Beaumont* 1462 Pat, *Thoresway Bayons alias Nevile* 1465 ib, -*waye* 1281 QW, 1310 ChancW, 1310 Pat, 1317 Ch *et passim* to 1405 RRep, -*waie* 1558 InstBen, 1576 LER, *Thorisway* 1285 Ipm, 1350 Fine, 1610 Speed, *Thorysway* 1428 FA

 Thyrisweye 1291 Tax

 Thursweye 1303 FA

 Thursway Beaumond 1467 Pat, *Thuresway* 1472 ib, 1517 ECB, *Thursway* 1526 Sub

 Thorseway 1652 *Rad*

 Thoreway 1242-43 Fees, -*wey*(*e*) 1249 Ipm, 1338 Pat, *Thorway* 1264 Cl

Thoresway is usually interpreted as 'Þori's road' from the ON pers.n. *Þórir,* ODan *Thorir, Thori* and OE **weg** 'a road', v. DEPN, SSNEM 222. ON *Þórir,* ODan *Thorir, Thori* is common enough in independent use in L, v. Feilitzen 393-94 and Fellows-Jensen, SPNLY 307-9, while the same pers.n. also occurs in North

Thoresby, some seven miles in a direct line from Thoresway, South Thoresby LSR, a lost *Thoresby*, in Revesby, LSR, DB lxix, as well as Thoresthorpe LSR. OE **weg** is rare in settlement names and Gelling, PNITL 83, takes the total number to be about 25. She points out that only three major names with this el. have a pers.n. as first el. -- Garmondsway Du, Hanwell (earlier *Haneweie*) O and Thoresway L. She does, however, note that **weg** is found frequently in minor names and f.ns, but there again it is very rarely compounded with pers.ns. Indeed, Dr John Insley points out that the only example with a Scand. pers.n. appears to be *Ravenildesweye* 1292 PN Db 754, 762, containing ON *Hrafnhildr* (fem.).

Dr Insley comments that "it would seem better to re-examine the etymology of Thoresway. The DB form could be taken to stand for a heathen name, ODan *Þorswǽ* 'the shrine dedicated to Thor', the second el. having subsequently been contaminated by OE **weg** as a result of semantic opacity. A direct parallel to an ODan *Þorswǽ* is provided by the Swedish p.n. Torsvi in Västergötland, for which *v.* Lundahl, NoB 43 (1955), 141. If we accept this etymology for Thoresway, the name must belong to the earliest phase of Scand. settlement, given that the re-conversion of the eastern Danelaw proceeded fairly rapidly in the 10th century".

The affixes *Beaumont* and *Nevill* are from families at one time holding the manor. Thoresway, as well as other places in L, was "held in dower by Joan of the endowment of John, late viscount Beaumont" 1460 Pat, while Roger *de Nevyl* held *ij partes i.f.* in Thoresway 1301 FA.

BLACK SPRINGS. CHALK PITS. GLEBE FM is apparently *ye low farm* 1697, - *Low Farm* 1707, *the Low Farm* 1712 all *Terrier*; in 1707 it is referred to as *ye sd Glebe farm*. HIGH STREET is *the streate* 1579, - *streete* 1625 both *Terrier*; this is the name of the ridgeway from Horncastle to Caistor, known as Middlegate Lane northwards from Caistor to South Ferriby. HILLS BROUGH, *burthe* 1579, *Brough* 1625, *Brough Hill* 1725 all *Terrier*, probably from ME **burgh** (OE **burh**) 'a dwelling, a mansion', *v.* MED s.v., section 3b; *Hills* may be from a local family, cf. Richard *Hill* 1604 *BT*. THE HOLT. MANOR HO, 1842 White. THE MILL (lost, TF 166 967), 1828 Bry, cf. *myll gatt* 1579, *milngayt* 1601 both *Terrier*, *v.* **myln**, **gata** 'a road'. MOUNT PLEASANT, presumably a complimentary nickname. PETER'S SPOUT. RECTORY is *the* *parsonage, A mansyon howse* 1579, *the Parsonaige* 1601, - *Parsonage* 1625 all *Terrier*, - *house* 1633 *Rad*, y^e *parsonage house* 1679, *the Parsonage House* 1690 all *Terrier*. RECTORY FM, 1839 *Dudd* and is presumably y^e *Parsonage Farm* 1724, *The* - 1781 both *Terrier*. It is *Top Farm* 1830 Gre. THE ROOKERY, a wood. SIMON'S POND. STONE FM. SWEED BED PLANTATION, cf. *Swettebye hole* 1579, *sweetabed hole* 1625, *Sweety Bed Hole* 1724, 1781 all *Terrier*, obscure. THORESWAY GRANGE, *the graung'* 1579, y^e *grainge* 1601 both *Terrier, the Grange* 1636, 1652 *Rad*; this is apparently an example of the later use of **grange** 'a homestead, small mansion or farm-house', quoted from L in EDD s.v. 2, cf. Croxton Grange PN L 2 100, but note that the Priory of St Leonard, Grimsby, held land in the parish. THORESWAY NORTH WOLD, cf. y^e *wouds* (sic) 1601, *wold furlands* 1625 both *Terrier, v.* **wald, furlang**; *furland* is a common variant of *furlong* in L. THORESWAY TOP FM, 1830 Gre and is *Lodge* F^m 1828 Bry. THORESWAY WARREN (lost), 1824 O, 1828 Bry, 1830 Gre, *The Warren, the Lings al's the Warren* 1652 *Rad*, cf. *all that Game of Conies planted on part of the Warren Walke* 1652 *ib*, y^e *Warren-Walk* 1724, *the Warren Walk* 1781 both *Terrier, v.* **walk**; for *the Lings*, cf. *lyng banke* in f.ns. (b).

Field-Names

Forms dated 1579, 1601, 1625, 1671, 1679, 1690, 1697, 1707, 1709, 1712, 1718, 1724 and 1781 are *Terrier*; 1633, 1636, 1652, 1653 are *Rad*; 1846 are *GDC*.

(a) Acre Rood Hill 1781 (*Acarwode hyll* 1579, *acre wood hill* 1601, *Akerwood hill* 1625, *Acre Rood hill*, - *hill furlong* 1652, - *Hills* 1653, *v.* **æcer, wudu**; forms in *Rood-* seem to have arisen by the loss of the *w* sound in *wood*); the late Blackbourns House Cl 1781 (*Black bourns home close* 1724, from the surn. *Blackbourn*, the earliest reference so far noted is to William *Blackburn* 1811 *BT*); The Brook 1781 (*v.* **brōc**, cf. *the becke* in f.ns. (b) *infra*); Bucknalls Pingle (1724, from the surn. *Bucknall*, cf. *Ligitt Bucknall* 1734 *BT*, and **pingel**); Bull Bank 1781 (*v.* **banke**); Caistor Gate 1781 (*castergatt, caystergatt* 1579, *Cayster gate, -gait* 1601, *Castor gate* 1625, *Caister Gate* 1636, *Caster gate Furlong* 1652, 'the road to Caistor', *v.* **gata**); Cottage Fd 1846; the Cow Pasture 1781 (*the* ... *Comon*) *Cowpaster* 1636, *the Cowpasture* 1652, y^e *Cow Pasture* 1724); Dunsgate 1781 (1625, 1653, 1724, *dunsgatt* 1579, *Dunsgate Furlong*, - *wonge* 1651 (*v.* **furlang**,

vangr 'a garden, an in-field'), (v. **gata**; the first el. is probably the surn. *Dunn*, v. Reaney s.n.); East Fd 1846 (*The est feyld* 1579, *the eastfield* 1625, - *Eastfeild* 1652, - *East Feild* 1653, v. **čast**, **feld**, one of the open fields of the parish, cf. the South Fd, the West Fd *infra*); Gibdale Wong 1781 (*gypedale, gypdalle* 1579, *Gibdale* 1625, *Gipdale, Gibdale wonge furlong, - Bottom* 1652, *Gib-Dale Wong* 1724, v. **dalr, vangr, furlang, botm**; the first el. is the surn. *Gibb* or *Gipp*, both from ME pet-forms of *Gilbert*, v. Reaney s.n.); Hall Garth, the Hall Garth Farm 1781 (*the hall garthe* 1579, *the Hall Gares* 1636, - *gares* 1652, *Hall Gares* 1653, (y^e) *Hall Garth* (*Farm*) 1724, v. **hall, garðr** 'an enclosure', with some confusion with **geiri**); Home Cl 1846; the Horse Pasture 1781 (y^e *Horse Pasture* 1724); the Ings 1781 (1652, *the Inges* 1625, *the Greate Ings of Thoresway, the Lower Ings, The Ings meadow* 1652, *the Lower Ings* 1653, y^e *Ings* 1724, v. **eng** 'meadow, pasture', in the pl., as elsewhere in the parish); Lincoln Street Road 1781 (1724, cf. *lyncoln' gatt* 1579 (v. **gata**), 'the road to Lincoln', v. **strǣt**); (the) Little Fd 1781 (1724); Long Btm 1846 (v. **botm**); the Long Hedge Walk 1781 (y^e *Long Hedge Walk* 1724, v. **hecg, walk**); New Cl 1781 (1652, 1653); North Fd 1846; Old Seeds 1846; The Orchard Cl 1781 (*the Orchard* 1652, 1653, 1690, y^e *Orchard* 1679, 1697, y^e *Orchard Close* 1724); the Ox Pasture 1781 (*the Oxpasture* 1652, y^e *Ox Pasture* 1724); Parsons Btm, - Hill 1846 (v. **botm, hyll**); the Parson's Platt Hedge 1781 (*Parsons platt* 1652, *the Parsons Platt* 1671, - *platt* 1679, y^e *parsons plat* 1697, v. **person(e), plat**[2] 'a small piece of land', with **hecg**); The Parsons Twelve Acre 1781 (y^e *Parsons Twelve Acres* 1724); The Pen Cl, Pen Lane 1781 (y^e *greater -*, y^e *lesser pen lan close* (sic) 1707, *the great -, The little pen close* 1712, y^e *Penn Close*, y^e *Little Penn Close* 1724, *Penn Lane* 1724, v. **penn**[2] 'a fold, an enclosure for animals'); Reed House Fm 1781 (*Reed house Farm* 1724); Saint Foin Bottom Cl, Sant Foin Hill Cl (sic) 1846 (v. **sainfoin**, the leguminous plant, *Onobrychis viciifolia*); the South Fd 1781, South - 1846 (*Thoresway South field* 1718, v. **sūð, feld**, one of the open fields of the village, cf. East Fd *supra*, the West Fd *infra*); South Hall Garth 1781 (v. **sūð, hall** with **garðr**); Southing Btm 1781 (v. **sūð, eng** with **botm**); The Spring Cl 1781 (1724, v. **spring**); Stainton Cl, - Corner 1846 (cf. *Staynton gatt, - meare* 1579, *Staynton gayt, - meare* 1601, *Stainton gate* 1625, *stanton streate* 1636, *Stanton dale* 1652, named from Stainton le Vale with **gata, (ge)mǣre, strǣt** and probably **deill**); Waingate 1781 (1724, *Wayn gatt* 1579, *wayne gayt* 1601, *wainegate* 1625, *Wainegate Furlong* 1652, *Wain gate* 1653, *Waynegate furlong alias Furth hill Furlong* 1652, *Furth hill* 1625, 'the waggon road or way', v. **wægn, gata**. The meaning of *Furth hill* is uncertain); the West Cl 1781 (*the West Close abutting on the West feild* 1652, *the West Close* 1653, y^e *west Close* 1707, *The West Close* 1712, y^e - 1724); the West fd 1781 (*The west feyld* 1579, - *Westfeild* 1625, - *West Feild* 1652, 1653, y^e *West Field* 1724, v. **west, feld**, one of the open fields of the parish, cf. East Fd, the South Fd *supra*); Willow Seeds

1846 (cf. *the Wyllowes* 1579, *the willowes* 1601); Wood Hill 1781 (1724, *wood hyll* 1579, *woodhill* 1625, *v.* **wudu, hyll** and cf. *wood dale* in f.ns. (b) *infra*).

(b) *asdall* 1579, *asdayl dayll* 1601 (perhaps 'the ash valley', from **askr, dalr** with **deill**); *battersdall* 1579, *batters dayle* 1601, *Battersdale* 1625 (*v.* **dalr** or **deill**; Dr John Insley suggests that the first el. is probably the ME (OFr) occupational term (*le*) *batur* 'a beater of cloth, a fuller, or a beater of metals, a coppersmith', or a surn. derived from it, cf. MED, s.v. *bēter*, Reaney, s.n. *Bater*); *the becke* 1579, *the beck(e)* 1601, 1625, *Beck Lane* 1652 (*v.* **bekkr** and cf. The Brook in f.ns. (a) *supra*); *the Buts* 1625 (perhaps 'the archery butts', *v.* **butt**^{2}); *bysgardalle* 1579, *Bisterdale* (sic) 1625; *one closse called the lyttle churchyarde* 1579, *a little spot of ground ... called the little churchyard* 1625; *the clake* (sic) 1579, *Clacke* (a place) 1625 (perhaps from **clacc** 'the top of a hill' or **klakkr** 'a hill'); *cowslack* 1579, *Cowe slake* 1601, *Cowslacke* 1625 (from **cū** and **slakki** 'a shallow valley', an OWScand word common in Norwegian-settled areas, but rare in the East Midlands); *coxwold gatt* 1579, *Coxwold gayt* 1601, *Cuckswold gate* 1625 ('the road to Cuxwold', *v.* **gata**); *one Marfaurr Cuckold' headland'* (sic) 1652, `*Marfars Cuckold head-land* (sic) 1653 (*v.* **marfur** 'a boundary furrow'; names in *cuckold* seem to refer to land favoured for illicit love-making, *v.* Field 56, and also **hēafod-land**); *croxbye gatt, - me(e)re* 1579, *- gate, - mayre* 1601, *Croxby gate, - hedge* 1625, *- meere* 1652 ('the road to Croxby', 'the boundary of Croxby', *v.* **gata, (ge)mære, hecg**); *the cryngylls* 1579 (from the pl. of **kringla** 'a circle', with reference to something circular in shape); *Dales walke it being likewise a sheepe walke* 1652 (*v.* **walk** and cf. *Ould close Walke infra*); *depedale, depedall(e) gatt, depedale hyll'* 1579, *deepdayll* 1601, *deepedale* (*hill*), *- gate* 1625, *Deepedale Bottom, the valleys called East depedale bottom* (sic) 1652 (*v.* **dēop, dalr** with **gata, hyll** and **botm**); *a Furlong called Dimsy Marfarr* 1652 (*v.* **marfur**; the reading *Dimsy* is uncertain, but we may be concerned with the surn. *Demps(e)y*); *doghull* (sic) 1579, *dogdhul* (sic) 1601 (*v.* **dogga, hyll**); *the est closse* 1579, *the Eastclosse* 1601; *farsdalle* 1579, *-dayll* 1601, *farze dale* 1625, *furze dale* 1625 (obscure); *Furth hill* (*v.* Waingate in f.ns. (a) *supra*); *the gares* 1579, 1625, *Gares furlong* 1652, *- Furlong* 1653 (from **geiri** 'a triangular plot of ground'); *middle -, upper greenedikes* 1625 (*v.* **grēne**^{1}, **dīk**); *grymsdalle* 1579, *-dayle* 1601, *Grimsdale gappe* 1625, *Grimsdale* 1652 (from the pers.n. or surn. *Grim* and **dalr**); *Haberhalf Furlong, Haverbaff* (*furlong*) 1652 (obscure); *ye High Close* 1652; *the Homes* 1579, *- homes* 1601, *- holmes* 1625 (*v.* **holmr**); *Horne Castle streete* 1652 ('the road to Horncastle', *v.* **strǣt**, probably an alternative name for High Street *supra*); *the toppe of the hyll* 1579 (*v.* **hyll**); *on knyuer* 1601, *Kniuers* 1625 (obscure); *Lady wood slacke* 1625, *Lady wood Lath* 1652 (from *lavedi* (OE **hlǣfdige**) 'a lady, Our Lady', **wudu**, with **slakki** (cf. *cowslack supra*), **hlaða** 'a barn'); *the Larth Close*

1652, 1653; *Lead aker furlong* 1652 (*Lead* may be a surn. here, with **æcer, akr** and **furlang**); *Leaworth* -, *Loworth bottom* 1652; *Ledawkin Bottome* (sic) 1652 (the first el. is probably the Fr article *le* plus the surn. *Dawkin,* for which *v.* Reaney s.n. *Dawkins,* as Dr John Insley points out, the second being **botm** 'a (valley) bottom'); *a furlong* ... *called Lincolne hill* 1652; *the little Closes* 1652, *the Litle* - 1653; *louthe gatt, louthedalle, lowth hyll* 1579, *lowthe gayt, Louthedaile,* *lowth hill* 1601, *Lowth gate, Lowthdale, Lowth hill* 1625 ('the road to Louth', *v.* **gata,** but the exact significance of *da(i)le* and *hill* here is uncertain); y^e *low pasture* 1707; *lyng banke* 1579 (*v.* **lyng** 'ling, heather', **banke**); *lyttyl dalle* 1579, *little dale* 1625, *littledale* 1652; *makames* (sic) 1579, *mawkames* 1601 (obscure); *the medowe* 1579, 1601; *the moore, the great* -, *the lyttle* - 1597, *the more, the great* -, *the lytle* - 1601 (*v.* **mōr**[1]); *Mawddales* 1625, *hither mowdales, farther Mondales* (sic) 1652 (the forms are too diverse to suggest a plausible etymology); *the New inned close* (*lately Inclosed*) 1636 (*inned* is from dial. *in* vb. 'to enclose, to reclaim waste land', *v.* NED s.v.; neither *New* nor *lately* should be taken to indicate even an approximate date of enclosure); *nordale,* - *gatt,* - *end* 1597, *nordayll,* - *gate,* - *end,* - *syd* 1601, *Northdale,* - *gate,* - *end* 1625, *North Dale bottom, East Northdale Bottom, west Northdale* 1652, *Northdale Mouth* 1652, *North dale Mouth* 1653 (*v.* **norð, dalr**); *normanby gatt* 1597, *Normanby Holgate* 1652 ('the road to Normanby le Wold', *v.* **gata,** 'the sunken road', 'the road in the hollow leading to Normanby le Wold', *v.* **hol**[2], **gata**); *nunngarthe syde, nunn headland* 1579, *none headland,* - *pytes* 1601, *Nunclose side, Nunpits,* - *head* 1625, *Nunland, nuns Close* 1652, *Nuns Close* 1653 (by which time it is said to be *belonging to the Collidge,* i.e. Trinity College Cambridge; these f.ns. commemorate the holdings in the parish of The Priory of St Leonard, Grimsby, *v.* 1406 Pat, 1535 VE ii, 68a and 1539-40 Dugd iv, 546b, *v.* **nunne** with **garðr, hĕafod-land, pytt, clos(e)** and **land**); *Old Close* 1652, *Ould close Walke* it being a sheep walke, 3 Sheepe walks commonly called Oldclose Walk 1652 (*v.* **walk** and cf. *Dales walke supra*); *pannow how,* - *hyll,* - *steyghe* 1579, - *howe,* - *hill,* - *steyghe* 1601, *Pannow stigh* 1652, *Panny hill,* - *stigh* 1625 (*v.* **haugr, hyll, stigr**; *pannow* is obscure); *the Parke Close* 1652 (*v.* **park**); *the parson breeke* 1579, - *breaks* 1601 (*v.* **persone, brēc** 'land broken up for cultivation'); *Parsonage Close* 1633 (cf. Rectory *supra*); *Parsons bush* 1625; *pease landes* 1579, *peas landes* 1601, *Peaselands, Long* -, *Short peaselands* 1625, *pease land furlong* 1652 (*v.* **pise, land** with **furlang**); *valley called the Prieste dale* 1652 (*v.* **prēost, dalr** and cf. Parsons Btm in f.n. (a) *supra* and *Parsons bush* in (b) *supra*); *Prousedale furlong* 1652 (from the surn. *Prouse,* Reaney s.n. *Prowse,* and **furlang**); *on quyn* 1579 (obscure); *rasyn gatt* 1579 ('the road to Market Rasen', *v.* **gata**); *Rothwell meere* 1625, - *Meere* 1652 ('the boundary with Rothwell', *v.* **(ge)mǣre**); *the runnyll* 1579, *the rundle* 1601 (*v.* **rynel** 'a small stream, a runnel'); *rydsdale gatt* 1579, *Risdale* 1625, 1652; *Scotter thornes*

1625 (probably from the surn. *Scotter,* from Scotter LWR, and þorn); *shepedyke hyll* 1579, *sheapdycke hyll* 1601, *sheepdike hill* 1625 (*v.* scēap, dīk with hyll; presumably this was for washing sheep, though the usual name for this is *Sheepwash*); *the shepegates* 1636 (*v.* shep-gate 'pasturage, or the right of pasturage for sheep'; for an early example *v.* PN L 2, 244); *the Shorte Furrs* 1652 (*v.* furh in the pl.; *furr* is a common form of *furrow* in L); *silver hill, Silver hill furlong* 1652 (*v.* seolfor, hyll; *silver* presumably refers to the colour of wild plants growing here); *lower -, upper Sleagth(es) sleyghtes, sleyght busshe* 1579, *fare* (i.e. far) - *low sleightes, sleight bush* 1601, *the upper slights* 1625 (perhaps from the pl. of slæget 'a sheep pasture', but the pl. of slétta 'a smooth, level field' is certainly possible); *the sleede* 1579, - *slead* 1601, - *sleds* 1625, - *Slead'* 1636, - *Sled* 1652, cf. *Slade -, Slead Close* 1652, *the Sleade Close* 1653 (*v.* slæd; Dr John Insley notes that ME *slâde, sled(e), sledde* can denote low lying ground, a valley, a flat grazing area, a stream, *v.* MED s.v.); *sleipes* 1601 (perhaps a noun, in the pl., derived from ON sleipr 'slippery', applied to places with this characteristic); *Smith feild walke* 1652 (*v.* walk); *South Bottom* 1652 (*v.* botm); *the southe hyll'* 1579, *Southhill side* 1625, *South-hill* 1652 (self-explanatory); *Spell bush* 1625 (literally 'the speech bush', *v.* spell, busc, presumably the site of a local assembly); *Sprowsdale (side)* 1625, *Sprows dale* 1652 (the first el. appears to be the OE pers.n. *Sprow,* though the forms are too late for certainty, the second is dalr 'a valley'); *strypes* 1579, *stripes* 1625 ('the narrow plots of land', *v.* strīp); *suddalle* 1579, *suddaill* 1601, *suddale* 1625, *Suddale Bottom* 1652 (*v.* sūð, dalr and cf. *nordale supra*; the 1601 form suggests at least the influence of deill 'a share of land'); *swyth nolles* 1579, *Swethnowles* 1625 (*v.* cnoll 'a knoll, a hillock' in the pl., with an uncertain first el.); y^e *teddering pasture* 1707 (i.e. a pasture for tethering beasts); *Thoresway field* 1709 (*v.* feld); *tombanke* 1579 (*v.* banke; again the first el. is uncertain); *the Town Closes* 1652; *the Townes end* 1579, *Thorsway Towne end* 1652 (self-explanatory); *Toynton gate, Taynton-gate* 1652 (presumably the first el. is a surn. from one of the Toyntons LSR, since High and Low Toynton and Toynton All Saints and St Peter are too far away for the name to mean 'the road to Toynton', *v.* gata); *Turfdale* 1652 (presumably 'the valley where turves are got', *v.* turf, dalr); *Wharledikes* 1625 (perhaps from hwerfel 'a circle', denoting something circular, dīk); *whyttgatt* 1579, *Whitegate foot,* - *hill* 1625, *White-gate,* - *Furlong* 1652 (*v.* hwīt, gata, perhaps a reference to the chalky nature of the road and note Chalk Pits *supra*); *wood dale* 1579, *certaine valleys or Bottoms called wodale* 1652, *wood furlong', woodfurland hole* 1579, *wode furlands (hole)* 1625 (*v.* dalr, furlang and cf. Wood Hill in f.ns. (a) *supra*; *furland* is a common spelling for *furlong* in L); *wrong meere* 1625 (*v.* vrangr 'crooked', (ge)mǣre 'a boundary, land on a boundary'); *wykam gatt* 1579, *Wikam gate* 1625 ('the road to *Wykeham* (in Nettleton, PN L 2, 238-39)', *v.* gata).

Thorganby

THORGANBY

Turgrebi (sic) (6x) 1086 DB

Turgrimbi 1086 DB, c.1148 YD vii (p), 1166 RBE (p), 1166-80
YCh xi (p), 1205 Cur, *-b'* 1185 Templar, *-by* 1202 FF,
Turgrimebi 1205 Cur, *Turegrimbi* Hy2 LN

Torgrembi 1086 DB

Torgrimbi c.1115 LS, 1186-1200 Dane (p), *-by* Hy2 (1409) Gilb,
Torgrimebi c.1115 LS

Turgramebi Hy2 Dane (p), *Turgrambi* 1202 Ass (p), 1206 ib,
-by 1206 FF, 1207 ib (p), 1242-43 Fees

Turegramesbi 1193, 1194 P both (p)

Turganneby (sic) 1205 FF

Torgrambi 1212 Fees, 1218 Ass, 1273 RRGr

Thurgrimbi a1183 Dane (p), *Thurgrymeby* 1331 Ch

Thorgrimbi lHy2 Dane (p), *Thorgrimebi* 1314 Ch

Thurgrambi 1212 Fees, *-by* 1242-43 ib, 1338 Pat

Thorgramby 1226 FF, c.1240 IB, 1242-43 Fees, 1245 FF, 1272
Ass, 1275 RRGr, 1294 RSu, 1303 FA, 1306 *NCot*, 1311
YearBk, 1312 Pat, 1313 Orig, 1316 FA *et freq* to 1431,
1440 *Yarb,* *-bi* Hy3 HarlCh

Thorgranbi 1268 RRGr, *-by* 1375 Peace, 1472 Fine, *Thorgraneby*
1346 Pat

Thorgamby 1254 ValNor, 1277 RRGr, 1291 Tax, 1292 *NCot,*
1303 FA, 1311 YearBk, 1317 Ipm *et passim* to 1482 *Yarb,*
Thorgaumby 1268 Ch

Thorganby 1296 *AD*, 1503, 1506 Ipm, 1526 Sub, 1558, 1560,
1565, 1618 *Yarb et passim,* *-bye* 1555 Yarb, 1576 Saxton,
1601, 1606 *Terrier*, 1610 Speed, *-bie* 1556, 1566 InstBen,
1558 *Yarb*, 1611, 1639 *Terrier, Thorgonby* 1535 Val iv,
-bye 1564 Pat

Thurganby 1529 Wills ii, 1666 VL

Thirganbie 1576 LER

Torngrebi (sic) 1199 P

Torngrenbi 1199 OblR

Torngramby 1242 RRG

Thorngramby 1280 Ch, 1324 Pat

Thorngranby 1281 QW

Thornganbie 1572 *Yarb*

'Thorgrim's farmstead, village', v. **bȳ**, the first el. being the ON pers.n. *Þorgrímr*, ODan *Thorgrim*. The development of the name is paralleled by that of Thorganby PN YE 263-64. Forms in *Torn-*, *Thorn-* appear to have been influenced by **þorn** 'a thorn-bush'.

CORN MILL. CORONATION PLANTATION. DEEPDALES WOOD, cf. *les Depdales, le Fardepdales* 1413, *Depedale, Middle Depedale* 1638, *Deepedale* 1639 all *Yarb, Deep Dales* 1804 *MiD*, 1805 *Yarb, - dales* 1842 *TA*, self-explanatory and topographically appropriate, v. **dēop, dalr.** THORGANBY COVER (lost, approx. TF 224 979), 1824 O, 1830 Gre. THORGANBY GORSE. THORGANBY HALL, 1804 *MiD*, 1828 Bry, 1830 Gre, *atte Hall* 1341 NI (p), v. **hall,** described by P&H 399 as "Late Georgian, there are remains of a former rebuilding of 1648". It is probably referred to as *the mannor howse of Mʳ Thomas Danbie* 1611 *Terrier*. THORGANBY SCREED, from ModE *screed* 'a narrow strip of land'. VALLEY PLANTATION.

Field-Names

Forms dated 1307, 1413, 1561, 1565, 1601[1], 1602, 1606[1], 1625, 1629, 1638, 1639[1], m17, 1655, 1661, 1664, 1665, 1670, 1671, 1672, 1673, 1674, 1675, 1676, 1677, 1679, 1684, 1805, 1833 are *Yarb*; 1601[2], 1606[2], 1611, 1619 (Croxby), 1639[2], 1671 (Croxby), 1692 (Croxby), 1694-95 (Croxby), 1700, 1709, 1712, 1715, 1718 are *Terrier*; 1804 are *MiD*; 1842 are *TA*.

(a) Judith Bell's Bank and Btm 1804, Tudy Bell Holt, - Btm, - Top 1805 (cf. *Tudeball garth* (sic) 1671, 1672, 1674, *Tudiball - * 1676, *Tudyball Garth* 1672, *Tudiballgarth* 1677, v. **banke, botm, holt, garðr**; Dr John Insley suggests that the first el. may possibly be the Old Breton pers.n. *Tutwal, Tudal,* with the second el. of the name modified through the influence of OE *-bald,* the 1804 form perhaps arising by coincidence or popular etymology, though no certainty is possible); Blacklocks Ings 1804 (from the surn. *Blacklock,* cf. John *Blacklock* 1753 *BT,* with **eng**); Bottoms, - North End, - Oziers 1805 (v. **botm**); Brick Holes 1805; Brick kiln Cl 1804, - Kiln Pce 1833, - pce 1842 (v. **pece**), Brickkiln Walk 1805 (v. **walk** 'a tract of pasture land for cattle or sheep', as elsewhere in this parish; cf. *le Kilnehowse (yeard)* 1601[1], v. **kiln-house**); Brick Yd 1804 (cf. *yᵉ Brick close* 1664, *the - * 1665, *- Close* 1670); Cabbage Pce 1833, - pce 1842 (v. **pece**);

Caistor Road pce 1804 (v. **pece**; Caistor is about eight miles NW of Thorganby);
Cash Cl 1804, Cash Hill (Holt) 1805, Cash Hill Plantn 1833 (named from the
Cashe family, cf. Michael *Cashe* 1641 LPR); Church cl 1842 (cf. *the church wonge*
1638, 1639[1], v. **vangr**); Cottage Cl 1805, 1842, - cl 1833, - Garths 1804 (v. **garðr**
'an enclosure', as elsewhere in this parish, and cf. *the Cottagers Close* 1674, 1677);
Cover Plat 1833, - plat 1842 (v. **cover(t), plat**[2] 'a plot of land', as elsewhere in
the parish); Cow Cl 1804, 1805, 1833, 1842 (*the Cow Close* 1664, 1670, -
Coweclose 1665), Cow Walk 1804 (v. **walk**), - Hill Holt 1805 (v. **cū, hyll**, with
holt); Croxby Walk 1833, 1842 (v. **walk**; named from the neighbouring parish of
Croxby *supra*); East Dale 1804 (cf. West Dale *infra*); Fish Pond 1805, 1833, Fish
Pond Walk 1805, Pond Walk 1833, - walk 1842 (v. **walk**), Fish Ponds &
Plantation 1842 (the 1833 and 1842 forms refer to the same feature); The Forty
Acres 1804, Forty Acres 1833, 1842, - acres 1842; Fox Cover 1805, 1833, 1842 (v.
fox, cover(t)); The Furze Cover 1804 (v. **fyrs, cover(t)**); Garth 1805 (v. **garðr**);
Goose Hills 1804 (v. **gōs, hyll**); Gravel Pce 1833, - pce 1842 (v. **pece**); Great Ley
1804 (from ModE **lea** (or *ley*) (from OE **lēah**) 'pasture, a meadow'); Green Cl
1804; Green Hills 1804, 1805 (1718, *Greenhills* 1672, *the Green hill* 1675, - *hills*
1676, 1677, *little Green hill* 1671, *the little Greenhill* 1672, v. **grēne**[1], **hyll**); High Cl
1804, 1833, - cl 1842, - Close Cover 1804 (v. **cover(t)**); Home Cl 1833, - cl 1842;
Home Hill 1805; Home Walk 1805 (v. **walk**); (Top) Horse Cl 1833, - cl 1842;
House Cl 1804 (*the house close* 1675, described as *adjoyning ye said Messuage*);
Ings 1805, 1833, 1842 (*ye Ings* 1664, *the -* 1665, 1670, *the Inge hedge* 1601[2], *the
Inge Gate, the Inges Hedge* 1606[1], *the Inge gate, The Inges hedge* 1606[2], *the Ynges
hedge* 1611 (v. **eng** 'meadow, pasture', as elsewhere in this parish); Ley Cl 1804,
1805, Lea - 1833, - cl 1842 (cf. *Lea-Croft* 1674, *Lea Croft* 1676, from ModE **lea**
'meadow, pasture' (from OE **lēah**) with **croft**); Line Lands 1804, 1833, - lands
1842 (*the -* 1675, *Linelands* 1684, 'the strips on which flax was grown', v. **līn,
land**); Louth Road 1804 (self-explanatory); Low Cl 1804 (*the Low Close* 1672,
1674, 1677, 1679, *the low -* 1673, 1676); Low Inges 1833, - Ings 1842 (*the low
Ings or Milne Acre* 1673, cf. "a mill in" *Thorgramby* 1226 FF and Ings *supra*);
Low Paddock 1805; Marl Pce 1833, - pce 1842 (v. **marle**); Meadow Cl 1805;
North -, South Middle Walk 1805, Middle Walk 1842 (v. **walk**); North Hills 1804
(1715, 1718, y^e *North Hills* 1664, *the -* 1665, 1670, v. **norð, hyll**); North Ings 1804
(m17, *lez Northenges* 1413, *the North Inges* 1625, 1638, 1639[1], *the Milne acre or
North Ings* 1674, v. **norð, eng**, cf. Ings, Low Inges *supra* and South Ings *infra*);
First -, Second -, Third or Farthest North Walk 1804 (v. **norð, walk**); Nursery
1804; Orchard 1804, 1805, 1833, 1842 (cf. *the Low Orchard* 1673; this is close to
Thorganby Hall), Orchard Cl 1833, - cl 1842; Paddock 1804, 1833, 1842;
Plantation 1833, 1842; Pond Walk (v. Fish Pond *supra*); Saddle Garth 1804 (v.
garðr); Sainfoin Plat 1833, Sanfoin pce 1842 (v. **sainfoin, plat**[2], **pece**); Sand Hills

1804, 1833, 1842 (*v.* **sand, hyll**); Second Plot 1833; Seventy Acre 1833, - Acres 1842 (the actual acreage of the field); Smithfield 1804; South Fd 1833, - fd 1842 (*the Southfeild* 1674, one of the open fields of the village); South Ings 1804 (1655, *les Southenges* 1413, *Sowth ynge* 1565, *the South inges* 1625, - *Inges* 1638, 1639[1], - *Ings* 1673, 1674, 1677, *South Inge close* 1639[1], *the South Ings or Great Ings* 1674, *the Great Ings* 1673, 1674, *v.* **sūð, eng**); South Walk 1805 (*v.* **sūð, walk,** cf. First North Walk *supra*); Stack Yard Cl 1833, - yard cl 1842 (*v.* **stakkr, geard**); Swinhop Plat 1833, - plat 1842 (from the adjoining parish of Swinhope and **plat**[2]); Lr -, Upr Sykes 1804, Middle -, Top - 1833, - sykes 1842, The sykes 1833, 1842 (*the sykes* 1625, *the Sikes* 1639[1], 1676, 1677, *that Close of pasture ground comonly called* - 1674, *v.* **sík** in the pl.); Thorganby New Fm 1804; Three Corner Pce 1833, - corner pce 1842 (*v.* **pece**); Tom Tongue 1804 (*Tom Tong* 1675, *tom* - 1676, *The Tom* - 1679, *v.* **tunge** 'a tongue of land' or **tang** 'a tang, a spit of land'; *Tom* is obscure); Town end pce 1804 (*v.* **pece**), - End Cl 1805, - end Cl 1833, - cl 1842 ('(land) at the end of the village', *v.* **tūn, ende**[1], cf. *the Towne end Gate or lane Commonly called the Watery lane* 1672, *the Town end gate* 1677); Town side 1804, Townside Walk 1805, Town Side - 1833, Town side - 1842; Tween Becks 1805 (*v.* **betwēonan** 'between', **bekkr**); Twenty Acres 1833, 1842; Twenty five Acres 1833, - acres 1842; Twenty six Acres 1833, - acres 1842; Upper pce 1804; Far Walk 1833, 1842, First - 1833, - walk 1842, Great Walk, Lt - 1833, 1842, Middle - 1833, West - 1833 (*v.* **walk**); Watergarth 1804 (*Watergarth wong* 1625, *the watergarth wonge* 1638, 1639[1] (*v.* **vangr** 'a garden, an in-field', as elsewhere in the parish), *Watergarth, Water garths* 1671, *Watergarths* 1672, 1677, *the* - 1673, 1674, *the Watergathe* (sic) 1676, *v.* **wæter** probably in the sense 'wet', **garðr**); West Dale 1804, 1833, - dale 1842 (cf. East Dale *supra*); West Garth 1805 (*v.* **west, garðr**); The White Hart Public House Farm 1804 (it is uncertain whether this is in Thorganby); Wold Newton Corner 1833, 1842 (cf. *newton gayte* 1601[2], *Newton Gate* 1606[1], - *gate* 1606[2] (*v.* **gata**) and *newton meere* 1565 (*v.* **(ge)mǣre** 'a boundary'), referring to the neighbouring parish of Wold Newton).

(b) *le Ambree landes* 1601[1] (*Ambree* is either ME *aumenerī, aumonerie, a(u)mbry* 'a place where alms are distributed, an almonry', or ME *almarie,* ModE *ambry, aumbry* (NED s.v.) 'a storehouse'; the second el. is **land**); *Akar nowke* 1601[2], *acre nouke* 1606[2], *Acar noucke* 1611, *the acre nook land* 1625 (*v.* **æcer, nōk**, described in 1625 as *one land*); *Atterby Close* 1671, 1672, - *Closes* 1674, 1677, - *garth* 1675 (named from the *Atterby* family, cf. William *Atturbe* 1641 LPR); *y*[e] *back side* 1700; *le beck* 1565, *the common beck* 1601, *the* (*comon*) *becke* 1606[2], *the Becke* 1606[1], 1611 (*v.* **bekkr**); *Belesbygate* 1413, *bylysbie gate* 1601[2], *Beelsbie* - 1606[2], 1611, *Belesbie garthe* (sic) 1602 ('the road to Beelsby (an adjoining parish)', *v.* **gata**; the 1602 form is perhaps an error); *les bondgardins*

1413 (this is probably from ME *bond(e)* (cf. **bondi**) and **gardin** in the pl., denoting 'cultivated pieces of land, gardens held by tenure of bond service or rent'); *ad Pontem de Thorgamby* 1307 (p), *atte Brig* 1372 Misc (p), y^e *Bridge* 1709 (*v.* **brycg**, the 1372 spelling in a Scandinavianized form); *Bull peece* m17 (*v.* **pece**); *one Land Called the Butland* 1625, *the But land* 1638, 1639[1] (*v.* **butte, land**); y^e *Calfe Close* 1664, *the Calf Close* 1670; *Chapple Close* 1655, *the Cappell Close* 1673, *the Chappell Close* 1674, 1676, *Chappell lane* 1674, 1676; *a furlonge cauld the clays* 1601[2], *the Claies* 1606[1], 1606[2], 1611, *the clayes* 1638, 1639[1], *Clay wong* 1625 (*v.* **clæg, vangr**) ('the clayey places', *v.* **clæg** in the pl.; there are many claypits in the parish); *Claytons Close* 1655 (*Clayton* (no forename) was a tenant m17); *the Corn Close* 1671, 1674, *the Corne* - 1672; *diuisa de Croxby* 1413, *Croxbie meere* 1565, *Croxbie meare* 1606[1], *Croxby* - 1606[2], *Croxbie mare* 1611 ('the boundary with Croxby (an adjoining parish)', *v.* **(ge)mære**); *a place called Dodswell* 1671 (from the pers.n. or surn. *Dodd*, with **wella**); *in orie*[li] *campo* 1413, *in campo orien'* 1565, *in Campo orientali* 1601[1], *the easte feild* 1601[2], *the East fielde* 1606[1], *the east feilde* 1606[2], 1611, *the Eastfeild* 1625, *the East feild* 1638, 1639[1], *The East feild* m17 (*v.* **east, feld**, one of the open fields of the parish); y^e *feildes meare* 1601[2] (*v.* **mære** and cf. *metas campi de thorganby* 112 (13) *Alv*); *terram Johis del Fewe* (sic) 1413; *Geyres* 1307 (*v.* **geiri**); *Gonnerby gate* 1625 ('the road to Gunnerby', *v.* **gata**); *terram Ric' de la Hawe* 1413 (*v.* **haga**[1] 'an enclosure, etc.'); *the hempyard* 1625 (*v.* **hænep, geard**); *the Hill close* 1664, *the Hillclose* 1665, *the Hill Close* 1670; *Hollywell garth* 1674, *Holy well garth or the Lathegarth* 1676, *Holly well garth, Holliwell Garth or Lathe Garth* 1677 (*v.* **hlaða** 'a barn') (*v.* **garðr,** *Hollywell* may be 'holy well', *v.* **hālig, wella**); *Holme furres* 1625 (*v.* **holmr, furh** 'a furrow'); *the Holmes* 1625, 1639[1], *the holmes* 1638, *the holme wonge* 1638 (*v.* **holmr, vangr**); *one land ... called the Land at Lanes* 1625, *the land att the Laynes* 1638, - *at the Layns* 1639[1] (*v.* **lane**); *le langethorne* 1413 (*v.* **lang**[1], presumably in the sense 'tall', **þorn**); *Lincoln' gate wong* 1625, *Lincolne gate wonge* 1638, 1639[1] (*v.* **gata** 'a road', **vangr** 'a garden, an in-field'; the road is presumably that from Grimsby to Lincoln); *the Lordes Closse* 1606[1], 1606[2]; *the Lowe howse* 1629; *Maltebie garthe* (described as *sometymes pcell of the Manner of Maltbie nere Louthe*), *Maltebie garthe* (described as *One little Close of pasture ... sometimes parcell of the Manner of Maltebie nere Lowthe*) both 1602, *Malteby close* 1629 (*v.* **garðr**); *Mandevelcrof* (sic) 1413 (from the surn. *Mandeville*, with **croft**); *Milne Acre* (*v.* Low Ings in (a) *supra*); *Missadow wong, Mossadow wong* (sic) 1625, *one wong called Mossodow* (sic) 1638, *One wonge called* - 1639[1] (*v.* **vangr**); *the new close ende* 1638, 1639[1], *new close hedge* m17; *Noe Mansion or parsonage house* 1611 (there are no other references to a Parsonage in *Terriers*, but note *the parsonage pingle infra*); *the North Fielde* 1606[1], *the Northefeilde* 1606[2], *the Northfeild* 1625, *the North feilde* 1638, 1639[1], *the North Feild belonging unto the*

East Feild m17, *the Northfeild* 1672 (one of the great fields of the village); *the parsonage pingle* 1611, *the Parsons Pingle* 1672, *parsons pingle* 1712 (*v.* **pingel**. Mr John Field points out that this is probably a piece of glebe land rather than one adjoining or pertaining to the parson's house); y^e *peas-close* 1664, *the Pease Close* 1664, 1670, *the Peasclose* (sic) 1665 (*v.* **pise**); *the pitts* 1639[1] (*v.* **pytt**); *pudding Hole, the Puddin Close* (sic) 1664, *the Pudding hole* 1665, - *Hole* 1670 (probably an area of soft, boggy soil); y^e *River side* 1709, y^e *Greens by* y^e *River Side* 1715; *Ronudilly willowes* (sic) 1625, *Ronndy willowes* (sic) 1638, - *Willowes* 1639[1] (obscure); Mr *Rothwels Sheepwalk* 1674 (*v.* **shepe-walk**); *the Sea-Croft* 1664, *the Sea Croft* 1665, 1670, *Seacroft* 1671, *Sea Croft* 1672; *the Seaven leas* 1638, *one plott of Swarth called the Seaven leas* 1639[1] (from **seofon** with ModE **lea** (from OE **lĕah**) 'pasture, meadow' in the pl.; note the use of the appellative in *eight leas* 1625); *the Sheep Dike* 1677 (*v.* **scĕap**, **dīk**); *shepe gatte* 1561 (*v.* **scĕap**, **gata**; the areal unit, the *sheepgate*, is defined: *one shepe gatte or the gatte of one hundrythe shepe* 1561); *the Sheepe Walke* 1661, *Walk or sheep ground*, - *Sheep walk* 1664, - *Sheep-walk* 1665 - *Sheep Walk* 1670, 1671, 1684, *the Sheep-walke* 1677, - *Sheepewalke* 1679 (*v.* **shepe-walk**); *Sigdele* (sic) 1206 FF; *toftum quod fuit Siwat* 1202 FF ('Sigwat's curtilage', *v.* **toft**, the first el. being the ON pers.n. *Sighvatr*, well-evidenced in L); *le Sowthall garthe wonge* 1601[1], *South halgarth wong* 1625 (*v.* **vangr**); *South Hall garth* 1638, 1639[1], *Southwells Gards Closes* (sic) 1684 (*v.* **garðr**; *Sowthall* etc. is probably 'the southern hall', from **sūð**, **hall**, cf. *West hall garth infra*; the 1664 form is probably an error); *South hill Carrs* 1655 (*v.* **kjarr**); *le Stonepitt, - Stonepitte wonge* 1601[1], *Stone pitt wonge* 1638, 1639[1], *the stonepitt land* 1625 (*v.* **stān-pytt**, **vangr**); *campis de Thorgamby* 1413, *in campis de Thorganbie* 1565, *Campos de Thorganbye* 1639[2], *the fyeldes of thorganby* 1561, *Thorganby feild* 1619, 1692, - *field* 1671, 1694-95 (*v.* **feld**, and cf. y^e *feildes meare supra*); *the thornes* 1625, 1638, 1639[1] (*v.* **þorn**); *Urforde gate* 1606[2] ('the road to Orford (in Stanton le Vale)', *v.* **gata**); *the ward crosse land* 1638, 1639[1] (perhaps from the surn. *Ward*, though the earliest reference to such a family, so far noted, is Mary *Ward* 1724 *BT*; *crosse land* is perhaps 'the selion lying across, intersecting, etc.' from ModE *cross* adj., adv. and **land**); *in occ^{li} campo* 1413, *in campo occiden'* 1565, *in Campo* - 1601[1], *the weeste Feild* (sic) 1601[2], *the west Fielde* 1606[1], - *feilde* 1606[2], *the West* - 1611, *the westfeild* 1625, *the west feild* 1638, *the West* - 1639[1] (*v.* **west**, **feld**, one of the great fields of the village, cf. *in orie^{li} campo supra*); *West hall garth* 1638, 1639[1] (*v.* **hall**, **garðr**, cf. *le Sowthall garthe wonge supra*); y^e *Wheate close* 1675, *the Wheat Close* 1677; *the Would close* 1661, y^e *Would Close* 1664, *Would Close* 1665, *the Wold Close* 1670 (*v.* **wald**); *Wrangdykes* 1307 (*v.* **vrangr** 'crooked', **dīk** in the pl.); *Wyth gat* 1413 (probably *v.* **viðr** 'a wood', **gata**).

Thornton le Moor

THORNTON LE MOOR (now in Owersby parish)
Torentun 1086 DB, -tune 1086 ib, -tone 1086 ib
Torntuna c.1115 LS, -ton 1208 FF, -ton in Mora 1343 Pap
Thorenton' 1221 Welles, 1254 ValNor, Thorinton' 1242-43 Fees
Thorneton 1236 Cl, 1263 FF, 1275 RH, 1281 QW, 1291 Tax,
1428 FA et passim to 1606 Monson, - in the More 1341
NI, in Mora "next" Suthkelleseye 1357 Cl, - in le More
1397 Pat, 1529, 1576 AD, - in the more 1526 ib, - in lee
More 1539 LP xiv, - in le more 1540 LindDep 82, - in le
moore 1583 Monson, - in the moore 1583 ib, - in le more
next Southe Kelsey 1609 ib, - in le Moore 1640 Foster
Thornton 1303, 1316 FA, 1331 Ch, 1346 FA et passim to 1541
AD, - in Mora 1291 RSu, 1381 Peace, - in mora 1327, 1332
SR, - in La More 1318 FF, - in la More 1409 RRep, - "in
the" More 1382, 1394 Pat, 1421 WillsPCC, - "in the Moor"
1472 ib, - in the Moore 1519 AD, - in the more 1581
Monson, - in the More 1645 Holywell, - in le More 1506
Cl, 1594-96 MinAcct, 1622 Monson, 1638, 1662 Foster, - in
le more 1542 Monson, 1553 AD, - in le mowre 1543 ib, - in
le moore 1579 ib, - "near" Kelsaie 1558 InstBen, - iuxta
Kelsey 1576 Monson, - next South Kelsay 1581 ib, - "by"
Ankholme 1444, 1445 Pat, -tone on lamoure 1345 Monson,
-tonne a1567 LNQ v, 1585 SC
Little Thornton alias Thornton in le More 1664 Monson
Thornton le Moor 1824 O

'The farmstead or village where thorn-trees grow', v. þorn, tūn,
cf. Thornton Curtis, PN L 2, 279-80. It is distinguished as on the
Moor, near South Kelsey and near the R. Ancholm.

BEASTHORPE

Bestorp Hy3? (1632) Dods 135, -thorp(') 1318 FF, 1331 Ch,
1348, 1373 Cor (p), 1384 FF, 1402 FA, -thorpe 1335 Pat,
-trope 1542 Monson, Bessthorp 1526 AD
Beesthorp(') 1336 FF, 1581, 1606 Monson, -thorpe 1594-96, 1609
Monson, 1660 Cragg, 1762 Foster, -thropp 1662 ib
Bysthorp(e) 1525, 1526 AD, Bistrop a1567 LNQ v, - laine

1625 *Terrier*
Beisthorp(e) 1529, 1578 *AD, Beysthorp* 1529 *ib*
Besethroope 1543 *AD*
Beasthorpe F.^{m} 1828 Bry

This would appear to be identical in origin with Besthorpe, PN Nt 183-84 and 201, for which the editors point out that, in view of the absence of forms with medial -*e*-, the first el. is unlikely to be the ODan pers.n. *Besi*, recorded once independently from L in DB (Feilitzen 201), They reject, therefore, Ekwall's interpretation, DEPN, that the names are in fact derived from the ODan pers.n. Their suggestion, accepted by Fellows-Jensen, SSNEM 123, is that it is OE **bēos** 'bent grass', as in numerous Beestons, *v.* DEPN s.n., and that the name would mean 'the outlying settlement where bent grass grows', *v.* **þorp**. Fellows-Jensen points out that Beasthorpe lies in Thornton Carrs, where bent grass would be liable to grow.

CAUTHORPE or CANTHORPE (lost, approximately TF 055 958)
    Campthorp' 1234-39 *Foster* (p), *Campthorpp'* 1338 *Monson*
    Camthorp 1236 Cl, - *iuxta thorneton* 1517 *Monson, Camthorpe*
        juxta thorneton' 1518 *ib*
    Camthorpbek^1 1280-85 *Foster, -beck'* 1299 *ib, Camthorpebeke,*
        -beck c.1570 *Monson*
    Cauthorp' 14 *Monson, -thorp* 1559, 1581, 1606, 1664 *ib, -thorpe*
        1598, 1602, 1606, 1609, 1626, 1647, 1655, 1660 *ib, -thrope*
        1650 *Inv, -thropp* 1662 *Foster, Cauthrop furres* 1601 *Terrier*
        (*v.* **furh** in the pl.)
    Cauthorppe Bekke 1498 *Monson, Cauthorpebecke* 1578 *AD*
    Cawtrop a1567 LNQ v, *-thorpe* 1594-96 *MinAcct*
    Couthorp 1664 *Monson*
    Canthorp 1715 *BT* (South Kelsey, St Nicholas), *Canthorpe* 1763
        Foster (-*u*- changed to -*n*- in different ink), *Canthorp Close*
        1839 *TA, Canthorpe Field* 1846 *Dixon*

There seems little doubt that these forms all refer to the same place, but they are so varied as to make an explanation of the first el. difficult in the extreme. It seems that spellings in *Cau*- are regular from the late 15th till the 17th century, after which the -*u*- has been changed to -*n*-. It is, in fact, impossible to be sure that this statement is correct for -*u*- and -*n*- are very frequently almost indistinguishable, though the two spellings in *Caw*- seem to support such a conclusion. All that can be said for certain is that the second el. is **þorp** 'an outlying, dependent settlement', presumably from Thornton le Moor, *v.* further DB lxxxvi.

CARR FM. THORNTON CARRS, 1824 *O*, 1830 Gre, - *Carr Ho.* 1828 *Bry*, *the Carrs* 1664 *Monson*, 1753, 1758 *LPC*, cf. *Thorneton car* 1591 *Monson*, *the Carr* 1638 *Foster*, 1755 *LPC*, *the Common Carr* 1797 *Dixon*, *Com^m Carr* 1803 *ib*, *Common Carr* 1839 *TA*, *the South Carr* 1638 *Foster*, *the North Car*(*r*) 1601, 1625 *Terrier*, *v.* **kjarr** 'brushwood', later 'a bog, a marsh'. CATER LANE, 1645 *Holywell*, 1700 *Terrier*, 1778 *LPC*, 1824 *O*, 1830 Gre, - *lane* 1622 *Monson*, 1846 *Dixon*, *Cater als Carter Lane* (sic) 1663 *TLE*, cf. *the Catter Lane Close* 1674 *Terrier*, *Cater Lane Close* 1758, 1787 *LPC*, - *lane Close* 1839 *TA*. Cater Lane is partly in Owersby, though most of the references are from Thornton le Moor documents. The surn. *Cater*, however, is recorded from Owersby, cf. John *Cator* 1445 AASR xxix, *Jone Cater* 1583 *BT*, *Thomas Cater* 1591 *ib*. GIPSEY LANE. GRAVEL HILL FM, *Gravel Hill* 1797 *Dixon*, 1824 *O*, 1828 Bry, 1830 Gre, *Gravil hill* 1839 *TA*; the farm is marked on *TAMap*. MANOR HO. SHIFTY NOOKING, 1824 *O*, 1830 Gre.

Field-Names

Forms dated 1327, 1332 are *SR*; 1348 are *Cor*; 1540, 1549, 1550, 1551, 1552, 1553, 1558, 1559, 1560, 1564, 1570, c.1570, 1573, 1576, 1580, 1581, 1583, 1591, 1609, 1622, 1653, 1664[1], 1721 are *Monson*; 1578 are *AD*; 1583 are ChancP; 1594-96 are *MinAcct*; 1601, 1625, 1664[2], 1674, 1679, 1700, 1703, 1706, 1707, 1788 are *Terrier*; 1638 are *Foster*; 1645 are *Holywell*; 1723, 1731, 1739, 1744, 1753, 1755, 1778, 1787 are *LPC*; 1797, 1803, 1822, 1840, and 1846 are *Dixon*; 1839 are *TA*; 1847 are *TAMap* (South Kelsey).

(a) Black miles 1839, - Miles 1846 (*Blacke Myles* 1609, *Blackmyles* 1625, 'the black soil', *v.* **blæc, mylde**); B^k Kiln Cl 1803, Brick kiln Cl 1839, - Kiln fd 1846 (*ye brickell closse, the brick kyll closse* 1601, *Brickkill close* 1625); Broxholm Cls 1753 (1700, 1731), Broxholme Cl 1839, - fd 1846 (*Broxholme Closes* 1739, 1744, named from the family of Robert *Broxholme* 1636 *Monson*); Calf Cl 1797, 1803, 1839, 1846; Close 1839; Common Car(r) 1839, 1846 (*v.* **kjarr**); Corn Cl 1797,

1839, 1846, - platt 1797, - plat 1839, - Plat 1846, - Sykes 1797, 1839, - Syke 1846 (v. plat^2, 'a small plot of ground', sík); Cottagers Cl 1797, 1846, Cotchers - 1803, Cocher - 1839 ('land allotted to cottagers', the 1803 and 1839 forms being common variants); Cow Cl 1839, 1846 (the Cow close 1625, - Close 1674, Cowclose 1679), - plat 1839, - Plat 1846, - Sykes 1797, 1839 (v. plat^2, sík); Furze Cl 1839, Great far furze Cl, High -, Low -, Middle -, Near furze Cl 1839, G^t - (sic), High furze cl 1846 (all in the extreme SE corner of the parish) (le newe Fur close 1578, old furr Close 1700, the furre Close 1622, ye great fur-closse commonly called Canthrop furres 1601, the great fur close 1625, ye great Fur Close 1703, probably from furh 'a furrow', cf. Le Furres infra); Grass Plat 1839, 1846 (v. plat^2); Great Cl 1839, 1846 (the great closse 1601, the great close 1609, 1625); Hall Croft 1839, 1846 (v. hall, croft); Hallow Briggs 1839, 1846 (v. brycg, perhaps 'causeway' here); Low -, Top high Mere 1797, - High mere 1839, - high mere 1846 (the furland called High Meire 1601, the High meare furlong 1625, v. (ge)mære 'a boundary'; furland is a common variant of furlang in north L); Holme Cl 1839 (v. holmr); the Home Cls 1797, Home Cl 1803, 1839, 1846; the Ings 1788, Great Ings 1797, G^t - (sic) 1803, Little - 1797, 1839, 1846, L^t - (sic) 1803 (the Inge hedg 1601, the Fore Inges 1622, the fore Inges hedge 1625, the fore inges 1674, inges 1664^2, ye Ings 1700, 1703, 1706, 1707, v. eng 'meadow, pasture', fore 'in front of'); Land Sykes 1797, 1839, - Syke 1846 (v. land, sík); Ley Fd 1846, Leyfield 1839 (v. læge 'fallow, untilled'); Little Fd 1797, - fd 1839; the Long Car 1778, the Long-Carr 1787 (Long Carr 1723, v. kjarr); Long Leys 1797, 1846, - leys 1839 ('pasture, meadow-land'; and v. the discussion under Carr Leys Wood in Brocklesby, PN L 2 66); Low fd, - Cl 1846; Middle fd 1846; Miles Cl 1797, 1839, 1846 (possibly from the surn. Miles); Near fd 1846; new Cl 1797, New cl 1839, - Cl 1846; Pinfold 1839 (v. pynd-fald); Pingle Cl 1797, 1839, Pingle 1846 (v. pingel); Plantation 1822, 1839, 1840; Punch Garth 1797 (Pounch garth 1674, one Close caled ponc gath (sic) 1679 (v. garðr, Punch is probably a byname based on ME paunch(e) 'the human stomach, belly'); The Rectory 1797, Rectory Cl 1839, 1846, - fd 1846; Robinsons Fd 1839, - fd 1846 (named from the Robinson family, cf. John Robinson 1703 Terrier); Rush Carr 1797, 1846, - Carrs 1839 (v. rysc, kjarr); Sheep syke 1839, - Syke 1846 (v. sík); South fd 1839, 1846 (the South fielde 1609); Spalding Cl 1839, - fd 1846 (from the name of a local family, cf. John Spalding 1690 BT); Stewhills 1797, 1803, 1839, Stew Hills 1846 (perhaps cf. the North Stewell 1638); Sykes Cl 1803, 1846, Syke - 1839 (the sick closse, sike closse 1601, Syke close 1625, v. sík); Thatch Carrs 1822, 1840 (Thack Carr 1622, the thach carre 1664, Great Thack Carr 1700, doubtless alluding to reed-beds from which thatching material (þak) could be cut, v. kjarr); Thornton le Moor Beck 1846 (the beck 1601, Thornton Beck 1721, v. bekkr); Thoroughfare 1846 (the Thurefare Close 1674); Thows 1803, Thows Cl 1797, 1839, 1846 (the

Middel Thawes, the upper Thawes 1601, *the uper Thowes, upper -, the middle Thowes hedg* 1625, obscure); Townend Fd 1797, Town end Cl 1846, Townend pce 1839 (*ye townes end* 1601); Gt willow Cl 1797, Gt -, Lt willow Cl 1839, - Willows Cl 1846, Lt Willow fd 1839; Wilson Cl 1839, Wilsons - 1846 (named from the family of Mathew *Wilson* 1839 *TA*).

(b) *the comon acredyk, ye acr' dyke* 1550, *le aker dyke* 1551, *le acredyk* 1553, *le acredike* 1558, *le Acre dike* 1560, *le Acredykhedg* (sic) 1564, *le acredikhedge* 1573 (*v*. **æcer, akr, dīk** and the old Acre-Dike Fence in Middle Rasen f.ns. (a) *supra*); *Atkinsons Houe* 1550 (named from the *Atkinson* family, cf. Thomas *Atkinson* 1591 *BT*); *Barfurres* 1601, *the barfurres* 1625 (*v*. **furh**, with perhaps **bær**[3] (**bere**) 'barley'); *Barkers close* 1625 (named from the family of John *Barker* 1625); *Beaneland furlonge* 1601, *B...landes* 1625 (*v*. **bēan, land, furlang**); *the buttes* 1601 (probably archery butts, *v*. **butt**[2]); *Camplyn headland* 1622, *Camlin head land* 1674, *Camplan headland* 1679 (*v*. **hēafod-land**; *Camplyn* may be a surn.); *Le Croft* 1327 (p) (*v*. **croft**); *dam sick, Dam sick well* 1601, *Damsyke well, the Dam syke* 1625, *the dame sike* 1664[2], *the dam sike* 1674, *dam sike* 1679 (from ME *damme* (**dammr**) 'a dam, a pond formed by damming' and **sīk**); *in campo oriental;* 1576, *in orientali compo* (sic) *de Thorneton* 1578, *ye east feild* 1601, *the east field* 1625 (one of the open fields of Thornton, cf. South Fd in (a) *supra*); *Le Furres* 1549, *Le Furrs* 1559 ('the furrows' from the pl. of **furh**); *atte Grene* 1327 (p), "The Greens" 1583; *Hardynges* 1560 (*v*. **heard, eng**); *Houlton Ground* 1700 (named from the adjoining parish of Holton le Moor); *Kente garthes* 1609, *kent garthe* 1625 (*v*. **garðr**; *Kent* is probably a surn.); *my ladyes Close* 1581, *the Ladies -, le Ladyes Close* 1583, *the Ladies Close* 1594-96, *our Ladies closse, the Ladies closse, my Ladies closse* 1601, *the ladye close* 1625, *the Ladyes Close* 1645, *Ladyes -* 1653 (*v*. **hlǣfdige** 'a lady', one of the 1601 forms suggesting it refers to 'Our Lady'); *the longe Lares* 1601, *the Long Laires* 1625 (perhaps from **leirr** 'mud, clay' or **leira** 'a muddy, clayey place'); *the littell closse by the town sid* 1601; *the Lower Parrok* 1581, *le -, the lower Parke* 1583, *the lower p'ke* 1594-96, *ye low Parke* 1625 (*v*. **pearroc, park**); *ad fossam Thom' mariot* 1348; *Mydelborough* 1601, *middlebrough* 1625 (*v*. **middel, be(o)rg** 'a mound, a hill'); *the middle stigh* 1625 (*v*. **middel, stīg**); *North Carr* 1664[1] (*v*. **norð, kjarr**); *Nue feld* 1674, *the Nufeld* 1679 ('the new field', *v*. **nīwe**); *the Ote Close* 1581, *le -, the Ote Close* 1583, *Oteclose* 1594-96, *the Oate Closse pingle* 1601, *the Oate Close* 1653 (*v*. **āte, clos(e), pingel**); *the padway* 1601; *the parsing Lane* (sic) 1674; *Peul-, Pewlgate* 1601; *potterflett* 1552 (perhaps from **flēot** in the sense of 'a stream' or the like, with **pottere** 'a potter', or the derived surn.); *the Prest close* 1674 (probably from **prēost** 'a priest' and **clos(e)**); *le stuble fyelde* 1570, *the stuble Felde* 1580 (denoting a field that has been reaped and not yet ploughed again, *v*. **stuble**

NED s.v. 4); *Thorneton dike* c.1570 (*v.* **dík**); *campos de Thornton* 1540, 1622, *Thornetonn feilde* 1591 (*v.* **feld**); *a Certaine close called Thorneton South Corne* (sic) 1609 (perhaps for - *South Corner*); *West Close* 1723; *in campo occidentali de Thornton* 1576, *the west feild* 1601, - *field* 1625 (one of the open fields of Thornton, cf. *in orientali campo supra*); *Westiby* 1327, *Westyby* 1332 both (p) ('west in the village', *v.* **vestr, í, bý**).

Usselby

USSELBY (now in the parish of Kirkby cum Osgodby)

> *Osoluabi, Osoluebi* c.1115 LS, *Osolvebi* 1156-57 (Ed2) YCh i, R1 (1308) Ch
>
> *Osolfby* 1209-35 LAHW, c.1221 Welles
>
> *Oselby* Hy2 (p1269) *Bard,* Hy3 AD, 1275, 1276 RH, 1281 QW, 1287, 1293 Ipm, 1298 AD, 1298 *Ass,* e14 AD, 1316 FA, 1324, 1327 AD, 1332 *SR,* 1331 Ch, 1344 AD, 1349 Fine *et freq* to 1622 *Td'E,* - *alias Usselby* 1606 *ib,* -*bye* 1545 LP xx, 1555 *Td'E,* -*bi* c.1200 RA iv (p), *Oseleby* 1223 Welles, *Osilby* 1386, 1392 *Foster,* 1546 Dugd iii, 1619, 1623 *Td'E,* *Osylby* 1433 AD, 1474 *LCCA, Hoselby* 1331 Ch, 14 AD
>
> *Osselby* 1267, e14, 1371, 1379, 1410 AD, 1536 LP xi, -*bye* 1538-9 Dugd vi, *Ossellby* 1460 *Foster, Ossilbye* 1537-38 *AOMB 409, Ossylby* 1559 Pat
>
> *Uselby* 13 AD, 1428 ASSR xxix, 1534, 1655 Monson, - *als Ulceby* (sic) 1647 *ib,* -*bie* 1589 *Td'E,* - *otherwise Oselbie* 1611 *ib, Uselebie* 1611 *ib,* 1612 *Monson, Usylby* 1557 Pat
>
> *Vsleby* 1539 *Monson,* 1618 *Foster,* 1655 *Monson,* 1766 *Td'E,* - *otherwise Uslaby* 1765 *ib,* - *otherwise Usselby* 1789 *Td'E,* -*bie* 1602 *Terrier,* 1613 *Td'E, Uslaby* 1629 *ib,* 1687 TLE
>
> *Usselby* 1324, 1330, 1345 AD, 1357 *Cor,* 1357, 1358, 1385, 1407, 1412, 1424 AD, 1445 ASSR xxix *et passim* - "next" *Kaster* 1509-10 LP i, - *alias Ulceby* (sic) 1571 Pat, -*bee* 1602 *Monson,* -*bye* 1610 Speed, *Usselbie ais Oselbie* 1611 *Td'E, Vssylby* 1530 Wills ii, *Ussilby* 1623 *Td'E, Ussolbye* 1576 Saxton, *Ussleby* 1586, 1788 *Td'E*
>
> *Owssylby* Hy7 ASSR xxix, *Ousleby als Usoleby* 1721 *Td'E,* - *als Useleby* 1721 *ib*
>
> *Ozelby* 1535 VE iv, *Uzleby* 1651 WillsPCC, - *otherwise Usselby* 1721 *Td'E*

'Oswulf's farmstead or village', *v.* **by**. Ekwall, DEPN s.n., assumes that the first el., the OE pers.n. *Ōswulf,* is here in an Anglicized form, *Ōsulf,* from ON *Ásulfr,* ODan *Ásulf,* but *Ōswulf* is well-attested in OE sources, cf. John Insley, *Studia Anthroponymica Scandinavica* 3, 26-27, 1985. There is, therefore, no reason to doubt that the first el. is the OE pers.n., especially in view of the fact that Scand. *Asulfr* is characteristically Norwegian and is rare in Denmark.

It is noteworthy that there is no trace of the OE gen.sg. -*es* in the forms of Usselby. It is possible that the 12th century -*a*- and -*e*- represent a Scand. gen.sg. -*a* < -*ar,* as proposed by Ekwall, IPN 62, suggesting that the p.n. had been given by Scandinavians. For a similar development cf. Audleby, Barnetby le Wold and Worlaby, PN L 2 88, 8-10 and 302-3 respectively.

FISH POND FM, 1894 *Td'E* (Plan), cf. *Fish Pond* 1828 Bry, *Usselby Fish Pond* 1839 *TA* (Owersby). FISH POND PLANTATION, 1894 *Td'E* (Plan). GIPSEY LANE, *Gibsy Lane* (sic) 1894 *ib.* THE GRANGE, 1894 *ib.* This is probably an example of the later use of **grange** 'a homestead, a small-mansion or farmhouse' recorded in EDD s.v. from L. KILN (lost, approximately TA 100 939), 1824 O, 1830 Gre, *Kilns* 1828 Bry, cf. *Kiln Close* 1829 *MiscDep 43.* NEW BRIDGE. THE ROOKERY, 1894 *Td'E* (Plan). THE SCREED, 1894 *ib,* from *screed* 'a narrow strip of land', *v.* NED sb I, 1b and cf. Long Screed in Cadney, PN L 2 78; it is in fact a narrow plantation. USSELBY HALL, *Hall* 1828 Bry, 1930 Gre, *Usselby House* 1835 *Td'E.* USSELBY MILL (lost, approximately TA 104 937), 1828 Bry, 1830 Gre, 1894 *Td'E* (Plan), *molendino aqua'* a1537 *MiscDep 43, Osselbye, firma molend'* 1538-39 Dugd vi, *Vslabie water-mill* 1609 *BT, the water mille in Uslabye* 1626 *Td'E, the water Mill* 1762 *Terrier, Water Mill* 1833 *Yarb,* cf. *le Milnedam* 13, 14 AD. USSELBY MOOR, 1811 *Td'E,* 1824 O, 1830 Gre, *inter moram* Hy2 (p1269) *Bard, le More* 13 AD, "the moor" 1330 ib. It is also *a certain Common ... known by the name of Usselby Common* 1811 *Td'E.* WEST FM, 1894 *Td'E* (Plan).

Field-Names

Undated forms in (a) are 1829 *MiscDep 43.* Forms dated c.1115 are LS; Hy2 (p1269) and 112 (p1269) are *Bard;* 13, Hy3, 1267, e14, 14, 1324, 1327, 1330, 1345,

1358, 1372, 1379, and n.d. are AD; 1386, 1631 are *Foster*; 1389-90 are *AD*; Hy7 are AASR xxix 41; 1537-8 are *AOMB*; 1603, 1606, 1611, 1613, 1619, 1622, 1623, 1626, 1629, 1637, 1638, 1639, 1711, 1728, 1793, 1811 are *Td'E*; 1657, 1658, 1660 are *TLE*; 1707 and 1762 are *Terrier*; 1828 are Bry.

(a) the beck 1762 (*ultra bec* c.1115, *le Bec* 13, *le Beke* e14, (land lying) "bisowt the beck" 14, *le Bech* 1330, *le Bek* 1371, *v.* **bekkr**); Brats (*le Brattes* 1637, *v.* **brot** 'a small piece of land', in the pl.); Brick Kiln Plat (*v.* **plat²**); Carpenters Cl; Claxby cls 1762 (1711); Clay Hill; Coney Yard (*v.* **coni, geard**); Cow Cl; Gt -, Lt Cridikes (*v.* **dík**); Cringlings (*v.* **kringla** 'a circle, something circular', **eng** 'meadow, pasture', as elsewhere in this parish); Croft Cl, Croft Hill; the East feild 1762 ("the east field" 13, 1327, 1345, *the east feild of Uslabye* 1626, *the Eastfeild* 1658, *the East feild of Uselby* 1660, y^e *east field* 1711, one of the common fields of the parish, cf. the West Fd of Usselby *infra*); Far Pce; Bottom -, Middle Grudge (possibly an uncomplimentary nickname); W. Hargraves -, Hargraves Cl; Hollows, Hollows Pce (cf. *Holow* 1371, *v.* **hol¹**, cf. *Holongate* 13 (with *n* for *u*), *Holougat, Binortholougat* 1386 ('the road running in a hollow', *v.* **hol²**, **gata**, with **bī, norð** 'to the north'); *Holousik'* 1386 (*v.* **hol²**, **sík** 'a small stream, a ditch'); Home Cl; Horne Cl; Horse Cl, - Plat (*v.* **plat²** 'a plot of ground', as elsewhere in this parish); Gt -, Lt Ings (cf. *Forhenges* 1358, *Fornheng'* (sic) n.d., *le Estynges de Osilby* 1386, *the East Inges of Uslabie* 1626, - *Ings of Uslaby* 1629, *the East Ings* 1631, *le East Inges* 1637, *the* - 1638, *the East ynges* 1639, *Lees west Ings* (sic) 1603, *the ynges Closes* 1638, 1639, *v.* **eng**); Kirkling Cl; Long Cl; Middle Cl; Middle Plat (*v.* **plat²**); North -, Old -, South Moor Plat 1829, the south Moor 1762 (*v.* **mōr¹**, **plat²**, cf. Usselby Moor *supra*); Mounsey Cl (this probably commemorates the name of John *Mounsey*, Curate of Usselby from at least 1755 to 1800 *BT*); New Cl (*the new close* 1537-8); Orchard; Paddock; Pars^e 1828 (i.e. "Parsonage", and note the statement in 1707 *Terrier no Vicaridge house*); Peacock Hole (named from the *Peacock* family, cf. George *Peacock* 1766 *Td'E*); the pinfold cl 1762, Pinfold Cl 1829 (*v.* **pynd-fald**); Bottom -, Middle -, Top Plat, Plat Holt Cl (*v.* **plat²**, **holt**); Poor House Gardens; Rasen Road Cl (selfexplanatory); Round Cl; Rye Cl; Sand Hills; the south Moor 1762 (*v. supra*); Stack Yd; Top Cl; Town-end Cl; Usselby Common 1811; the West Fd of Usselby 1793 ("the west field" 13, 1324, *the west feild* 1606, 1619, *the West* - 1622, *the west feild of Usleby* 1657, *in campo occidental' de Osilby* 1386, *in occident' campo de Uslebie, in Occident' campo de Uselbie* 1613, one of the open fields of the parish, cf. the East feild *supra*); Wet Furrows.

(b) (*le*) *akerdic* Hy2 (p1269), 112 (p1269), *Acre Dyke Close* 1606, 1623, *the acre dyke* (*close*) 1619, *the Acre dyke Close* 1622 (*v.* **æcer**, **akr**, **dík** and the old

Acre-Dike Fence in Middle Rasen f.ns. (a) *supra*); *Bayilgarth* 1386 (*v.* **garðr**, the first el. being probably the OE fem. pers.n. *Bēaghild*); *Cat' thyng* Hy7 (from the surn. *Cater*, cf. John *Cator* 1445 AASR xxix, and þing 'property, premises'); *Cevenlhol* (sic in transcript) 1324; *Chaxchrokes* (sic in transcript) 1371; *ad Crucem* Hy3 (2x) both (p), (*v.* **cros**); *Disnayland·* 1386 (from **land** and the surn. *Disnay*, Thomas *Dysenay de Kyngerby* 1386 is a witness to the charter); *in campo* -, *in campis de Usselby* 1358 (both checked from original deeds), *in campis de Oselby* 1389-90, - *de Ussleby* 1637, *Usselby feilde* 1611, 1619, *the Feildes of Osilby* 1619 (*v.* **feld**); *ad fimem* (sic) *occidental' de Ossilby* 1386 ('Usselby West End'); *Foxheugh* (probably with -*e*- for -*o*-) 1358 (*v.* **fox**, **hōh** 'a spur of land'); *Hauscho* 1386 (uncertain); *Hegilgat* 1372 (*v.* **gata**; the first el. may perhaps be an unrecorded OE fem. pers.n. **Hēahild*); *Hengcroft dike* Hy3 (probably from **eng** (with prosthetic *H*-) and **croft**, with **dīk**); *les Houstelanges* 1324, *les Oustlanges* 1327 ('the eastern long strips', *v.* **austr**, **lang**²); *le Hyll* 1345, *le hyl* 1386 (*v.* **hyll**); *Jedwynackyr* 1386 (the first el. is probably a Scandinavianized form of OE *Ēadwine*, as suggested by Dr John Insley, who compares ON *Iatgeirr* from OE *Ēadgār*, ON *Iatvarðr*, OSwed *Iædwardh* from OE *Ēadward*, and Yaddlethorpe LWR, *Iadulf(es)torp* 1086 DB, whose first el. is a Scandinavianized form of OE *Ēadwulf*, the second el. is ON akr 'a plot of arable land'); *Linelandes, Linhevedlande* Hy3 (*v.* **līn** 'flax', **land**, **hēafod-land**); *Northercroftes* Hy2 (p1169), *Nordercroftes* 112 (p1269), *North Croftes* 13, *Northe croftes* Hy3, *North croftes* e14 (*v.* **norðor**, **norð**, **croft**); *Oxheplaice* 1379 (*v.* **place**); *the Pond close als mill close* 1639 (cf. Mill Ings (a) *supra*); *tr' Prioris de Elsham* 1386 ('land of the Prior of Elsham'); *pulley Close* 1537-8; *Redholm'* 1386 (*v.* **hrēod** 'reed', **holmr**); *Scronxton thyng* 1537-8 (from þing 'property, premises' with the surn. *Scronxton*); *South Croftes* 13, *Suthecroftes* Hy3, *Suthcroftes* e14 (*v.* **sūð**, **croft**); "the south field" 1358 (only a single reference to this has been noted, while East & West Field are well-evidenced); *Sperri hevedlande* Hy3, *Speriholes* 1371 (presumably from a pers.n. or surn. with **hēafod-land** and **hol**²); *Stanfurlags* (sic) 1267, *Stainfurlanghes* Hy3, *Staynfurglas* (sic) 1371, *Staynfurlangs* n.d. (*v.* **stān**, **steinn**, 'a stone', **furlang**); *Thirnberth* 13, *Chirnbergh* (sic in transcript) e14 (possibly 'thorn hill', *v.* **þyrne**, **beorg**); *Thyrstpit* 1372 ('the demon-haunted pit', *v.* **þyrs**, **pytt**); *Usselby croftes, Osselby Croftes* 1358 (*v.* **croft** in the pl.); *Usselby Thorne, Osselbythorn* 1358 (presumably from a prominent thorn-tree); *yᵉ water hill homestead* (sic) 1711; *le Wath* 13, e14 (*v.* **wað** 'a ford'); *le Westlangs* 1324, *le Westelangges* 1386 ('the western long strips', *v.* **west**, **lang**², cf. *les Houstelanges supra*); *tr' Will'i Attewell'* 1386 ('the land of William at the Well', *v.* **wella**).

Walesby

WALESBY
> *Walesbi* (3x) 1096 DB, c.1115 LS, 1188 P (p), Hy2, l12,
>   1196-1203 RA iv, 1202 Ass (p), 1204 P, 1204, 1205, 1206
>   Cur, c.1210 RA iv, *-bia* 1212 Fees, *-by* 1187 (1409) Gilb,
>   1198 (1328) Ch, John RA iv, 1223 Cur, 1240 FF, c.1240
>   RA iv, 1242-43 Fees, 1249 FF, 1254 ValNor, 1263, 1268
>   Ipm, 1269 FF, 1272 *Cor*, 1275 RH, 1281 QW, 1291 Tax,
>   1303 FA *et freq*, *-b'* 1240-50 RA iv, *-bie* 1576 LER, *-bye*
>   1576 Saxton, 1610 Speed, *Walisby* 1295 RSu, *Walysby*
>   1509 Ipm, 1526 Sub, *-bye* a1537 *MiscDep 43*, 1547 Pat
> *Walesbi Hundred* 1086 DB
> *Whalesby* 1414 Cl
> *Waylesbie* 1576 *FLIrnham*, *Wailsby* 1708 *Td'E*
> *Wallesby* 1467-72 ECP, *-bie* 1601 *Terrier*
> *Walebi* 1188, 1190, 1191, 1192, 1193, 1194, 1195, 1197 all P
>   (p), 1196 ChancR, 1196 Cur, *-by* 1220 ib, 1239 RRG,
>   1246 Pat, 1247 RRG

'Val's farmstead, village', *v.* **by**. The pers.n. is ON *Valr,*
ODan *Val,* as in Walshcroft Wapentake *supra,* the same man
presumably giving his name to the district and to the settlement.
Walesby, PN Nt 63-64, has an identical etymology. It should be
noted, however, that Dr John Insley includes the formal possibility
that the first el. of these names may be the OE pers.n. *Walh.* He
points out that "OE *Walh* occurs in 8th and 9th century charters,
and is attested as late as the early 11th century (in Ker 176c:
Durham) as the name of a manumitted serf."

OTBY
> *Otesbi* 1086 DB
> *Otebi* 1086 DB, Hy1 LN (p), Hy2 Dane (p), *-by* 1166 RBE (p),
>   1177 P (p), 1183 Dane (p), 1566 Pat, 1666 VL
> *Ottebi* c.1115 LS, 1141-54 RA vi (p), Hy2 ib iv (p), Hy2 Dane
>   (p), 1170-98 Revesby (p), lHy2 Dane (p), l12 RA iv (p),
>   1200 P (p), 1212 Fees, *-by* 1140-47 RA i (p), Hy2 (1409)
>   Gilb (p), l12 *MiD* (p), 1242-43 Fees, 1269 FF, 1272 *Ass,*
>   1273 Misc, 1275 RH, 1281 QW, 1303 FA *et passim* to 1507

Lanc, *-beya* 1198 (1328) Ch, *-bie* 1576 *FLIrnham*
Othebi 12 *RevesInv*, *-by* Hy3 RevesbyS
Otby 1327 *SR*, 1386 Peace, 1618 *Foster et passim*

'Otti's farmstead, village' *v.* **by**, the first el. being the ODan pers.n. *Otti*, which Fellows-Jensen, SSNEM 62, points out had been thought to be an early loan from Continental Germanic, but has been shown to be a hypocoristic form of *Ottarr*. *Otti* is found independently in DB in Sf and Y.

RISBY
  Risebi 1086 DB, 1191, 1194, 1195 P, 1196 ChancR, 1197, 1198,
    1199 P, 1202 FF, 1205 P, 1206 Ass, 1206 Abbr, 1219 Ass,
    -b' 1185 Templar, *Risabi* c.1115 LS, *Riseby* 1193 P, 1208
    FF, eHy3 (1409) Gilb, 1242-32 Fees, *Ryseby* 1242-43 ib,
    1256 FF
  Risseby 1385 Peace, 1416 Ormsby, *Rysseby* 1467-72 ECP
  Risby 1268 Ch, 1276 , 1310 RH, 1310 *MM*, - *iuxta Walesby*
    1373 *FF et passim*, *Risby Hall* 1824 O, 1830 Gre, *Rysby*
    1275 RH, 1287 Ipm, 1467 WillsPCC
  Reysby 1519 DV i, 1547 Pat, *Reisby in Walesby* 1542-43 Dugd
    vi

'The farmstead, village among brushwood', from ON **hrís** 'brushwood, shrubs' and **by**, identical with Risby Sf, YE and LWR, as well as Rejsby and Risby in Denmark, as Fellows-Jensen points out, SSNEM 65. It is possible that Risby, found four times as a settlement-name in this country, is, in fact, a name transferred from Denmark, but, if this were so, it was presumably as topographically appropriate for the English names as for the Danish.

CATSKIN LANE, cf. *the Catskin* 1792 *Td'E*, *High Catskin* 1795 *EnclA* (Tealby), *Catskins close* 1724 *NW*, *Long Cat Skin Close* 1795 *EnclA* (Tealby), *Catskin Close* 1829 *Padley*, obscure. HIGH STREET, *the hye streete* 1579, *the streete, the highe streete* 1611, cf. *streete furlonge* 1579, 1611, *Street furlong(e)* 1638, 1664, 1697, 1700 all *Terrier*; this is the name of the ancient trackway from Horncastle to Caistor, which becomes Middlegate Lane from Caistor to South

Ferriby. LOYD'S HO (lost, approximately TF 125 932), 1828 Bry, presumably from a local surn. MIDDLE MOOR (lost), 1828 Bry, 1852 *TA*; it is shown as that part of Walesby Moor, south of the road running west from Walesby to the parish boundary. NOVA SCOTIA BRIDGE, a nickname of remoteness, being situated on the boundary with North Willingham. RECTORY, 1842 White, *Rect^y* 1828 Bry and is *the parsonage* 1579, 1601, 1664, *the Parsonage house* 1638, *ye Parsonage house* 1709, 1718, 1724 all *Terrier*, *a Parsonage House* 1829 *Padley*, and cf. *the parsonage leyes, leyes belonginge to the parsonage* 1579, *the parsonage lees, Leaes belong to the parsonage* (sic) 1601, *the parsonage closes* 1638, *the Parsonage closes* 1664 all *Terrier. leyes*, etc. is from the pl. of **lea** 'meadow, pasture', itself a later development of OE **lēah**. RED HO. RISBY MANOR is *Risby Hall* 1824 O, 1830 Gre. RISBY MOOR, 1709 *Terrier*, 1743 *Td'E*, 1804 *NW*, *the moores of Risbye* 1638, *Risby Moore* 1671, 1697, *ye Moore of Risby* 1700, *ye Moors of Risby* 1724 all *Terrier*. SNOBS AWL (lost, approximately TF 128 929), 1828 Bry. TOP FM. WALESBY GRANGE, cf. "The Grange lands" 1558-79 ChancP. WALESBY HALL (Kelly), *Aula* "of" *Walesby* 1268 Ipm (p), *atte Halle* 1327 *SR* (p). WALESBY MOOR, 1700 *Terrier*, *the more of Walesby* 1556, 1557 *Td'E*, *the common mo(o)re* 1579, 1601, *Wailesbye common moore* 1638, *Wailsby common Moore* 1664, *the moor* 1697, *ye Moor of Walesby* 1724 all *Terrier*, and cf. *Walesby Inmoore* 1671 ib. WALESBY TOP FM, - PLANTATION.

Field-Names

Forms dated lHy2 are MCD 290; l12, c.1210, and 1240-50 are RA iv; 1327, 1332 are *SR*; 1558-79 are ChancP; 1579, 1601[1], 1638, 1664, 1671, 1697, 1700, 1709, 1718, 1724, and 1793 are *Terrier*; 1601[2] are *NW*; 1768 and 1852[1] are *Yarb*; 1829 are *Padley*; 1852[2] are *TA*.

(a) Atholt Cl 1829 (perhaps from the ME surn. *atte Holt*); East -, North -, South -, West Barn Wold 1852[2]; Barr's Paddock 1852[2] (from the surn. *Barr*, cf. William *Barr* 1810 *BT*); Gt -, Lt Beanfield 1852[2] (*v.* **bēan, feld**); Biliff Cl (sic) 1829 (presumably an error for Bilcliff, a surn. well-evidenced in the parish, cf. Edward *bylcliff* 1563 *BT*); Bottle Ings 1829 (*v.* **eng**); Breathings 1829 (*bredhenges* (sic) lHy2 'the broad meadows', *v.* **breiðr, eng**, a Scandinavian compound, cf.

brethinge becke or river 1579, *breything beck* - 1601[1], and for another instance of the name *v.* the f.ns. (b) Brocklesby, PN L 2 70); Broad Sykes 1829 (*Brode sicke* 1579, *Broadsike* 1601[1], *brode syke* 1638, *Broad sike* 1664, *v.* **bräd, sik**); Brick Kiln Moor 1852[2] (*v.* **mōr**[1]); Cabbage Cl 1829; Calf Cl 1829; Great -, Little Chapel Cl 1829, Gt -, Lt Chapel 1852[1] (*Chappell Dale* 1579, *Chappel* - 1601[1], *Chappell daile* 1638, 1664, - *dale* 1671, 1697, - *Dale* 1700, *Chapple dale* 1709, - *Dale* 1718, *Chapple-dale* 1724, *v.* **deill**); Church Cl 1829 (*the Churche Closse* 1579, *the churche close* 1601[1], *the Churchclose* 1638, *the Church close* 1664, - *Close* 1671, 1697, *y*[e] - 1700, - *Church-closes* 1709, - *Church Closes* 1718, - *closes* 1724, cf. *Kirkhill* in (b) *infra*); Church Yd 1852[2]; Far -, Near Cliff Cl 1829 (*a Close call'd Cliff* 1700, *y*[e] *Cliff* 1718, 1724, *v.* **clif**); Cottage Moor (sic) 1829 (*v.* **mōr**[1]); Cow Cl 1829, - Paddock 1852[2]; Cream poke Cl 1768 (a complimentary term for rich pasture, cf. *Creame Poake Nooke* in the f.ns. (b) of Bigby, PN L 2 55); Cuckoo Cl 1829; Eight Acre 1852[2]; Fitch Close Btm, - Close Hill 1829 (named from the *Fitch* family, cf. Joshua *Fitch* 1700); Five Acre 1852[2]; Garth 1829 (*v.* **garðr** 'an enclosure'; it is located near a house); Garth End Cl 1829 (*the garthe endes* 1579, *the garth* - 1601[1], *v.* **garðr, ende**[1]); Gravel Rib Fd 1852[2] (perhaps alluding to a gravelly soil, *v.* **gravel**; *Rib* is from ME *rib(be)* (OE **ribbe**) "a medicinal herb; ribwort (*Plantago lanceolata*), hound's tongue (*Cyonoglossum officinale*), ? costmary (*Chrysanthemum balsamita*)" MED s.v. (2), cf. Rib Moor *infra*); Great Cl 1829 (area 37a. 2r. 8p.); Grimplings 1829 (*grimplinge* (*Closse*) 1579, *Grimpeling* (*Close*) 1601); Little Hardy 1852[2]; Hewson Cl 1852[2] (from the surn. *Hewson*); Hill Cl 1829, 1852[2] (*hill close* 1671, *the Hill Close* 1671, 1697, *y*[e] - 1700, 1718, 1724, - *Hill-close* 1709, cf. *subtus montem* lHy2, *v.* **hyll**); Hoisier Holt 1852[2] (*v.* **oyser, holt**); Home Cl 1829; Horse Cl 1852[2], Horse Paddock 1852[2]; In Wold 1852[2] ('the inner wold', *v.* **in, wald**, cf. North -, South Out Wold *infra*); Ings 1829, The Ings 1852[2] (*the Ings* 1664, *litleinge* 1579, *little Inge* 1601[1], *v.* **eng** 'a meadow'); Knott Cl 1829 (*Knotte* 1579, *Knote* 1601[1], the knott close 1638, 1664, *v.* **knottr** 'a hillock'); Far Ling Cl 1829, Ling Moor 1852[2] (*v.* **lyng, mōr**[1]); Middle Low Barn Pce 1852[2]; North -, South Long Wold 1852[2] (*v.* **wald**); Lord's Cl 1829 (*the Lordes Close* 1579, - *lordes cloase* 1601[1], - *Lords close* 1638, - *Close* 1671, 1697, *y*[e] *Lord's* - 1709, 1718, 1724, probably enclosed demesne land; the early forms with *the* seem to exclude the possibility that *Lord* is a surn. here); East & North Low Fd 1852[2] (Risby) (*Rysbie northe lowe feilde* 1579, *Risbie north low feild* 1601[1]), South & West Low Fd 1852[2] (Risby) (*the south lowe felde* 1579, *Risbie South low feild* 1601[1]; this is *the little lowe field* 1638, *little lowe feild* 1664, *the little Low* - 1671, self-explanatory); Mill Cl, Water Corn Mill 1829 (on the northern boundary of the parish, cf. *le Mulnegate* in (b) *infra*); New Long Acre 1829 (cf. *Long Acres* 1579, *Longe* - 1601[1], *long acres* 1638, *v.* **lang, æcer**); Normanby Side Cl 1829 (referring to the neighbouring parish of Normanby le Wold); Little Northings 1829 (*northinge* 1579, *North Inge* 1601, *Northinges* 1638, *Northinge* 1664, *Northing* 1671, *v.* **norð, eng,** and cf. South Ings *infra*); Noye (sic) 1829; North -, South Out Wold 1852[2] (*v.* **ūt** 'outer', **wald,** cf. In Wold *supra*); (Little) Ox Pasture 1829 (*the Oxe pasture close* 1638, *the Ox pasture Close* 1671, *the oxpasture close* 1697, *y*[e] *Ox Pasture Close* 1700, *y*[e] *Ox pasture-close* 1709, *the ox-pasture Close* 1718, 1724); Pasture Cl 1829; Paulins 1852[2] (from the surn. *Paulin*); Pingle and Orchard 1829 (*v.* **pingel**); Pit Cl 1829 (*v.* **pytt**); Rib Moor 1852[2] (*v.* **ribbe, mōr**[1], cf. Gravel Rib Fd *supra*); Rickpurse or North Moor 1852[2] (*v.* **norð, mōr**[1]; *Rickpurse* may be a derogatory name alluding to lack of profit from the land, from *rick* vb. 'to strain, to wrench'); Rickyard Barn etc. etc. 1852[2]; Sand's Cl 1829 (*Sandes* 1579, 1601[1], *the sands* 1638, *the Sands* 1664, *v.* **sand**); South Ings 1829 (*Southinge* 1579, *Southinges* 1601, *the lowe southinge* 1664, *the low southing* 1671, *v.* **sūð, eng** and cf. Little Northings *supra*); Square Cl 1829 (*the square close* 1638, *the Square* - 1664); Sunrise Cl 1829 (*Sunryse* 1579, *Sunnrise* 1601[1], *Sunrise in* (sic) 1638, presumably land facing the east); Ten Acre Moor 1852[2] (*v.* **mōr**[1]); Town End Cl 1829; Wakelandes 1829 (*Wa*[...]*e Landes* (blot on MS) 1671, *Wakelands* 1709, 1718, 1724, *wateland Spring* (sic) 1697, 'lands where wakes or festivals are held', *v.* **wacu, land**); Wardales 1852[2]; Well Becks 1829 (*Wellebec* 1240-50, *Welbecke* 1579, *Welbeck* 1601[1], *a close called* - 1638, - *called Welbecks* 1664, *v.* **wella, bekkr** and cf. Welbeck Spring PN L 2 234); Whittakers East -, Whittakers West Moor 1852[2] (from the surn. *Whittaker*, cf. Edwin *Whittaker* 1851 *Census*, with **ēast, west, mōr**[1]); Plantation of Willows 1829; Wold Cl 1829 (cf. *the woldes* 1638, - *Woldes* 1664, - *Wold* 1697, *y*[e] *Wolds* 1700, 1709, 1718, *Wailesbye Would* 1671, *Walesby Wold* 1724, *the hedge of the woldes* 1638, *v.* **wald**).

(b) *Battbondes* 1579, *Batt bondes* 1601[1], *Baf bonds* (sic) 1638, *Baffbonds* (sic) 1664, *Baff bonds* 1671, - *bondes* 1700 (obscure); "close called" *Belfreygarth* 1558-79 (*v.* **garðr,** the first el. is dial. *belfrey* 'a lean-to or shelter-shed'); *Beornes Welle* lHy2 (from the OE pers.n. *Beorn* or ON *Bjǫrn* and **wella**); *Bradeen Closes* 1652 *Rad* (probably from the surn. *Bradden*); *Braymbushe* 1579, *Brame bushe* 1601[1], - *bush* 1664, *Bromebush* 1638, *Bramebush* 1671, *Brame bush* 1700 (*v.* **brōm** 'broom', **busc** and cf. *Nettelbustmar* PN L 2 277); *Brodale* 1579, *Brodalle* 1601, *broad daile* 1638, - *dayle* 1664, *broad dale* 1671 ('broad portion of land', *v.* **brād, deill**); *Bromecliffe* 1579, 1638, *Brome clife,* - *Lees* 1601[1] (*v.* **lea** (OE **lēah**) 'a meadow'), *Brome cliffe* 1664 (*v.* **brōm, clif**); *Brothland* 1240-50 (perhaps from **brot** with **land**); *Burt landes* 1579, *Burtlandes* 1601[1], *Burck* - 1638, *Burtlands* 1664 (*v.* **land**, the first el. being probably the surn. *Burt*); *Bushe hill* 1579, - *Hill* 1601[1] (*v.* **busc, hyll**); *buttermilk style* 1671; *byschopbryge lane* 1541-42 *Inv*, *Bishop brigge Lane* 1747 *BT* (presumably leading to Bishop Bridge in Glentham LWR); *Castregate* lHy2,

Cayster gate 1579, 1601[1], *Caister* - 1638, 1664, - *Gate* 1697, 1700, *Caister road* 1709, - *road* 1718, 1724 ('the road to Caistor', *v.* **gata**); *Catt stone* 1579, 1601, *Catstone* 1664 (*v.* **catt, stān**); *the lane goeing to the church* 1638, *the Church Lane* 1671, - *way* 1697, y^e *lane* y^t *lead to* y^e *Church,* y^e *Ch'ch Lane* 1700, y^e *Church-Lane* 1709, - *Church Lane* 1718, 1724 (self-explanatory, but cf. *Kirkhill infra*); *Claypitt furlonge,* - *gate* 1579, - *Furlong, clay pitt gate* 1601[1], *Clay pitt field* 1638, - *feild* 1664; *Cockrell hill* 1579, *Cockerell Hill* 1601[1], *Cockerill* (*hill*) 1638, 1664, *Cockerill Close* 1700, - *close* 1709 (possibly from the surn. *Cockerell*); *durdall, Durdall hill* 1579, *Duredalle Hill* 1601, *dardaile* 1638; *Flatebandis* 1240-50, *Flattlandes* 1579, 1638, *Flatt landes* 1601[1], *Flattlands* 1664 (*v.* **flatr** 'flat, level', **land**; the 1240-50 spelling appears to be corrupt); *Forth Garthe* 1544-7 ECP (*v.* **forð, garðr**); *Grimbaudeland'* 1240-50 (from the ME pers.n. *Grimbald,* itself from the Frankish per.sn. *Grimbald,* well attested in northern Gaul and Flanders, as Dr John Insley points out, with **land**); *the gripp* 1579, - *grippe* 1601[1], *the gripe* 1664 (*v.* **grype** 'a drain, a ditch'); *Grymsbie gate* 1579, *Grimsbie* - 1601[1] ('the road to Grimsby', *v.* **gata**); *Gyrsplates* 1579, *Gresplattes* 1601 ('grass plots', *v.* **gærs** or **gres, plat**[2]); *Hardcrofte* 1638 (*v.* **heard** in the sense 'hard to till', **croft**); *Harecroft* 1240-50, *Haracrofte* 1664 (*v.* **hara** 'a hare', **croft**); *terram Simonis de Herdegate* 1240-50; *heuedlanddeiles* lHy2 ('headland allotments', *v.* **hēafod-land, deill**); *Hiring'* 1240-50; *Hogden* 1579, 1601[1], 1638, 1664; *Holme* 1240-50 (*v.* **holmr**); *Home dale* 1579, - *Dale* 1601[1], *Homesdaile* 1638, *Homesdale* 1664, 1700; y^e *Home-stall* 1709 (*v.* **hām-stall**); *Howtofte*(*s*) 1579, *Howtoft* 1601, 1638, 1664, 1671 (*v.* **haugr, toft**); *Hundehil* 1240-50 (*v.* **hund, hyll**); *Hurninge* 1579, 1601[1], 1638, *Hurne inge* 1664, *Hurnein* 1671, *Hurning* 1700, *Hurninge Gate* 1579, - *gate* 1601[1] (*v.* **hyrne** 'a nook, a corner of land', **eng**); *Kickrill hole* (sic) 1579, - *Hole* 1601; *Kirkhill* 1579, *Kirke hill* 1601[1], *Kirkehill hole Close* 1638, *Kirkhill close* 1664 (the persistence is noteworthy of *Kirk-* (from **kirkja** or the Scandinavianization of **cirice**) in all forms, compared with *Church* in *the lane goeing to the church supra* and the early spellings of Church Cl (a) *supra*); *Lameing close* 1638 (named from the *Laming* family, well-evidenced in the parish, cf. John *Laminge* 1579); *Langhehoues* lHy2 (*v.* **lang**[1]; the second el. may be the pl. of **haugr**); *lincoln' Leyes* 1579, *Lincolne Leas* 1601[1], - *leyes* 1638, *Lincoln Leyes* 1664, - *leys* 1671 (possibly from the surn. *Lincoln*; *Leyes* is from **lea** (OE **lēah**) 'meadow, pasture' in the pl.); *Littelcroft* Hy3 RevesbyS (*v.* **lytel, croft**); *atte Loft'* 1327, 1332 both (p) (*v.* **lopt** a loft'); *Longe Acres* 1664; *Long crofte* 1664; *lyne toftes* 1579, - *tofts* 1638, *Line toftes* 1601[1], *Lynetofts* 1664, *Lynetofts* 1671 (*v.* **lin** 'flax', **toft**); *Mason Garthe* 1544-7 ECP (*v.* **garðr**), *Mason Dale* 1579, 1700, - *dale* 1601[1], 1697, *Maison daile* 1638, 1664, *Mayson dayle* 1638, *Mason-dale* 1709, *Mason-Dale* 1724, *Mason Dale Bottom,* - *Dale hill* 1671 (from the surn. *Mason,* though the earliest noted reference for this surn. in the parish is William *Mason* 1738 BT, with

deill 'an allotment, a share of land'); *May dailes* 1638, *Maydailes* 1664, 1671, *May dales* 1697, *Maydale* 1700, *May-dales* 1709, 1724, *Maydales* 1718, *Maydaile pasture* 1664 (*v.* **deill** 'a share of land', the first el. is uncertain); *Medale furlonge* 1579, *Midle furlonge* 1579, 1601[1], the *Middle furlong* 1664, 1671, y^e *middle furlong*' 1700 (self-explanatory; it would appear that *Medale* reflects ON **meðal** 'middle'); *Methel croft* Ed1 *AddCh* (from ON **meðal** 'middle' and **croft**); *midle gate* 1579, *Middle gate* 1601[1] (*v.* **gata** 'a road'); *le Mulnegate* 1240-50, *milnegate, milgate* 1579, *milne gate* 1601[1], *Milne gate feild* 1664, 1671, *Milne gate* (*furlonge*) 1579, *Milnegate furlonge* 1601[1], *milne gate furlong* 1671 (cf. *totum molendinum* 1601[2], *v.* **myln**); *Milne hill furlonge* 1579, *milne* -, *Milne Hill* 1601; *Mylnehall furlong* (sic) 1638, *Milnehall* - 1664, 1671, *Miln Hall Furlong* 1700; *Nether Barghe* 1579, *Neither barghe* 1601[1], the *nether barthe* 1638 (*v.* **neoðera** 'lower', **beorg** 'a mound, a hill'); *newe Closse* 1579, *new close* 1601[1]; *noddgreine* (sic) lHy2 (presumably for *nord-*, *v.* **norð**, **grein**, cf. *sudgreine infra*); *Norddall, Norddale subtus montem, Norddale˙ lid* lHy2 (*v.* **hlíð**), *Northdale* Hy3 RevesbyS, *nordale* 1579, 1601[1] (probably 'the northern valley', *v.* **norð**, **dæl**, **dalr**)' *in campo de Otteby* Ed1 *AddCh*, 1364 MCD, - *de Ot'by* 1276 RH, *otbye fieldes* 1638, *Otby Field* 1700 (self-explanatory); ˙*Otby Gate* 1671 (*v.* **gata**); *Otbye grange* 1558-79 (*v.* **grange**); *Otby hedge* 1671; *otby Lands* 1709, 1718, *ottby* - 1724; *diuisam de Hutteb'* 1240-50, *Otbye meere* 1579, *Otby Meere* 1697, 1724, *Otby Meare bancke* 1671 (*v.* **(ge)mǣre** 'a boundary, a boundary-strip'); *Otbye north feelde* 1579, *Otbie north Feild* 1601[1], - *field* 1638, *Otby North feild* 1664, 1671, - *field* 1700, *Otbye South-field* 1638, *Otby South field* 1700; *Otbye pastures* 1638, *Otby pastures* 1664, *otby* - 1709, 1718, *ottby* - 1724 (all named from Otby *supra*); *paster hill* 1579, 1601[1] (i.e. *pasture*); the *Pye Closse* 1579, - *Pie Cloase* 1601[1], *pye hole* 1579 (named from the surn. *Pye*, cf. *Willelmus Py de eadem* (i.e. Walesby) 1364 MCD 846); *Rhetthe* (sic) 1240-50; *the river betwixt Wailsbie & Normanbie* 1601; *le Routhelandes* 1240-50, *Long Rowthinge, Shortrowthinge* 1579, *Longe* -, *Short Rowthing* 1601[1], *long Roathing, short Roathinge* 1638, *longe Rowthinge, Short Routhinge* 1664, *Long Rowthing, Short Rowthinge* 1671 (*v.* **land**, **eng**, the first el. being perhaps the ON pers.n. *Rauðr, Rauði*); *Rush hill* 1579, *Rushill* 1601[1], 1638, 1664, *Rush hill* 1671, - *Hill* 1697, - *hilles* 1700, *Rush-hills* 1709, *Rushills* 1718, *rush-hills* 1724 ('the rush-covered hills', *v.* **risc**, **hyll**); *Rydale* 1579, 1601[1], *Ryedaile* 1638, 1664, *the Rye close* 1671, *the Rie Close* 1697, y^e *Rye* - 1700, y^e *Rye-close* 1709, 1718, - *Rye close* 1724 (from **ryge** 'rye', with **deill** and **clos(e)**); *Rysbie acre dicke* 1579, *Risbie acre dike* 1601[1] (a fairly common minor name in L, probably 'the field ditch', cf. the old Acre-Dike Fence in Middle Rasen f.ns. (a) *supra*); *Rysbie Easte wold feelde* 1579, *Risbie east would feilde* 1601[1], *Risbye East Wolde* 1638, *Risby east wolde* 1664, *Risby East Would* 1671 (*v.* **wald**); *Rysbie lowe south feelde* 1579, *Risby lowe south field* 1638, *Risbye lowe South feild* 1664, - *low South feild* 1671 (one of the open fields of Risby);

Rysbie meare 1579, *Reisbie* - 1601[1] (*v.* **(ge)mære**); *Rysbie West Wolde* 1579, *Risbie west would* 1601[1], *the west fielde or west wolde* 1638, - *feild or west wolde* 1664, *the westfield or west Would* 1671 (*v.* **west, wald** and cf. *Rysbie Easte wold feelde supra*); *Scam landes* 1579, *Scamlandes* 1601[1] (perhaps 'short lands', *v.* **skammr** 'short', **land**); *Scar hill* 1579, 1638, - *Hill* 1601[1] (the forms are late, but the first el. may be ON **sker** 'a rock, a scar', dial. *scar* 'a rocky cliff, a bed of rough gravel'); *Shortbutes* 1579, *Shorte Buttes* 1601[1], *Short butts* 1638, 1671, *Shorte* - 1664 (*v.* **sceort, butte**); *Shortcauckes* 1579, *Short Cauks, long Cauks* 1601[1], *Short Caulkes* 1638, *Shorte Cauks* 1664, *Short* - 1671 (*v.* **calc** 'chalk', in the pl., probably the sense is 'the chalky places' and cf. First ... Chalks PN L 2 11); *Les Sikes* 1240-50 (*v.* **sik**); *Smale thorne* 1579, 1638, - *Thorne* 1601[1], *Small thorn* 1700 (*v.* **smæl** perhaps in the sense 'thin', **þorn**); *Snardale* 1240-50 (Dr John Insley suggests that this is perhaps 'the valley where birds are snared', from ME *snåre* (OE **snearu**, ON **snara**) 'a snare for catching birds' and **dalr**); *the southe feelde* 1579, *Otbie the South Felde* 1601[1], *Otbye south-field* 1638 (one of the open fields of Otby); *atte Spout'* 1327 (p) (*v.* **spout** 'a spout, gutter'); *Stainton gate* 1671, *Stainton road* 1709, 1718, 1724 ('the road to Stainton', alluding to the neighbouring parish of Stainton le Vale, *v.* **gata**); *atte Stanes* 1332 (p) (*v.* **stān**); *Stanilandale* 1240-50 (possibly from the surn. *Staniland* or an earlier name 'the stony tract or strip of land', *v.* **stānig, land** with **deill**); *stonbrige furlonge* 1579, *stone bridge Furlong* 1601 (*v.* **stān, brycg**, with **furlang**); *Streete hurndale* 1579, *Street Hurne dale* 1601[1], *Streethorne dayle* 1638, 1664, - *dale* 1671 (*Streete* presumably refers to High Street *supra*); *sudgreine* lHy2 (*v.* **sūð** 'south(ern)', **grein** 'a branching valley', cf. *noddgreine supra*); *Taddale* 1240-50 (probably 'the valley frequented by toads', *v.* **tadde, dalr**); *Teavelby street* 1638, 1671, - *road* 1724, *Tevelby Street* 1664, *Tealby* - 1697, 1709, *Tealbye* - 1718 ('the road to Tealby', *v.* **stræt**); *tevilbie meere* 1579, *Teavelbie meare* 1601[1] ('the boundary with Tealby', *v.* **(ge)mære**); *Thoresway gate* 1601[1] ('the road to Thoresway', *v.* **gata**); *Thorp furlonge* 1579, *Thorpe furlong* 1601[1], 1671 - *furlonge* 1638. 1664 (*v.* **furlang**); *Toft* 1240-50 (*v.* **toft** 'a messuage, a curtilage'); *tofto Walteri de Holegate* 112, *toftum* - c.1210 (*Holegate* is 'the road running in a hollow', *v.* **hol**[2], **gata**); *the towne Close* 1579, - *cloase* 1601[1], - *close* 1638, 1664; *duas terras que uocantur tridinges* lHy2 ('two lands which are called *tridinges* i.e., third parts', *v.* **þriðjungr**); *Wailesbye feildes* 1638, - *feild* 1671, *Walesby field* 1700 (*v.* **feld**); *Walesbi daile* lHy2 (*v.* **deill**); *Walesbie furlonge* 1579, *Wailsbe Furlong* 1601[1], *Wailsbye furlonge* 1638, *Wailesby* - 1664, *Walesby* - 1671 (*v.* **furlang**); *Walesbie meere* 1579, *Wailsbie meere* 1601[1] (*v.* **(ge)mære** 'a boundary', probably that with Otby); *Walesbie North feelde* 1579, *Wallesbie north feilde* 1601[1], *Wailesby north field* 1638, *Wailesby North feild* 1664, 1671, *the North field of Wallesbe* 1697, *Walesby North Field* 1700, 1718, *Walesby North-Field* 1709, *Walesby North field* 1724, *Wailesby lowe North field* 1638,

- *feild* 1664 (one of the open fields of the village); *Walesbie Southe feelde* 1579, *wailsbie south Feilde* 1601[1], *Wailesby south feild* 1638, - *South feild* 1664, *Walesbye South feild* 1671, *South field* 1697, *Walesby South Field* 1700, 1718, y^e *South-field* 1724 (cf. the prec.); *Walesbie northe lowe feelde* 1579, *Wailsbie north low feilde* 1601[1], *Walesbie south Lowe feelde* 1579, *Wailsbie south low feilde* 1601[1] (self-explanatory); *Wallesby Common* 1697; *the wayn way* 1579, *the waineway* 1601 (*v.* **wægn** 'a waggon', **weg**); *iuxta fontem* lHy2, *atte Welle de Walesby* 1318 YearBk (p), *atte Welle* 1327, 1332 both (p) (*v.* **wella**); *molendinum del West, terram Ade West* 1240-50, *The West House* 1558-79, *the west house* 1569-70 *KRMB 38*, *Westfeld de Otteby* 1355 *Cor* (p), *the westfeild or West Would* 1671 (one of the open fields of Otby); *Westhorp* 1355 *Cor* (p) (*v.* **west**, **þorp**, but this may not be a local surn.); *Wodefurlang* 1240-50, *atte Wod'* 1327 (p), - *Wode* 1332 (p) (*v.* **wudu**, **furlang**); *Wrongstong* Ed1 *AddCh* ('the crooked rood', *v.* **vrangr**, **stong**); *Wycam Dale* 1579, *Wicam* - 1601[1], *Wyromdaile* (sic) 1638, *Wickham dayle* 1664, *Wickham Dale* 1700 (from the surn. *Wickam*, cf. Richard *Wicam* 1601[1], with **deill** 'an allotment, a share of land'); *wynter close* 1638, *the winter close* 1664, 1671, *Winter Close* 1697, y^e - 1700 (named from the family of Edward *Wynter* 1601[1]).

North Willingham

NORTH WILLINGHAM

 Wiuilingeham (2x) 1086 DB, *Wiuelingeham* 1193, 1194 P, 1209-35 LAHW, *Wivelingeham* c.1221 Welles, *Wyvelingeham* 1225 Cur, 1234 Welles

 Wiflingeham c.1115 LS

 Wiflingham 1086 DB, c.1115 LS, 1187 (1409) Gilb, l12 RA ii, 1210-15, a1224 ib iv, *-yngham* c.1150 (1409) Gilb, 1327 *SR*, *Wyflyngham* c.1150, 1156-61, Hy2, c.1160, c.1220 all (1409) Gilb, 1316 *Barne*, 1317 AD, 1341, 1357 *Barne et passim* to 1455 *ib*, *-ingham* c.1190 RA iv, 1276 RH, 1335 *Barne*, 1339 *Goulding*, 1346, 1347 *Barne*, 1385 AD *et passim* to 1560 Pat

 Wiffelingham 1326 *Barne*, *Wifflingham* 1470 *ib*, 1597 *NW*, 1602 *LindDep 42*, 1709 *NW*, *Wyfflyngham* 1429 *Barne*, 1468, 1483 AD, 1551 Pat, *-ingham* 1493 *Barne*, 1678 *NW*, - *alias North Willingham* 1555 Pat

 Wiuelingham 1196-1202 RA iv, 1198, 1199 P (p), 1203 Ass, *Wiulingham* eHy3 (1409) Gilb, *Wyuelyngham* Hy2 (1409) Gilb, 1298 Ass, *-ingham* 1240 FF, 1242-32 Fees, l13 *Barne*,

Wyuellingham eHy3 (1409) Gilb, *Wyvelingham* 1311 Ipm,
- "by" *Sixhil* 1316 Pat, 1328 Banco, *Wyvelyngham* 1250
(1446) Pat, 1291 Tax, 1303, 1316, 1325 FA, 1325, 1327 Pat,
1340 AD, 1428 FA, *-ingham* 1252 Ch, 1254 ValNor, 1295
AD, *Wivelingham* 1246, 1260 RRG, *Wiveligham* 1212 Fees
Weyflyngham c.1160 (1409) Gilb
Wullingham 1187 (1409) Gilb, 1242-43 Fees
Wylingham 1242-43 Fees, 1274 RRGr, 1322 Ipm, 1375 Peace,
-*yngham* 1332 *SR*, 1346 FA, *Wilingham* 1341 *Barne*
Willingham 1187 (1409) Gilb, 1526 Sub, - *als North Willingham
als North Wiflingham* 1650 *LindDep 42*, - *als North
Willingham als North Wifflingham* 1659 *ib*, - *alias North
Willingham alias North Wiflingham* 1769 *NW*, *Willigham*
1203 Ass, *Wyllyngham* 1343 NI, a1537 *MiscDep 43*, - *iuxta
Teuelby* 1369 *Barne*
Welyngham 1539-40 Dugd iv
North Willyngham 1502 Ipm, *Northwyllyngham* 1535 VE iv, 1561
Pat, *North Wyllingham* 1609 *NW*
Northwyfflyngham 1503 Ipm, *Northe Wiffflingham* 1559 Pat
Northwillingham 1547, 1553 Pat, 1564 *NW*, 1597 *MinAcct*, 1590,
1592 *LindDep 42*, 1610 Speed, - *alias Willingham alias
Wyfflingham* 1560 Pat, - *alias Northwiflingham* 1600 *Barne,
North Willingham* 1555 Pat, 1576 LER, 1596 *LindDep 42*,
1602 *Barne et freq*, - *alias Willingham* 1565 Pat,
Northwillinghame 1592, 1594 *LindDep 42*

'The homestead, estate of the family, people of Wifel', *v.* **hām**,
the first el. being OE *Wifelinga*, gen.pl. of OE *Wifelingas*, a group
name meaning 'Wifel's people'. The OE pers.n. **Wifel* is not
recorded independently, but must have been fairly common, as
Ekwall, DEPN s.n. Wilcote, points out. Identical with North
Willingham are Willingham by Stow LWR and Willingham PN C
173-74, while *Wifel* is also the first el. of Wilsford Kest as well as
Wilsford PN W 326 and 372. It should be noted that the
etymologies of Cherry Willingham LWR and South Willingham LSR
are different from that of North Willingham, though identical, each
meaning 'the homestead, estate of Willa's people'.

ASH GROVE. BELLA HILL, cf. *Bella Hill Moor* 1867 *NW*,
perhaps the same as *bell hill* 1697 *ib*. BLOATER HILL, 1824 O,

Bloatoe Hill 1606 *LindDep 42, bloto* 1697 *NW, Blowtay Hill* 1707 *Terrier, Blotoe hill closes* 1746 *NW,* obscure. BROOM COVERT. CHURCH HOLES, cf. *Church Hole Close* 1867 *NW.* THE CLUMP, *Clump Hill* 1828 Bry, 1830 Gre, cf. - *Close* 1867 *NW.* DOG KENNEL FM, cf. *Kennel* 1828 Bry. HIGH STREET, *the High Streete* 1682 *NW,* - *high Street* 1693 *ib,* the name of the ancient trackway from Horncastle to Caistor, called Middlegate Lane from Caistor to South Ferriby. LING PLANTATION, cf. *Lynghill Close* 1592 *LindDep 42,* - *hole* 1594 *ib, Lingell Close* 1657, 1668, 1670, 1696, (*the*) *lowe Linghills* 1684, 1687, 1690, *the Ling-hill close* 1709 all *NW,* 'the hill where heather grows', *v.* **lyng, hyll.** LOWFIELD FM, *the lowefeild* 1606 *LindDep 42,* - *lowfielde* 1676, - *Low feild* 1682, - *Low Feild* 1704, - *Lowfeilds* 1712, 1715 all *NW,* self-explanatory. MILL HILL, 1867 *NW,* and cf. *land wher the horse myln' standith'* (sic) 1532 *AOMB 409;* there is reference to *a mill* 1619 *NW.* ROBERT'S WOOD, named from the family of Arthur *Roberts* 1641 LPR, John *Robarts* 1769 *NW.* SNAKE HOLES. SPRINGBANK FM. TEALBY LANE PLANTATION, cf. *Tealby Lane Close* 1867 *NW,* leading to the neighbouring parish of Tealby. THORNEY COVERT, - *Cover* 1828 Bry, 1830 Gre. TOYNE MOOR, *Toyn Moor* 1867 *NW,* named from the family of *Jon Toyne* 1697 *NW,* Richard *Toyne* 1737 *BT.* VICARAGE (local), *the Vicaridge House* 1606, *A* - 1700, *the Vicarige House* 1706, *The Vicarage house* 1718 all *Terrier, ye said Rectorie or Parsonage house* 1658 *LindDep 42,* and note *There is no House belonging to the Vicarage of North Willingham* 1822 *Terrier.* WILLINGHAM CORNER, cf. *Corner Inn* 1828 Bry, *Willingham Corner Inn* 1830 Gre. WILLINGHAM HILL (lost), 1828 Bry, cf. - *Furlong* 1793 *Td'E.* It was the name of the road (the modern A 631) running east of the village towards Ludford. WILLINGHAM HO, 1824 O, 1830 Gre, cf. *the Hall* 1707, 1822 *Terrier,* and note the reference in 1822 *ib* -- *the old Hall, which Hall is now entirely taken down and rebuilt on a different site at the West End of the parish.* It was still called *Willingham Hall* 1828 Bry. WILLINGHAM PARK, *Park* 1867 *NW.* WISTERIA COTTAGE.

Field-Names

Forms dated c.1150 and 1187 (1409) are Gilb; 1210-15, and a1224 are RA iv;

1276 are RH; 113[1], 1326, 1341, 1346, 1347, 1357, 1395, 1396, 1436, and 1560[1] are
Barne; 113[2], e14[2], 1317, 1340, 1456, 1483, and n.d. are AD; e14[1] are *Goulding*;
1531-2, 1535-6, and 1537-8 are *AOMB 409*; 1553, 1555, and 1560[2] are Pat; a1537
are *MiscDep 43*; 1537-9 are LDRN; 1564, 1567, 1568, 1582, 1616, 1621, 1628,
1642, 1650[2], 1657, 1666, 1667, 1668, 1670, 1674, 1676, 1682, 1684, 1687, 1690, 1692,
1693, 1696, 1697, 1700[2], 1704, 1707, 1711, 1712, 1713, 1715, 1725, 1746, 1769, 119,
and 1867 are *NW*; 1579, 1590, 1592, 1594, 1596, 1600, 1602, 1606[1], 1633, 1635,
1637[1], 1648, 1650[1], 1653, 1658, 1659, 1680, and 1695 are *LindDep 42*; 1596-7 are
MinAcct; 116 are *Dep*; 1606[2], 1700[1], 1706, 1707, 1718, and 1822 are *Terrier*; 1625
are Heneage; 1637[2] are *Yarb*.

(a) Ash Holt Cl 1867; Bar Cl 1867; Barn Fd, - Wold 1867; Far Blacksmith
Cl 1867; Blotshill 1867; Bond Hills 1867 (perhaps 'bean allotments', if it is
identified with *Boundale* 113[1], *closes or pastures comonly called bawndell and
dovehill* 1567, *Bondles Bottom Dike* 1707, *v.* **baun, deill** and Dove Hills *infra*);
Brick Btm, - Cl, - Kiln, Brickkiln Cl 1867; Bull Pasture (Cl, - Gdn) 1867
(*Bullpasture Close* 1621); Calf Cl 1867; Clark Cl 1867 (named from the family of
John *Clerk* 1642 LPR); First -, Second Clover Plat 1867 (*v.* **plat**[2] 'a small plot
of land'); Collingwood Cl 1867 (from the surn. *Collingwood*; William *Collingwood*
is named in the same document); The Common or Moor 1822 (*the More* 1531-2,
the Morez 1537-8, *the more* 1579, *the Mores* 1594, *the moore of North Willingham*
1693, *le Moore* (*Close*) 1642, *the Moore* 1680, 1682, 1692, *the Moor* 1706, *the
common moore* 1606[2], *North Willingham Common Moor, that part of the moor
Called The Common* 1697, *the Moore Close* 1692, *Moor Close* 1697, *the upper
More close* 1667, *the upper Moore Close* 1674, *the upper Moore* 1684, *the upper
and Lower Moore Closes* 1687, *The* -, *the upper and lower Moore Closes* 1700[2],
the Upper Moore Close, the lower moore Close 1687, *the uper* (sic) & *lower moore
Closes* 1690, *the Low Moor* 1746, *One severall more grounde* 1606[1] (*v.* **severall**),
Moor ground 1700[1], *v.* **mōr**[1]); Coney Green 1822 (1707), Lt Coney Green 1867
(cf. *the conye garth* 1531-2, *le Conygarth(e)* 1555, *the conygarthes* 1568, *Cony garth
close* 1693, *v.* **coni, garōr** 'an enclosure'; *Coney Green* is a frequent variant of
coninger 'a rabbit warren'); Corner Cl, - Plantn 1867; Cottage Pasture 1867; The
Cow Cl 1822, Cow Cl 1867 (1707, *the Cow close* 1531-2, *Coweclose* 1560[1], *Mr.
Drury Cow Close* 1697); Far -, New Cowholds 1867 (*Cowhold Common* 1697);
Cowhouse & Garth 1867; Crook Cl 1867 (1707); Dales West End, Far Dales
East 1867 (*v.* **deill**); Deepdale 119 (*depedale* 1594, *v.* **dēop, dalr**); Dish bourn
Hill 1867; Dog Dale 1867 (*dockedayle* 1326, *docke dales* 1564, *v.* **docce** (the
plant), **deill**); Dove Hills 1867 (*Dovehill* 1555, *closes or pastures comonly called
bawndell and dovehill* 1567, *v.* **dūfe**, cf. Bond Hills *supra* and Dufeholm in (b)
infra); North East -, South East Walk 1867 (*v.* **ēast, walk** 'a sheep-walk', as

elsewhere in this parish, with **norð** and **sūð**); Etherington Cl 1867 (named from the *Etherington*, earlier *Everington*, family, cf. Thomas *Everinton* 1670 *BT*, Thomas *Everington* 1764 *NW*); Fanthorp hill 1867 (named from the *Fanthorp* family, cf. John *Fantrop* 1755 *BT*, John *Fannthorp* 1792 *NW*); Fish Ponds 1867; Fletty Btm 1867 (v. **botm**, cf. *Close called Flettie Close* 1606[1], *flety Closse* 1697, *Fletty Close* 1704, 1713); Flint Walk 1867 (v. **walk**; the first el. may be the surn. *Flint* or v. **flint**); Foot Road Cl, - Road Paddock 1867; Fowlsyke 1867 (*foul sike* 1697, v. **fūl**, **sīk**); Fox Cover Hills 1867; Greenland Walk 1867 (v. **walk**; *Greenland* may be a surn. or from **grēne**[1], **land**); Hare Holes 1867 (*Hareholles* 1537-8 *One close called Hareholes* 1606[1], *Haregoles Newe Close* (sic) 1693, v. **hara**, **hol**[1]); the Home Cl 1822 (1707); Hop Pasture Btm 1867; Horne Cl 1867; Hortons Garth 1867 (from the surn. *Horton*, with **garðr**; James *Horton* is named in the same document and the family is recorded earlier in James *Horton* 1815 *BT*); House Cl 1867 (1700[1], 1707, *the House Garden Close* 1706), House Pce 1867; How Croft 119 (v. **croft**); (Little -, Meadow) Ings, Ing's Pasture 1867 (*Le Inges* 1560[1], *lez ynges* 1590, *les -* 1592, *the -* 1594, 1602, 1606, *le Inges* 1642, *North Willingham Inges* 1682, 1693, *the Ings of North willingham* 1692, *the Ings* 1696, *The -, ye Ings* 1700[2], 1707, v. **eng** 'a meadow' in the pl.); (Far -, First -, Second) Legsley Plat 1867 (v. **plat**[2]); North -, West Leighton Barn & Fd 1867 (from the surn. *Leighton*, cf. John *Leighton* 1828 *NW*); Ley Cl 1867 (1555, v. **lǣge** 'fallow, untilled', **clos(e)**); First -, Second Lilley 1867 (from the surn. *Lilley*, cf. Richard *Lilly*, named in the same document); Limons Cl 1867 (perhaps from the surn. *Limon*); Little Ings 1867; Long Cl 1867; Long Lees 1822 (1707), Longleys 1867 (*long leys (end)* 1697, v. **lang**[1], **lea** (OE **lēah**) in the pl.); Machan Btm 1867 (from the surn. *Macham* or *Matcham*, with **botm**); Meadow Cl 1867 (stated to be "meadow"), Meadow Paddock 1867; Nan Turner Cl 1867 (named from a member of the *Turner* family, cf. John *Turner* 1564 *NW*, and Turner Cl *infra*); New Cl 1867 (*Neweclose* 1560[1]), New Delight 1867 (an arable close, 5a 1r 1p in area; cf. Great Delight PN Ch **3** 131, ib **5** 269, where a derivation is proposed from a postulated *lihte 'a light place, a place in a wood where the trees are more sparse'. The 1867 f.n. is, however, almost certainly a complimentary nickname); New Found Land 1867 (probably 'land newly brought into cultivation'); North Barn Cl 1867; Orchard 1867; padcroft 1822, Padcroft 1867 (1707, *Padcrofte* 1560[1], v. **padde** 'a toad', **croft**); Paddock 1867; Pinfold Square 1867 (v. **pynd-fald**); Pleasure Grounds & Plantation 1867; Red Chalk Hill 1867 (probably alluding to red ochre, cf. "Native red ochre is called red chalk and reddle in England" 1839 NED s.v. *ochre* (1); it would be the raw material of *raddle* for marking sheep); Road Walk 1867; Round Plantn 1867 (named from its shape); Sand Pit Cl 1867; Seventeen Acres 1867; Sixhills Lane Cl 1867 (from the neighbouring parish of Sixhills LSR); South Fd 1867 (*Southefeilde* 1555, *the*

South feilde 1579, *the sowthe feildes* 1594, *the south feild* 1602, 1680, *ye South feild* 1635, *the South Field* 1707, *in Australi Campo* 1596, 1600; one of the open fields of the parish, *v.* sūð, feld, cf. *Northfelde* in (b) *infra*); Stanhill 119 (*Stonehill close* 1602, *two Closes caled Stonehill* 1606[1], *ye 2nd Stone hill Bank* 1707, *v.* stān, hyll); Stone Pit (Cl) 1867 (*v.* stān, pytt); Stone Wall Pce 1867; Thistle Hill 1867; Town Cl, - Paddock 1867; Townsend Cl 1867, Townend Plott 119; Turner Cl 1867 (named from the *Turner* family, cf. Nan Turner Cl *supra*); Twelve -, Twenty -, Twenty five Acres 1867; Upton Cl 1867 (named from the *Upton* family, cf. George *Upton* 1817 *BT*); Walk 1867 (*v.* walk); Water Mdw 1867; Webster Btm 1867 (possibly from the occup.n. *Webster, cf. Robert West de Tiuelby Webster* 1396 *Barne*, with botm); Westhills 1867 (*West Hills* 1579, *v.* west, hyll); North -, South West Walk 1867 (*v.* walk); Wet Cl 119 (cf. *Wate landes* 113[1]); Wheatley Cl 1867 (probably from the surn. *Wheatley*); Wilkentree Plott 119 (perhaps alluding to the rowan or mountain ash, locally known as the *wickentree*); Far -, Lt Wold 1867 (*v.* wald); Yowdram (sic) 1867 (*Zollerdam hill* 1456 (with *Z* for *ʒ*) *Yowderdam* 1707, obscure).

(b) "a furlong called" *Aldresshes* 1456; *Armitage moor* 1697 (*v.* hermitage); *Aycerholme* 113[1] (*v.* æcer, holmr); *Aykkedykes* (sic) 113[1] (*v.* æcer, akr, dīk and the old Acre-Dike Fence in Middle Rasen f.ns. (a) *supra*); *Band* 113[2]; *Bardall Close* 1625, *Bardoll Close* 1650[2], *Bardale* 1658, *Bardale close, Bardall* - 1659; *Barrandes Landes* 1553 (the rent was given to the lights in the church, perhaps 'Baron's lands', cf. John *Baron de Wyflyngham* 1448 *Barne*); *common layne called blynd layn'* 1537-8 (a *blind lane* is a cul-de-sac); *boney moor* 1697 (possibly from the surn. *Bonney*); *Bramdaill* 1555 (*v.* brōm, deill); *Close called Broadwaters* 1606[1] (from the surn. *Broadwater*); *Browndale* 1596 (from the surn. *Brown* with deill; the earliest reference so far noted to the surn. is Richard *Browne* 1619 *BT*); *Butt Close* 1707 (*v.* butte); *Caldeweldale* e14[2] (*v.* cald, wella with dalr); *the Clay Pits* 1707 (*v.* clǣg, pytt); *Cokdayle* 1346, *Cockfurres* 1596, *Cockefurre Close* 1676, *Cock Furre Close* 1712, 1715 (possibly from the surn. *Cock*, cf. John *Cock* 1570 *Inv*, with deill and furh, clos(e), and cf. (*the*) *rock Fur close infra*); *Coker Croft* n.d. (from the surn. *Coker*, with croft); *Cooke laine* 1628 (named from the *Cooke* family, cf. Julian *Cook* 1570 *Inv* and William *Cooke* 1628); *Cowper Dayle* 1531-2, *Coperdalle alias leyclose* 1560[1] (*v.* deill; the first el. is probably the surn. *Couper*, cf. John *Couper* 1357, cf. Ley Cl in (a) *supra*); *the Kowpasture* 1676; *Crakemyre Furlong* 1596 (*v.* krāka 'a crow, a raven', myrr 'a marsh', a Scand. compound, with furlang); *dogdike Close* 1650[1], 1650[2] (*v.* dīk; the first el. may be docce, cf. Dog Dale in (a) *supra*); *Dufeholm* e14[2] (*v.* dūfe, holmr); *le Egg* 1596, *the Egge* 116 (*v.* ecg in a Scandinavianized form, and cf. *motten meadow infra*); *Eleueakrefurlang'* (sic) 113[1] (*v.* æcer, akr, furlang; the first el. may be the ON pers.n. *Eileifr*, as Dr John Insley suggests); *Engecroft* 1187 (*v.* **eng, croft**); "the" *Estend* 1483 (*v.* **ēast, ende**[1]); *estlanges* 1346 ('the eastern long strips of land', *v.* **ēast, lang**[2]); *le Fallowe yate* 1596 ('gate to the fallow (field)', *v.* **falh, geat**); *Fenwell'* 1210-15, a1224 (*v.* **fenn, wella**); *in campis de Wiflingham'* 1210-15, *in campis* e14[1], *in campis de Wyflingham* 1341, *in campis de* - 1347, *in campis de Wyflyngham* 1395, 1396, 1436, *in campis de Wyllyngham* a1537, *the feildes of North wiflyngham* 1606[1], *the common feildes* 1606[2], *the Field* 1706, 1707 (self-explanatory, *v.* **feld**); *le Fury Hyll* 1555, *furr hill* 1579, *fur hill Close* 1594, *le furhill* 1596, *lytle furre hill* 1564 (*v.* **hyll**); *the Graunge* 1531-2, *the Graunge of Wyfflyngham* 1535-6, "granges in Willyngham ..." 1537-9, "a messuage called" *Le Grange* 1560[2], *Le Fi...e close alias Graunge close* 1560[1], *illam grangiam nostram* 1560[1], *the Graunge* 1616 (this was a **grange** of Sixhills Priory); *Grenehyl* 113[1], *-hil* 113[2], *Grenhyll'* 1346, *Greene hill* 1600 (*v.* **grēne**[1], **hyll**); *halter Daile* 1579 (*v.* **deill**; the first el. may be a surn.); *Hawetoft* 1210-15, *Houtoft* n.d., "toft & croft called" *Houtoft* 1340, *Howtoft moor* 1697 (*v.* **toft**; the first el. is uncertain); *hell dore ...es* 1564, *Helldoor* -, *Hell doore Bottome* 1596, *Hell Doors* 1707 (probably a derogatory name, cf. Hell-Fire Gate, PN Gl 1 227 and Hell Gates Piece in Harlow Ess); *the high feild of North Willingham* 1693, *the High Field* 1707 (cf. *in campis de Wiflingham' supra*); *the heighe Inges* 1631, *the Heighe -* 1637[1], *yᵉ highe Ings* 1648, *yᵉ high Inges* 1653, *the high Ings* 1680 (*v.* **hēah, eng**); *horseleach noke* 1697 (*v.* **nōk**; a *horse-leech* may be either 'a farrier' or 'a large blood-sucking worm (*Haemopsis sanguisorba*)'); *horse platte* 1594 (*v.* **plat**[2]); *le Innomes* 1456, *the Innvmes, the Inham Close* 1531-32, *the Innvmes* 1531-32, 1537-38, *le Juvines* (sic) 1555 (*v.* **innām** 'a piece of land taken in or enclosed'; the form *Juvines* 1555 seems to be an error resulting from a misreading of the minims of *Invmes*); *the Intak'* 1537-8 (*v.* **inntak**); *kell moor* 1697 (named from the *Kell* family, cf. *late Kell* 1697, and probably *Barnard Keyll* a1537, *v.* **mōr**[1]); *the kerk Close* 1531-2, *Kirkeclose* 1560 (*v.* **kirkja** or a Scandinavianization of **ciricе**, with **clos(e)**); *lamcootehylles* 1346 (*v.* **lamb, cot**, with **hyll**); *langmargate* 113[1] ('the road to *Langmare* ('the long boundary'), *v.* **lang**[1], **(ge)mǣre**, with **gata**); *Langedalle Hoverhend* 1317 ('the upper end of *Langdale*', *v.* **lang**[1], **dalr** or **deill**, with **uferra, ende**[1]); *langlandes* 1346 (*v.* **lang**[1], **land**); *lare hill* 1697 (*v.* **leirr** 'clay'); *Leylandes* 1555 (*v.* **lǣge** 'fallow', **land**); *Linwood Platts* 1746 (from the neighbouring parish of Linwood and **plat**[2]); *Long greene* 1666, *Long Green* 1690 (*v.* **lang**[1], **grēne**[2]); *Longhill Closes* 1606[1]; (*Richard Lilly*) *long moor* 1697 (*v.* **lang**[1], **mōr**[1]); *the Lowe Ings* 1666, *the lowe Ings* 1690 (*v.* **eng**); *the lower close* 1684; *Ludfordgate* 113[2] ('the road to Ludford (the adjoining parish)', *v.* **gata**); *lytilhowe* 1346 (*v.* **lýtel, litill, haugr** 'a mound, a hill'); *Mansike* 113[2] (*v.* **(ge)mǣne** 'common', **sík**); *Marfurlonges* 1346 (*v.* **(ge)mǣre** 'a boundary', **furlang**); *a marstallo* (sic) 113[1] (cf. *le marstal'*, *- Marstal* 1Hy3 in Killingholme parish, PN L 2 209, where it is suggested that *marstal'* is from OE *_mær(e)stall_ 'a pool' or the like); *Mast dale moor* 1697; *maudeholmenes* 113 (from the ME fem. pers.n. *Maud* and **holmr**, with **nes**); *le Middelhill* 1555, *Middleshill* 1564 (*v.* **middel, hyll**); *Moign Close* 1594 (named from the *Moigne* family, cf. Henry *Moigne* 1455 *Barne*, Thomas *Moigne* 1483 AD, Thomas *Moyne* 1568 *NW*); *Moor Close, moor north side beck* 1697 (i.e., 'moor on the north side of the beck', *v.* **mōr**[1], **bekkr**); *motten meadow lying under the Egge* 116; *new plat* 1697 (*v.* **plat**[2] 'a small plot of ground'); *le Noble More* 1600; *Nordstoc* n.d.[2] (*v.* **norð, stocc** 'a tree-trunk, a stump' and cf. *Suthstoc infra*); *in Boriali Campo* 1592, 1596, *in boriali* - 1600, *Northfelde* 1555, *the northefyeld* 1564, *the northe fieldes* 1594, *the North feilds* 1633, - *feild* 1680, - *Field* 1707, *v.* **norð, feld**, one of the open fields of the parish, cf. South Fd in (a) *supra*); *cum tota le Northsokene de predicta villa* c.1150 (1409) (*v.* **norð, socn** 'an estate'); *Northwood* 1606[1], *the North Wood* 1682, *the Northwood* 1693 (*v.* **norð, wudu**); *the old House Garth* 1707 (*v.* **garðr**); *1 ley, 2 leys, 3 leys* 1697 (*v.* **lea** 'piece of meadowland or pasture'); *the Orcherde* 1531-2 (*v.* **orceard**); *Ambrose Osneys Close* 1707, *Osneys Close* 1725 (named from a member of the *Osney* family, cf. Robert *Osnaye* a1537, Richard *Osney* 1590 *Inv*); *le Overclose* 1555 (*v.* **uferra** 'upper', **clos(e)**); *Payde Closse* 1531-2; *pease Hole* 1707 (*v.* **pise, hol**[1]); *the Pinnell* 1657, *the pingle* 1668, *the Pingle* 1670, 1693 (*v.* **pingel**; *pinnel* is a dial. form of *pingle*); *Prior's walk* 1625 (pasture for 200 sheep; the reference is presumably to the *Prior* of Sixhills Priory); *Pyttes* 113[1] (*v.* **pytt**); *infra quatuor diuisas de Wyuelingham* 113[1] ('beneath the four boundaries (or divisions) of Willingham'); *a highway leadinge to Markett Rasin* 1693 (self-explanatory); *Close called Remepoke* (sic) 1606[1] (probably an error for *Cremepoke*, cf. Cream poke Noke in the f.ns. (a) of *Walesby supra*); *ad Ripam de Wyflingham* n.d. (p) (*v.* **banke**); (*the*) *rock Fur close* 1709 (possibly an error for *cock Fur* -, cf. *Cokdayle supra*); *Rugholme* 113[1] (*v.* **rūh, holmr**); *Russels Close alias Linghill Close* 1707 (with *alias Linghill Close* added in a different hand and cf. Ling Plantation *supra*), *Messuage ... formerly called or knowne by the name of Russells Farme* 1711, *Russells Farme* 1712, 1715 (named from the *Russell* family, cf. Thomas *Russell* a1539, and frequently in *BT*); *Salter Bank* 1707 (perhaps from the surn. *Salter*); *Sandgates* 1596 (*v.* **sand, gata**); *Setcop* 113[2] (*v.* **set-copp** 'a flat-topped hill'); *Sheffeild close* 1637[2], *Sheffeild Close* 1650[1], 1650[2], - *close* 1658, 1659, *Sheffields Close* 1707 (named from the *Sheffield* family; Vincent *Sheffeilde* was the owner of the close in 1637 and it was conveyed by him in 1650[2]); *le Shepegate* 1582, *pastur' vocat' Shepegate* 1596-7, *the sheepegate* 1616 (*v.* **shep-gate** 'a sheep-walk'); *Sixel Hill* n.d. (from the neighbouring parish of Sixhills and **hyll**); *skeldlandes* 1346 (Dr John Insley comments that this means 'the separated strips of land', from ME *skelled*, past part. of ME *skillen* 'to separate one thing from another', a Scand. loan word, and the pl. of **land**); *the Sodd Wall* 1746; *Soke Dales* 1555 (cf. *the Northsokene* supra); *Stanweldalbek'* 1346 (*v.* **stān, wella,** with **dalr** and **bekkr**); *Stanydale -, Stonydale Close* 1596, *Standing dale* 1680 (*v.* **stānig, dalr**); *in viam de Wflingham* 1276, *the Comon Streete* 1680, *the Common Street* 1695, *the Street* 1707 (*v.* **strǣt,** with reference to the village street); *Suth Langlandes* n.d.^{2} (*v.* **lang**[1], **land,** with **sūð**); *Suthstoc* n.d.^{2} (*v.* **sūð, stocc,** and cf. *Nordstoc* supra); *sweet hills* 1697 (no doubt a complimentary nickname); *le syke* e14^2 (*v.* **sīk**); *Tevilby meare* 1666 ('the boundary with Tealby (an adjoining parish)', *v.* **(ge)mǣre**); *Thrithidale* (sic) 113^2 (perhaps 'a one-third share', *v.* **þridda, deill,** or the first el. may represent **þritig** 'thirty'); *toft* 1210-15, *Toft* a1224 (*v.* **toft** 'a curtilage, a messuage'); *le Tong(e)* 1555 (*v.* **tunge,** in the topographical sense of 'a tongue of land'); *one litle pece of medow ... called the troughe* 1579 (*v.* **trōg,** possibly for some topographical reason, or because a drinking-trough was located there); *the tyle Cloise* 1531-2, *the tile Cloise* 1535-6, *Tyleclose* 1555, *Tyle close* 1560, 1616 (*v.* **tigel, clos(e)**); *vicarige moor* 1697 (probably part of the glebe, *v.* **mōr**[1]); *le Warlotes* 1346 (*v.* **warlot**); *Waynewelle* 113[1] (*v.* **wægn** 'a waggon', **wella** 'a spring, a well'); *Wellwonge Furlong* 1596, *Well wongs* 1680 (*v.* **wella, vangr** 'a garden, an in-field'); "the" *Westend* 1483 (*v.* **west, ende**[1]); *Westlandes* 1553 (*v.* **land**); *West moor* 1697 (*v.* **west, mōr**[1]); *Whaitelandhull'* 1210-15, *Whattelandhull'* a1224 (from **hwǣte** or **hveiti** 'wheat', **land,** with **hyll**); *Wlfstankumb'* 1210-15, *Wlstancumb'* a1224, *Wlstayncumb* 113^2, *Woltstoncombe* 1456 (*v.* **cumb** 'a valley'; the first el. is the OE pers.n. *Wulfstān,* of which *Wlstayn-* 113^2 is a Scandinavianized form); *Wordengland* 113[1]; *Wrangedale* 113[1] (*v.* **vrangr** 'crooked' with **dalr** or **deill**); *yornortehylles* 1346 (with *y* = *þ*), *yar ... tte hill* 1596 (the reading is uncertain) (cf. *þornotehille, þiornothil* in Immingham f.ns. (b), PN L 2 172); *Yowdalez* 1531-2, *Youdall Inges* 1560 (perhaps 'ewe lots', *v.* **eowu, deill** with **eng**).

INDEX

This Index is based on the following principles:

    (a) It includes all the place-names in the body of the work.

    (b) It covers only the main reference to each place and no cross-references are noted.

    (c) Street-names are included, but only those of Market Rasen have been treated separately in the body of the text.

    (d) "Lost" names are printed in italics.

    (e) In grouping names together no distinction has been made between those written in one word or two, e.g. Red Ho and Redhouse have been grouped together.

    (f) Only very few field-names (of special historical or philological interest) have been included. The spellings have sometimes been normalized under one head-form for convenience, e.g. *Acre Dike.*